Andrey A. Radionov · Vadim R. Gasiyarov
Editors

Advances in Automation IV

Proceedings of the International Russian Automation Conference, RusAutoCon2022, September 4–10, 2022, Sochi, Russia

Springer

Editors
Andrey A. Radionov
Moscow Polytechnic University
Moscow, Russia

Vadim R. Gasiyarov
Moscow Polytechnic University
Moscow, Russia

ISSN 1876-1100 ISSN 1876-1119 (electronic)
Lecture Notes in Electrical Engineering
ISBN 978-3-031-22310-5 ISBN 978-3-031-22311-2 (eBook)
https://doi.org/10.1007/978-3-031-22311-2

This Springer imprint is published by the registered company Springer Nature Switzerland AG
The registered company address is: Gewerbestrasse 11, 6330 Cham, Switzerland

Preface

International Russian Automation Conference (RusAutoCon) took place during September 4–10, 2022, in Sochi, Russian Federation. The conference was organized by Moscow Polytechnic University. The international program committee has selected 205 reports.

The conference was divided into eight sections, including

1. Process Automation;
2. Modeling and Simulation;
3. Control Theory;
4. Machine Learning, Big Data, Internet of Things;
5. Industrial Robotics and Mechatronic Systems;
6. Computer Vision;
7. Industrial Automation Systems Cybersecurity;
8. Diagnostics and Reliability of Automatic Control Systems;

The international program committee has selected totally 41 papers for publishing in Lecture Notes in Electrical Engineering (Springer International Publishing AG).

The organizing committee would like to express our sincere appreciation to everybody who has contributed to the conference. Heartfelt thanks are due to authors, reviewers, participants and all the team of organizers for their support and enthusiasm which granted success to the conference.

Andrey A. Radionov
Conference Chair

Contents

Process Automation

Mathematical and Software for Building the Rating of the Largest IaaS Suppliers by the Threshold Aggregation Method

S. V. Razumnikov(✉)

National Research Tomsk Polytechnic University, 30, Lenina Street, Tomsk 634050, Russia
demolove7@inbox.ru

Abstract. When making decisions about the choice of any alternative, it is necessary to find out which of the options under consideration will be better. The article presents a non-compensatory aggregation model for building ratings, which is based on the threshold aggregation rule. This method avoids compensating low scores for alternatives with higher scores. A scheme of evaluation stages for this model is proposed, and the software "Formation of an aggregate rating" has been developed. The non-compensatory aggregation model was programmed in C# in Visual Studio 2019. Based on the developed program, a rating of the largest IaaS providers was built, according to which you can choose the best option for an enterprise when using cloud services.

Keywords: Cloud services · Providers · IaaS · Rating · Threshold aggregation · Model · Program

1 Introduction

Cloud technologies have firmly entered our lives. More and more enterprises are considering the possibility of moving their IT infrastructure to the clouds completely or at least partially [1, 2]. In this regard, they have a question about choosing a cloud service provider, and what kind of services they need to work [3]. Partnering with a provider should be approached responsibly, as reliable operation, security and convenience will depend on it. When choosing a cloud service provider, customers can evaluate it according to various parameters. To conduct an analytical analysis of providers, one should study contracts, agreements, documents, reviews of Internet resources [4]. Such an analysis should be conducted by a formed enterprise cloud strategy team, consisting of relevant specialists of the organization and invited experts.

The management of enterprises pays special attention to the rating of the best suppliers offered on the market. This rating reflects the reputation of the provider and customer feedback [5]. However, it is important to take into account the fact that there may be superiority in some indicators, and thus they can compensate for lower values in other indicators. And given the importance of choosing a cloud provider, this is unacceptable.

A. A. Radionov and V. R. Gasiyarov (Eds.): RusAutoCon 2022, LNEE 986, pp. 3–11, 2023.
https://doi.org/10.1007/978-3-031-22311-2_1

To avoid averaging sets of assessments and at the same time take into account all indicators, you can apply the threshold aggregation method, which has proven itself well in multi-criteria assessment problems [6, 7]. Thus, the evaluation of providers providing cloud services for enterprises, without compensating for the best positions, is a modern urgent task.

The purpose of the article is to present a non-compensatory aggregation model for building a rating, in particular, for choosing an IaaS provider, as well as to develop software that implements this mathematical model and allows you to generate an aggregated rating.

2 IaaS Vendor Market

The introduction of cloud IT services has become massive, so the expectations from the transition have also grown, according to an analytical study by Gartner [8]. They predict that the market for public cloud IT services will grow in 2021. The total revenue of such services will grow to $266.4 billion in 2022 [9].

The most demanded segment is still SaaS. It will grow to $116 billion in 2021. The IaaS segment is up to $50 billion, which is 24% more than in 2020. Analysts attribute the accelerated growth of the IaaS segment to the needs of workloads and modern applications, as they need infrastructure that traditional data centers cannot provide [10–12].

With the increase in cloud technologies, the segment of managed cloud IT services is becoming more competitive and more complex. According to analysts, by 2022, up to 60% of enterprises will use the services of third-party providers to manage cloud IT services. This is twice as many as in 2018.

Now there is a clear trend towards market consolidation - there are large federal-scale suppliers that can compete with the scale of infrastructure, the quality of services provided and the capabilities of digital ecosystems. In the next 3–5 years, several large suppliers are expected to play an important role in the market [13, 14]. Let us illustrate the construction of a rating based on a non-compensatory threshold aggregation model.

3 Mathematical Model for Building a Raring by the Threshold Aggregation Method

Consider a mathematical model for building an aggregate rating. Let k be the number of alternatives to be evaluated according to n criteria. m is the number of gradations. Alternatives can be various subjects or objects, such as employees, suppliers, or projects.

Alternatives can be evaluated according to the chosen gradation scale, which can be from two or more. A five-gradation scale is often used, as this is the usual assessment scale for an expert (for example, in schools and universities). A 10-point scale is also convenient for grading. More than 10 is not appropriate to use, although it is possible. We will adhere to a ten-gradation scale for creating software when building aggregated ratings.

For the mathematical model of rating construction, we apply the threshold aggregation method and calculate the indices that will indicate the preference for each alternative, i.e. her position in the rankings.

This rule is that we find the number of each assessment, first ones, then twos, that is, assessments that are one higher, etc., up to the last given gradation [15–20].

According to the threshold aggregation rule, the desired preference index, which will characterize the alternative in the rating, will be equal to (1):

$$F(x) = \sum_{j=1}^{m} C_{n-V_j(x)+m-j-1}^{m-j} = \sum_{j=1}^{m} C_{n-(\eta(1)+\eta(2)+\ldots+\eta(j))+m-j-1}^{m-j} \tag{1}$$

F we find the normalized value by formula (2):

$$I_{Threshold} = \frac{F}{F_{\max}} \tag{2}$$

where $F_{\max}$ is the preference index, which can be obtained with the maximum estimates of the alternative for all criteria. The values that can be calculated using the formula will be in the range from 0 to 1. The closer to 1, the higher the indicator value. On Fig. 1 shows the scheme (algorithm) of the evaluation stages according to the non-compensatory aggregation model.

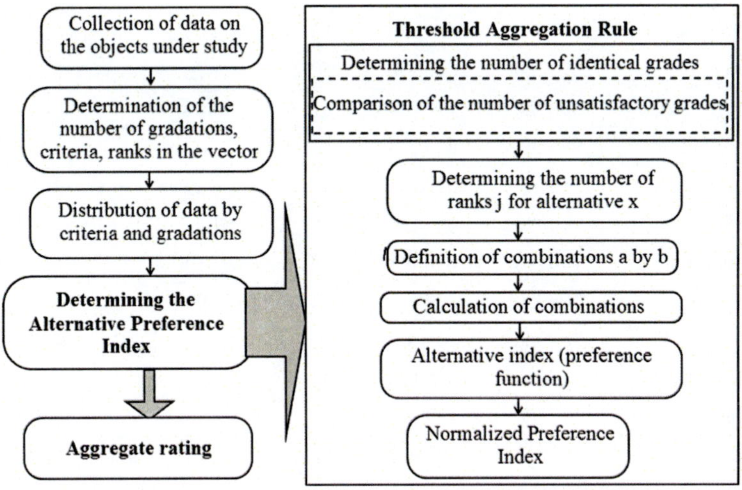

Fig. 1. Scheme of stages of evaluation according to the model of non-compensatory aggregation.

To obtain an aggregate rating of the objects being assessed, it is necessary to complete the stages presented in this scheme.

4 Formation of an Aggregated Rating of the Largest IaaS Providers

The paper presents data from the 18 largest IaaS providers. To identify suppliers, numbers from 1 to 18 are assigned based on the total number of alternatives considered.

When constructing the rating, specific IaaS service providers will act as alternatives. Consider the data of providers for 5 indicators presented in Table 1.

A 10-gradation scale was developed according to the given data (Table 2). The analyst will put down points (gradations) according to the given ten-gradation scale.

Since the indicators are quantitative, we will apply a ten-gradation scale (Table 2) and, according to the data from Table 1, we will assign points to the evaluated suppliers and enter them in Table 3.

Table 1. Supplier data.

№ п/п	Provider	Indicators				
		Revenue for IaaS services in 2018, thousand rubles	Revenue for IaaS services in 2017, thousand rubles	Revenue ratio 2018/2017 (growth), %	Share of IaaS revenue from the total revenue of the enterprise, %	Number of clients (IaaS customers)
1	Softline	2641147	1500652	76	3	1150
2	IT-Grad	2462189	360000	584	80	5361
3	Rostecom	2425000	1629430	49	-	н/д
4	DataLine	1975000	1427000	38	23	460
5	Selectel	1571760	1046660	50	76	more 20000
6	Krok	1560926	1194184	31	5	240
7	Servionika	999867	833233	20	-	60
8	OnCloud	912000	698000	31	39	304
9	IBS	774670	666494	16	-	-
10	M1Cloud	770540	610886	26	100	-
11	CorpSoft24	308000	322000	−4	90	123
12	Oxygen	281000	143000	96	-	400
13	Linxdatacenter	119084	88313	35	11	-
14	3data	117700	33000	257	35	230
15	Oblakoteka	109000	97000	12	100	1230
16	RUVDS.com	103391	64136	61	100	-
17	Caravan Aero	42745	36164	18	100	more 300
18	LanCloud	40158	26281	53	8	68

Using the developed software, we will enter the necessary data and perform calculations. The final results are presented on Fig. 2.

Table 2. 10-point rating scale for providers for five indicators.

Data ranges for five indicators				Point (gradation)	Description
Revenue for IaaS services in 2017 and 2018, thousand rubles	Revenue ratio 2018/2017 (growth), %	Revenue ratio 2018/2017 (growth), %	Number of clients (IaaS customers)		
Up to 50 million rubles	Up to 0	0–14	0–100	1	Minimum
50–100	0–10	15–24	100–250	2	The indicator is slightly above the minimum values
100–500	10–50	25–34	250–500	3	The indicator is higher than the minimum value
500–750	50–75	35–44	500–750	4	The value of the indicator is at a good level
750–1000	75–100	45–54	750–1000	5	Value above good level
1000–1250	100–150	55–64	1000–1500	6	Very good performance level
1250–1500	150–250	65–74	1500–2000	7	Excellent level
1500–2000	250–350	75–84	2000–4000	8	The indicator is lower than the maximum value
2000–2500	350–500	85–94	4000–8000	9	The indicator is slightly below the maximum values
Over 2500	Over 500	95–100	Over 8000	10	The highest score

Please note that as a result of model calculations, some positions of the vectors that were first presented to suppliers have changed. This suggests that simply ordering by score will not give an accurate result. Only after the calculations have been performed using the threshold aggregation model, one can assert the correct position in the ranking.

Selectel is the top IaaS provider in the model calculations. Its preference index is 1673. LanCloud, with a value of 34, ranks last in the ranking.

Table 3. Supplier data.

№ п/п	Provider	Indicators				
		Revenue for IaaS services in 2018, thousand rubles	Revenue for IaaS services in 2017, thousand rubles	Revenue ratio 2018/2017 (growth), %	Share of IaaS revenue from the total revenue of the enterprise, %	Number of clients (IaaS customers)
1	Softline	10	8	5	1	6
2	IT-Grad	9	3	10	8	9
3	Rostelekom	9	8	3	5	5
4	DataLine	8	7	3	2	3
5	Selectel	8	5	4	8	10
6	Krok	8	6	3	1	2
7	Servionika	5	5	3	5	1
8	OnCloud	5	4	3	4	3
9	IBS	5	4	3	5	5
10	M1Cloud	5	4	3	10	5
11	CorpSoft24	3	3	1	9	2
12	Oxygen	3	3	5	5	3
13	Linxdatacenter	3	2	3	1	5
14	3data	3	1	8	3	2
15	Oblakoteka	3	2	3	10	6
16	RUVDS.com	3	2	4	10	5
17	Caravan Aero	1	1	2	10	3
18	LanCloud	1	1	4	8	1

5 Software "Formation of the Aggregate Rating"

The threshold aggregation model is programmed in C# in Visual Studio 2019.

The program implements the following functions:

1. Accounting for data on alternatives, criteria and gradations for evaluation.
2. Calculation of parameters and number of combinations.
3. Calculation of the preference index for building an aggregate rating.
4. Formation of the rating chart.

The program provides 5 data outputs, including the final one - the preference index, on the basis of which the rating will be built.

When starting the program, you can enter the name of the project and the full name of the employee performing the rating. To get started, you must enter the number of criteria, alternatives and gradations. It is on the basis of these entered parameters that tables for data entry will be formed, and calculations will be made according to the non-compensatory model.

The program is divided into 6 tabs. After entering the parameters on the first tab "Data Entry", the values of the criteria for each alternative are entered. The alternatives themselves can be unloaded from the database or from a specially prepared file. You can also select criteria by clicking on the "Criteria" button. Filling in the database on alternatives and criteria can be unique for each enterprise, based on its specifics. The program is designed as universally as possible for any area where you can, one way or another, make ratings.

On the 2nd tab "Formation of the rating", the user is prompted to choose: "spread into gradations" or "leave unchanged". If the real values of the criteria were entered on the first tab, then it will be necessary to select "split into gradations", a clean table will be formed in which the grades will need to be posted. If expert assessments were initially entered immediately (for example, in the case of assessing quality indicators), then you must select "leave unchanged", and the entered data on the 1st tab will automatically be copied to the 2nd tab. On the 3rd tab "Threshold aggregation" intermediate calculations are made according to the rule of threshold aggregation. In the future, no input is required, only sequential pressing of the buttons that are responsible for the intermediate output of the results.

On the 4th tab "Index calculation", combinations of a and b are calculated, including for Fmax, as well as the calculation of the preference indices themselves together with normalized values (Fig. 2). Before calculating the index, you must click on the "Apply condition" button for the restriction to work $C_{-1}^{0} = 1$. When you click on the "Fmax" button, its value will be displayed, which will participate in the calculation of the normalized index. The "Generate rating" button will rank the evaluated alternatives according to the index from the smallest to the largest. The "Combination formula" button will open a form in which you can enter any values of a and b to check or simply calculate the value of the number of combinations.

On the 5th tab "Graph" you can create a rating graph for all evaluated alternatives. The x-axis shows the alternatives, the y-axis shows the value of the preference index.

On the 6th tab "Binding to the database" it is possible to save the results to the database and generate a different type of graph.

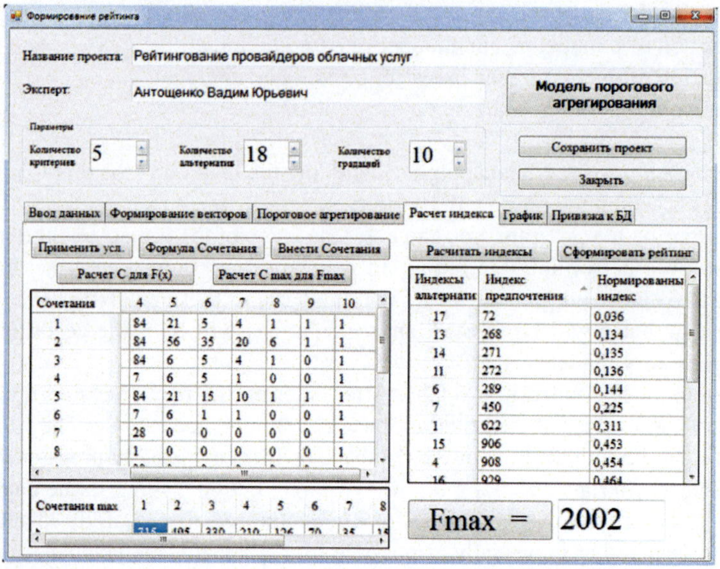

Fig. 2. Dialog box of the program "Formation of aggregated ratings", tab "Index calculation"

6 Conclusion

The article presents a non-compensatory aggregated rating building model for choosing the best provider that provides cloud services according to the IaaS model. The mathematical model is based on the rule of threshold aggregation, which is used in multicriteria decision-making problems. An algorithm (scheme) of assessment stages for non-compensatory aggregation was built and the software "Formation of an aggregate rating" was developed. The mathematical model was programmed in C# in the Visual Studio 2019 environment. Using the program, a rating of the largest IaaS providers was built, which reflects the comparative importance of various IaaS providers for the enterprise. According to the built rating, you can choose the best provider when using cloud services.

References

1. Paul, P.K., Ghose, M.K.: Cloud computing: possibilities, challenges and opportunities with special reference to its emerging need in the academic and working area of information science. Int. Conf. Modell. Optim. Comput. **38**, 2222–2227 (2012)
2. Reynolds, P., Yetton, P.: Aligning business and IT strategies in multi-business organization. J. Inf. Technol. **30**(2), 101–118 (2015)
3. Lin, W., Wang, H., Zhang, Y., et al.: A cloud server energy consumption measurement system for heterogeneous cloud environments. Inf. Sci. **468**, 47–62 (2018)
4. Razumnikov, S.V., Kurmanbay, A.K.: Models of evaluating efficiency and risks on integration of cloud-base IT-services of the machine-building enterprise: a system approach. IOP Conf. Ser. Mater. Sci. Eng. **124**(1), 012089 (2016)

5. Razumnikov, S.V., Prankevich, D.A.: Integrated model to assess cloud deployment effectiveness when developing an IT-strategy. IOP Conf. Ser. Mater. Sci. Eng. **127**(1), 012018 (2016)
6. Razumnikov, S.V.: Assessing efficiency of cloud-based services by the method of linear programming. Appl. Mech. Mater. **379**, 235–239 (2013)
7. Aleskerov, F.T., Khabina, E.L., Schwartz, D.A.: Binary Relations, Graphs and Collective Solutions, p. 344. FIZMATLIT, Moscow (2012)
8. Plekhanov, M.I.: The modern market of IT infrastructure. Int. Sci. Student J. **9**, 154–156 (2019)
9. Infrastructure as a Service, IaaS (Russian market). TAdviser. http://www.tadviser.ru/index.php/Infrastructure_as_a_Service,_IaaS. Accessed 04 Jun 2022
10. Lee, I.: Pricing and profit management models for SaaS providers and IaaS providers. J. Theor. Appl. Electron. Commer. Res. **16**(4) (49), 1–15 (2021)
11. Zheng, B., Pan, L., Liu, S.: Market-oriented online bi-objective service scheduling for pleasingly parallel jobs with variable resources in cloud environments. J. Syst. Softw. **176**, 110934 (2021)
12. Jia, R., Yang, Y., Grundy, J., et al.: A systematic review of scheduling approaches on multi-tenancy cloud platforms. Inf. Softw. Technol. **132**, 106478 (2021)
13. Tsai, W.-L.: Constructing assessment indicators for enterprises employing cloud IaaS. Asia Pac. Manag. Rev. **26**(1), 23–29 (2021)
14. Konjaang, J.K., Xu, L.: Meta-heuristic approaches for effective scheduling in infrastructure as a service cloud: a systematic review. J. Netw. Syst. Manage. **29**(2), 1–57 (2021). https://doi.org/10.1007/s10922-020-09577-2
15. Kalyagin, V.A., Chistyakov, V.V.: Axiomatic model of non-compensatory aggregation. House of the State University Higher School of Economics, p. 76 (2009)
16. Aleskerov, F.T., Kataeva, E.S., Pislyakov, V.V., Yakuba, A.I.: Evaluation of the contribution of scientists by the threshold aggregation method. Management of large systems. Special issue 44: Scientometrics and expertise in the management of science, pp. 172–189 (2013)
17. Aleskerov, F.T., Yuzbashev, D.A., Yakuba, V.I.: Threshold aggregation of three-gradation rankings. Automation and Telemechanics **1**, 147–152 (2007)
18. Aleskerov, F.A., Yakuba, V., Yuzbashev, D.: Threshold aggregation of three-graded rankings. Math. Social Sci. **53**, 106–110 (2007)
19. Aleskerov, F., Chistyakov, V.V., Kalyagin, V.A.: The threshold aggregation. Econ. lett. **107**(2), 161–162 (2010)
20. Aleskerov, F., Chistyakov, V., Kalyagin, V.: Social threshold aggregations. Soc. Choice Welfare **35**(4), 627–646 (2020)

Development and Software Implementation of the Design Documentation Verification Model Based on Association Rules

A. Kabirova(✉), I. Kulagin, and L. Emaletdinova

Kazan National Research Technical University Named After A.N. Tupolev-KAI, 10, Street Karl Marx, Kazan 420000, Russia
kabirovaaigul@mail.ru

Abstract. The article provides a method based on the construction and use of association rules. The method is used to verification the design documentation in order to determine the correct choice of design sets and related components. The algorithm of the proposed method is programmatically implemented as a subsystem of the automated information system for the projection of design documentation. The specification that enters the input of the information system is checked according to the association rules. If errors are found, a list of recommendations for finalizing the document for a specialist is formed. Following these recommendations makes it possible to reduce the complexity of checking design documentation at the stage of technical control by narrowing the set of analyzed specification sets. In order to build association rules the Apriori algorithm is used. As a result of approbation of the program module, were built 870 association rules which are used to determine incorrectly compiled specifications.

Keywords: Association rules · Design documentation · Information system · Verification model

1 Introduction

Today a lot of time is allocated for checking and identifying the incorrectness of the components of the specifications of design documentation in mechanical engineering, and this work is very laborious. Timely identification and elimination of errors in the specifications of design documentation can reduce the cost of their elimination during the construction of the facility. Therefore, the task of automating the process of identifying errors in specifications is an actual task.

The specification is related to the concept of a set. A set is understood as a list of products that accompany the main position of the document. For example, for a pipe these are flanges, gaskets and a mounting kit and so on.

The most common technical errors in the development of the specification is the incorrect choice of related elements by the specialist, that is the incorrect compilation of the set. For example, a designer can select a pipe, flanges to it, but not add gaskets or fasteners to assemble a flange connection or select these elements incorrectly.

A. A. Radionov and V. R. Gasiyarov (Eds.): RusAutoCon 2022, LNEE 986, pp. 12–22, 2023.
https://doi.org/10.1007/978-3-031-22311-2_2

Such errors are detected by a specialist in the department of technical control of design documentation, but manual verification of the document is highly laborious.

Currently, association rules are used to solve many problems in various fields [1–11]. In order to automate the process of checking the specifications of design documentation, this paper also proposes to apply the method of constructing association rules [12–21]. The use of association rules allows reducing the time of checking the specification and detecting errors in the most frequent sets that make up the specification.

2 Formulation of the Problem

Let there be a specification for the pipeline drawing, the analysis of which made it possible to identify the following transactions (Table 1):

In the future, to reduce the cumbersome presentation, we will use abbreviations instead of a long name. For example: instead of "Pipe 215 × 7 MNG5-1 GOST 17217-2018" we will use the abbreviation "Pipe". Thus, the information in Table 1 in abbreviated form will be as follows (Table 2):

Table 1. An example of transactions of the specification of design documentation

№	Transactions
1	Pipe DKRNT 258 × 5 MNG 5-1 GOST 17217-2018, Free pipe flange on a welded brass ring Du250 (555-3M.250), Gasket 310-275 Paronite VATI-TC 2,0 TU 2575-275-00149363-2006, Electrical insulating track connection unit Du250 (938-38.153-11), Bolt M16-6gx60.58.Spzc 15.xp GOST 7798-70, Screw M16-6H.5 Spzc 9.xp GOST 5915-70, Washer 16.65Г Spzc 24.xp GOST 6402-70
2	Pipe DKRNT 258x5 MNG 5-1 GOST 17217-2018, Free pipe flange on a welded brass ring Du250 (555-3M.250), Electrical insulating track connection unit Du250 (938-38.153-11)
…	…
9	Pipe DKRNT 215x7 MNG 5-1 GOST 17217-2018, Free pipe flange on a welded brass ring Du200 (555-3M.249), Gasket 249-220 Paronite VATI-TC 2,0 TU 2575-275-00149363-2006, Electrical insulating track connection unit Du 200 (938-38.153-10), Bolt M14-6gx70.58 Spzc 15.xp GOST 7798-70, Screw M14-6H.5 Spzc 9.xp GOST 5915-70, Washer 14.65Г Spzc 24.xp GOST 6402-70
10	Adapter cone

Table 2. Example of transactions with abbreviated product names

№	Transactions
1	Pipe 3, Flange 2, Gasket 2, Unit 3, Bolt 3, Screw 2, Washer 2
2	Pipe 3, Flange 2, Unit 3
3	Pipe 2, Flange 1, Gasket 1, Screw 1, Washer 1, Pin
4	Pipe 2, Flange 1, Gasket 1, Bolt 1, Screw 1, Washer 1, Pin
5	Pipe 2, Flange 1, Gasket 1, Bolt 1, Screw 1, Washer 1
6	Pipe 3, Flange 2, Gasket 2, Pipe 1, Flange 1, Bolt 4, Screw 2, Washer 2
7	Pipe 2, Flange 1, Gasket 1, Bolt 1, Bolt 2, Screw 1, Washer 1
8	Pipe 2, Flange 1, Gasket 1, Unit 1, Bolt 1, Screw 1, Washer 1
9	Pipe 2, Flange 1, Gasket 1, Unit 2, Bolt 1, Screw 1, Washer 1
10	Cone

3 Method of Constructing Association Rules

To build association rules: $A \rightarrow B$ or "If A, then B" in the work is used the Apriori algorithm as the simplest in terms of implementation and effective in application [22]. The method of searching of association rules according to this algorithm is to use "frequent sets" and consists of two stages:

1. Finding frequent subject sets.
2. Generation of association rules that satisfy the conditions of the minimum support threshold and reliability.

Stage 1. Finding frequent subject sets.
The algorithm for finding frequent subject sets is as follows:

1. The threshold α is set as the frequency (number) of occurrence of the subject set.
2. Transactions are presented in a formalized form of a table, where 1 is an indicator that this item is present in the transaction, 0 is not present.
3. The values in each line are summed up, the frequency (number) of the appearance of each item is obtained. Frequent one-subject sets F_1 are formed, for which the frequency is not less than the size of the threshold α.
4. Based on the one-subject set F_1, two-subject sets are formed by linking F_1 to itself. The frequency of each of the two-subject sets is calculated and frequent two-subject sets F_2 are formed, for which the frequency is not less than the size of the threshold α.
5. Given the number $k = 3$ subject sets.
6. Formation of k-subject sets:
7. To do this, a set of candidates F_k in k-subject sets is created by linking the set F_{k-1} with itself. If it is impossible to link subject sets, then go to step 8. Otherwise, F_k

is reduced in accordance with the antimonotonicity property, that is, if the set F_{k-1} is not frequent, then the set F_k will also not be frequent. The reduced set gives the final k-subject sets. Go to step 7.

8. $k = k + 1$. Repeated step 6.
9. To complete.

Stage 2. Generation of association rules.
For each frequent subject set F_k we apply a two-step procedure:

1. All possible subsets F_{k-1} of the set F_k are generated.
2. The association R: $F_{k-1} \rightarrow (F_k - F_{k-1})$ is considered, where $(F_k - F_{k-1})$ is a set of F_k without a subset of
3. F_{k-1}. An association R is considered an association rule if it satisfies the specified minimum support and/or reliability threshold.
4. Steps 1–2 are repeated for each subset F_{k-1} from the set F_k.

The support of an association rule is the frequency of occurrence of both the condition and the consequence of the association rule in the total number of transactions: $P(A \cap B)$.

The reliability of the association rule $A \rightarrow B$ is the frequency of the occurrence of the consequence B, provided that the "condition A" takes place: $P(B/A)$.

The significance of association rules is assessed using the Leverage indicator: $P(A \cap B) - P(A) \cdot P(B)$ and Lift indicator: $\frac{P(B/A)}{P(B)}$.

If the condition and the consequence are independent, then the Leverage is close to 0. This means that although the condition and the effect often occur together, they occur just as often separately, they are just popular.

If the Lift value of transaction $A \rightarrow B$ is greater than 1, then condition A appears more often in transactions containing consequence B than in others [22].

4 Example

In this example, due to the small number of transactions under consideration, we will consider a set to be frequent if it occurs in 30% or more transactions, that is, the minimum threshold for the considered example is $\alpha = 3$.

Let's present transactions in a formalized form (Table 3).

Table 3 shows that eight of the twenty unique items will be included in frequent one-subject sets F_1:

$$F_1 = \{\text{Bolt 1; Screw 1; Gasket 1; Pipe 2; Pipe 3; Flange 1;Flange 2; Washer 1}\}$$

Based on one-subject sets F_1 we build two-subject sets by connecting one - subject sets to each other and also determine the frequency of their occurrence α in transactions (Table 4).

As a result of the analysis of the frequency of occurrence of two-subject sets from Table 4, it was determined that 12 out of 28 two-subject sets will not be included in the F_2 set, since their frequency is less than the established minimum threshold of 3. Next, we use the set F_2 to form the set F_3- 3 subject sets. To do this, we link sets F_2 to each other if they have the first items in common (Table 5).

The set F_3 is analyzed from the point of view of antimonotonicity. For each three-subject set from the set F_3 subsets of size $k - 1 = 3 - 1 = 2$ are created and checked. Analysis of Table 5 showed that all three-subject sets are frequent. Further, according to the same principle, four and five subject sets F_4 and F_5 were formed. F_6 - 6 subject set was also formed (Table 6).

On this, the problem of searching of frequent subject sets on the initial set of transactions is solved. We pass to the second stage of the algorithm.

Let's build associations that contain one element in the condition. To do this, we apply the described method of generating association rules to the set F_2 (Table 7).

Table 3. Frequency of occurrence of one-item sets in transactions

№	Product names	Transaction number										Frequency α
		1	2	3	4	5	6	7	8	9	10	
1	Bolt 1	0	0	0	1	1	0	1	1	1	0	5
2	Bolt 2	0	0	0	0	0	0	1	0	0	0	1
3	Bolt 3	1	0	0	0	0	0	0	0	0	0	1
4	Bolt 4	0	0	0	0	0	1	0	0	0	0	1
5	Screw 1	0	0	1	1	1	0	1	1	1	0	6
6	Screw 2	1	0	0	0	0	1	0	0	0	0	2
7	Cone	0	0	0	0	0	0	0	0	0	1	1
8	Gasket 1	0	0	1	1	1	0	1	1	1	0	6
9	Gasket 2	1	0	0	0	0	1	0	0	0	0	2
10	Pipe 1	0	0	0	0	0	1	0	0	0	0	1
11	Pipe 2	0	0	1	1	1	0	1	1	1	0	6
12	Pipe 3	1	1	0	0	0	1	0	0	0	0	3
13	Unit 1	0	0	0	0	0	0	0	1	0	0	1
14	Unit 2	0	0	0	0	0	0	0	0	1	0	1
15	Unit 3	1	1	0	0	0	0	0	0	0	0	2
16	Flange 1	0	0	1	1	1	1	1	1	1	0	7
17	Flange 2	1	1	0	0	0	1	0	0	0	0	3
18	Washer1	0	0	1	1	1	0	1	1	1	0	6
19	Washer2	1	0	0	0	0	1	0	0	0	0	2
20	Pin	0	0	1	1	0	0	0	0	0	0	2

Table 4. Frequency of occurrence of two- subject sets in transactions

№	Two-subject sets	α	№	Two-subject sets	α	№	Two-subject sets	α
1	Bolt 1, Screw 1	5	10	Screw 1, Pipe 3	0	19	Pipe 2, Pipe 3	0
2	Bolt 1, Gasket 1	5	11	Screw 1, Flange 1	6	20	Pipe 2, Flange 1	6
3	Bolt 1, Pipe 2	5	12	Screw 1, Flange 2	0	21	Pipe 2, Flange 2	0
4	Bolt 1, Pipe 3	0	13	Screw 1, Washer 1	6	22	Pipe 2, Washer 1	6
5	Bolt 1, Flange 1	5	14	Gasket 1, Pipe 2	4	23	Pipe 3, Flange 1	1
6	Bolt 1, Flange 2	0	15	Gasket 1, Pipe 3	0	24	Pipe 3, Flange 2	3
7	Bolt 1, Washer 1	5	16	Gasket 1, Flange 1	6	25	Pipe 3, Washer 1	0
8	Screw 1, Gasket 1	6	17	Gasket 1, Flange 2	0	26	Flange 1, Flange 2	1
9	Screw 1, Pipe 2	6	18	Gasket 1, Washer 1	6	27	Flange 1, Washer 1	6
						28	Flange 2, Washer 1	0

Table 5. Three-subject sets in transactions

№	Three – subject sets	№	Three – subject sets	№	Three – subject sets
1	Bolt 1, Screw 1, Gasket 1	8	Bolt 1, Pipe 2, Flange 1	15	Screw 1, Pipe 2, Washer 1
2	Bolt 1, Screw 1, Pipe 2	9	Bolt 1, Pipe 2, Washer 1	16	Screw 1, Flange 1, Washer 1
3	Bolt 1, Screw 1, Flange 1	10	Bolt 1, Flange 1, Washer 1	17	Gasket 1, Pipe 2, Flange 1
4	Bolt 1, Screw 1, Washer 1	11	Screw 1, Gasket 1, Pipe 2	18	Gasket 1, Pipe 2, Washer 1
5	Bolt 1, Gasket 1, Pipe 2	12	Screw 1, Gasket 1, Flange 1	19	Gasket 1, Flange 1, Washer 1
6	Bolt 1, Gasket 1, Flange 1	13	Screw 1, Gasket 1, Washer 1	20	Pipe 2, Flange 1, Washer 1
7	Bolt 1, Gasket 1, Washer 1	14	Screw 1, Pipe 2, Flange 1		

Table 6. Six-subject set in transactions

№	Six-subject set
1	Bolt 1, Screw 1, Gasket 1, Pipe 2, Flange 1, Washer 1

Based on the set F_3 we build associations that contain two elements in the condition (Table 8):

Table 7. Examples of associations with one element in the condition and their indicators

№	Associations	Support (%)	Reliability (%)	Leverage	Lift
1	If {Bolt 1}, then {Screw 1}	5/10 = 50	5/5 = 100	5/10–(0*1/10) = 0,5	(5/5)/(6/10) = 1,66
2	If {Screw 1}, then {Bolt 1}	5/10 = 50	5/6 = 83	5/10–(1/10*0) = 0,5	(5/6)/(5/10) = 1,66
3	If {Bolt 1}, then {Gasket 1}	5/10 = 50	5/5 = 100	5/10–(0*1/10) = 0,5	(5/5)/(6/10) = 1,66
…	…	…	…		
29	If {Pipe 3}, then {Flange 2}	3/10 = 30	3/3 = 100	3/10–(0*0) = 0,3	(3/3)/(3/10) = 3,33
30	If {Flange 2}, then {Pipe 3}	3/10 = 30	3/3 = 100	3/10–(0*0) = 0,3	(3/3)/(3/10) = 3,33
31	If {Flange 1}, then {Washer 1}	6/10 = 60	6/7 = 85	6/10–(1/10*0) = 0,6	(6/7)/(6/10) = 1,42
32	If {Washer 1}, then {Flange1}	6/10 = 60	6/6 = 100	6/10–(0*1/10) = 0,6	(6/6)/(7/10) = 1,42

Table 8. Examples of associations with two elements in the condition and their indicators

№	Associations	Support (%)	Reliability (%)	Leverage	Lift
1	If {Bolt 1, Screw 1}, then {Gasket 1}	5/10 = 50	5/5 = 100	5/10–(0*0) = 0,5	(5/5)/(6/10) = 1,66
2	If {Bolt 1, Gasket 1}, then {Screw 1}	5/10 = 50	5/5 = 100	5/10–(0*0) = 0,5	(5/5)/(6/10) = 1,66
3	If {Screw 1, Gasket 1}, then {Bolt 1}	5/10 = 50	5/6 = 83	5/10–(0*0) = 0,5	(5/6)/(5/10) = 1,66
…	…	…	…		
57	If {Flange1, Washer 1}, then {Gasket 1}	6/10 = 60	6/6 = 100	6/10–(0*0) = 0,6	(6/6)/(6/10) = 1,66
58	If {Pipe 2, Flange1}, then {Washer 1}	6/10 = 60	6/6 = 100	6/10–(0*0) = 0,6	(6/6)/(6/10) = 1,66
59	If {Pipe 2, Washer 1}, then {Flange1}	6/10 = 60	6/6 = 100	6/10–(0*0) = 0,6	(6/6)/(7/10) = 1,42
60	If {Flange1, Washer 1}, then { Pipe 2}	6/10 = 60	6/6 = 100	6/10–(0*0) = 0,6	(6/6)/(6/10) = 1,66

By analogy based on the sets F_4 and F_5 we build associations that contain, respectively, three elements and four elements in the condition. Based on the set F_6, we build associations that contain five elements in the condition (Table 9):

Table 9. Examples of associations with five elements in the condition and their indicators

№	Associations	Support (%)	Reliability (%)	Leverage	Lift
1	If {Bolt 1, Screw 1, Gasket 1, Pipe 2, Flange1}, then {Washer 1}	5/10 = 50	5/5 = 100	5/10–0 = 0,5	(5/5)/(6/10) = 1,66
2	If {Bolt 1, Screw 1, Gasket 1, Pipe 2, Washer 1}, then {Flange1}	5/10 = 50	5/5 = 100	5/10–0 = 0,5	(5/5)/(7/10) = 1,42
3	If {Bolt 1, Screw 1, Gasket 1, Flange1, Washer 1}, then {Pipe 2}	5/10 = 50	5/5 = 100	5/10–0 = 0,5	(5/5)/(6/10) = 1,66
4	If {Bolt 1, Screw 1, Pipe 2, Flange1, Washer 1}, then {Gasket 1}	5/10 = 50	5/5 = 100	5/10–0 = 0,5	(5/5)/(6/10) = 1,66
5	If {Bolt 1, Gasket 1, Pipe 2, Flange1, Washer 1}, then {Screw 1}	5/10 = 50	5/5 = 100	5/10–0 = 0,5	(5/5)/(6/10) = 1,66
6	If {Screw 1, Gasket 1, Pipe 2, Flange1, Washer 1}, then {Bolt 1}	5/10 = 50	5/6 = 83	5/10−(0*1/10) = 0,5	(5/6)/(5/10) = 1,66

Thus, when considering 10 transactions, 188 associations were built. Of these 93 associations were included in the number of association rules, the support of which is at least 50%, the reliability is at least 80% and the leverage is at least 0,3, the lift is greater 1.

5 Results

Verification of specifications based on the formulated association rules is as follows:

1. Select transactions and determine their number L. Set transaction number l = 1.
2. Form a set a^*, consisting of one element of the transaction l.
3. $k = 1$.
4. Consider rules whose condition contains k elements. If there is a rule containing in the condition a^* and in the conclusion the element $b^* \in \{b_1, b_2, ..., b_n\}$ then go to step 5. Otherwise, go to step 7.

5. Combine the elements of the condition and conclusion: $\tilde{a} = \{a^*, b^*\}$, where $b^* \in \{b_1, b_2, ..., b_n\}$ and check for the presence of this set in transaction l. If the set is present then go to step 6. Otherwise, issue a recommendation to the specialist (in the AIS, the transaction is highlighted in red) about the need for additional verification of this transaction for correctness.
6. $k = k + 1$, $a^* = \tilde{a}$. Go to step 4.
7. $l = l + 1$. If $l \le L$, go to step 2, otherwise go to step 8.
8. To complete.

The algorithm of the proposed method is implemented as a software module of an automated information system for the projection of the design documentation "Mercury" for JSC "Zelenodolskoye PKB". The result of applying the algorithm to one of the specifications is shown in the Fig. 1.

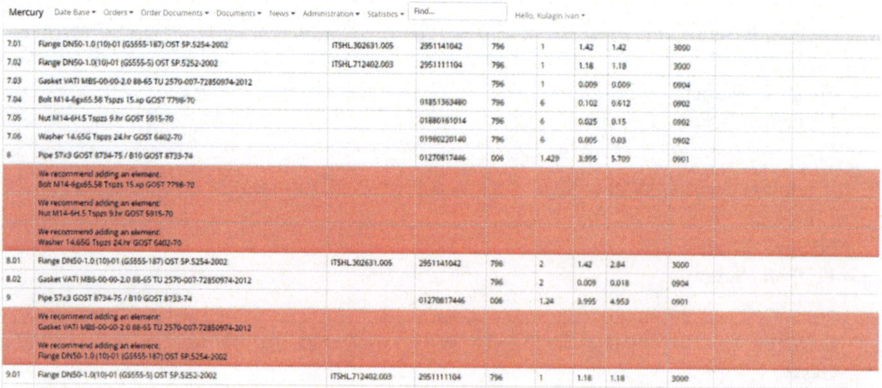

Fig. 1. The result of applying the method.

6 Conclusion

The results of the study showed the effectiveness of using association rules to verification design documentation for the correct choice of design sets and related components. If there are errors in the specification, then a list of recommendations for finalizing the document is formed for the specialist. Following these recommendations makes it possible to reduce the complexity of checking design documentation at the stage of technical control by reduction the set of analyzed specification sets.

This approach is not used to check typical transactions, since their correctness can be directly programmed in the software package.

References

1. Ordonez, C., Ezquerra, N., Santana, C.A.: Constraining and summarizing association rules in medical data. Knowl. Inf. Syst. **9**, 1–2 (2006)

2. Chiu, S.H., Chen, C.C., Yuan, G.F., et al.: Association algorithm to mine the rules that govern enzyme definition and to classify protein sequences. BMC Bioinformatics **7**, 304 (2006)
3. Carmona-Saez, P., Chagoyen, M., Rodriguez, A., Trelles, O., Carazo, J., Pascual-Montano, A.: Integrated analysis of gene expression by association rules discovery. BMC Bioinformatics **7**, 54 (2006)
4. Vinodh, S., Prakash, N.H., Selvan, K.E.: Evaluation of leanness using fuzzy association rules mining. Int. J. Adv. Manuf. Technol. **57**, 343–352 (2011)
5. Rizaev, I.S., Charnin, L.M., Yakhina, Z.T.: The search for regularities between zimovalnymi events based on association rules. Bulletin of the Kazan State Technical University named after A.N Tupolev **4**, 141–145 (2014)
6. Suchacka, G., Chodak, G.: Using association rules to assess purchase probability in online stores. IseB **15**(3), 751–780 (2016). https://doi.org/10.1007/s10257-016-0329-4
7. Zhou, H., Hirasawa, K.: Evolving temporal association rules in recommender system. Neural Comput. Appl. **31**(7), 2605–2619 (2017). https://doi.org/10.1007/s00521-017-3217-z
8. Bova, V.V., Kuliev, E.V., Shcheglov, S.N.: Estimating the effectiveness of the method for searching the associative rules for the tasks of processing big data. Izvestiya SFEDU. Engineering Sciences **2**(212), 66–78 (2020)
9. Obidallah, W.J., Raahemi, B., Ruhi, U.: Clustering and association rules for web service discovery and recommendation: a systematic literature review. SN Computer Science **1**(1), 1–33 (2019). https://doi.org/10.1007/s42979-019-0026-8
10. Peng, M., Lee, S., D'Souza, A.G., et al.: Development and validation of data quality rules in administrative health data using association rule mining. BMC Med. Inform. Decis. Mak. **20**, 75 (2020)
11. Mirhashemi, S.H., Mirzaei, F.: Using combined clustering algorithms and association rules for better management of the amount of water delivered to the irrigation network of Abyek Plain. Iran. Neural Comput. Applic. **34**, 3875–3883 (2022)
12. Zaki, M.J., Parthasarathy, S., Ogihara, M., et al.: Parallel algorithms for discovery of association rules. Data Min. Knowl. Disc. **1**, 343–373 (1997)
13. Coenen, F., Goulbourne, G., Leng, P.: Tree structures for mining association rules. Data Min. Knowl. Disc. **8**, 25–51 (2004)
14. Hämäläinen, W.: StatApriori: an efficient algorithm for searching statistically significant association rules. Knowl. Inf. Syst. **23**, 373–399 (2010)
15. Luna, J.M., Romero, J.R., Ventura, S.: Design and behavior study of a grammar-guided genetic programming algorithm for mining association rules. Knowl. Inf. Syst. **32**, 53–76 (2012)
16. Agrawal, J., Agrawal, S., Singhai, A., Sharma, S.: SET-PSO-based approach for mining positive and negative association rules. Knowl. Inf. Syst. **45**(2), 453–471 (2014). https://doi.org/10.1007/s10115-014-0795-2
17. Luna, J.M., Romero, C., Romero, J.R., Ventura, S.: An evolutionary algorithm for the discovery of rare class association rules in learning management systems. Appl. Intell. **42**(3), 501–513 (2014). https://doi.org/10.1007/s10489-014-0603-4
18. Oliinyk, A.A., Subbotin, S.A.: A stochastic approach for association rule extraction. Pattern Recognit. Image Anal. **26**(2), 419–426 (2016). https://doi.org/10.1134/S1054661816020139
19. Palmov, S.V.: Review of search algorithms for associative rules. Science and Business: Ways of Development **10**(64), 77–80 (2016)
20. Ait-Mlouk, A., Gharnati, F., Agouti, T.: An improved approach for association rule mining using a multi-criteria decision support system: a case study in road safety. Eur. Transp. Res. Rev. **9**(3), 1–13 (2017). https://doi.org/10.1007/s12544-017-0257-5

21. Olyanich, I.A.: Comparison of algorithms of construction of associative rules on the basis of the data set of customer transactions. News of the Samara Research Center of the Russian Academy of Sciences **6–2**(86), 379–382 (2018)
22. Paklin, N.B., Oreshkov, V.I.: Business Analytics: From Data to Knowledge (+CD): Textbook, 2nd edn. revised and supplemented. Peter, St. Petersburg, p. 704 (2010)

The Theory of Corporate Information Systems

D. Yu. Stepanov(✉)

MIREA – Russian Technological University, 78, Vernadsky Prospekti, Moscow 119454, Russia
mail@stepanovd.com

Abstract. The paper introduces a new discipline that combines the design of information systems, databases, programming, as well as project and change management, called the theory of corporate information systems. The content of each area correlates with the levels of software implementation: processes, data, application, technology as well as the project and changes. The theory assumes the formalization of strategies for analysis, design, roles and authorization, technical and business cutover, change management, build, data migration, training, testing and post go-live support, each of which is decomposed into subtasks, consistent with other activities, and then planned for execution. Planning plays a key role during the implementation of corporate information systems, since most of the project tasks involve receiving feedback from future users of the system. Thus, the proposed theory combines the existing technical disciplines and expands the scope of their application to larger-scale IT projects.

Keywords: ERP-systems · Corporate information systems · Implementation

1 Introduction

It is impossible to imagine our today's life without the use of modern information technologies. Smartphones and voice assistants, the Internet of Things and data mining, online cinemas and virtual reality [1]. The situation is similar in the workplace: we are surrounded by the industrial Internet of Things, robotics and predictive analytics, as well as many various automation tools [2]. Thus, digitalization affects almost all areas of our life. The unifying link in all these technologies is the need for analysis, development and implementation of software. The latter, depending on the application area and the scale of implementation, are characterized by the generally accepted terms information systems and corporate information systems [3]. Despite the fact that both definitions, in principle, say the same thing, the process of implementing these systems varies greatly. Ways, methods and approaches for deployment of information systems by default are trying to be applied to implement a more complex corporate information systems, which is not always applicable, rational and effective [4]. The practice of implementing software solutions related to ERP, ERP2 or ERP3 standards [5–7] confirms the existing knowledge gap in the theory of deployment of corporate information systems, which will be discussed in this paper.

Everyone who is interested in the field of information systems and technologies has the opportunity to easily find literature to their liking. Here we can highlight the works

A. A. Radionov and V. R. Gasiyarov (Eds.): RusAutoCon 2022, LNEE 986, pp. 23–34, 2023.
https://doi.org/10.1007/978-3-031-22311-2_3

devoted to software development [8, 9], information systems design [3, 10–13], as well as project management for the implementation of a similar class of software systems [14, 15]. The works [8, 9] are rightfully considered classics, thoroughly describing the ways, methods and algorithms of building software development. However, the issues of the implementation of big projects are practically not covered here. A greater emphasis on the design of software systems is made in [3, 11–13], where implementation methodologies, methods of modeling business processes and organizational structures, as well as typical realization problems are demonstrated. In these works, information and corporate systems are represented by synonyms. Finally, the monographs [14, 15] provide a foundation for managing any implementation projects, regardless of their subject orientation. Summarizing all these, we can say following: information and corporate systems are considered in most literature sources as software products of equal scale both in terms of their realization and implementation.

The purpose of this paper is to propose a theory of corporate information systems that demonstrates the distinctive features of the analysis, development and deployment of corporate systems as opposed to information systems. Consideration of the differences will allow us to approach the implementation of corporate information systems more rationally and ensure a more likely success of the project.

2 Computer Programs, Information Systems and Corporate Information Systems

Depending on the coverage of the business processes of the enterprise by the programs, several types of software can be distinguished (Fig. 1). The first of which is a minimal unit: a computer program. A computer program is a set of data and ordered commands designed to obtain a certain result by a computer [16]. Application development requires structural thinking, which makes it possible to present the logic of the program in the form of a sequence of operations, i.e. an algorithm, on information. Thus, this class of software almost always involves data processing and consists of three elements:

- ordered commands;
- structured data;
- algorithms.

It should be clarified, despite the fact that data is mentioned in the definition of a computer program, they are mainly having a secondary role: considered as input parameters. Examples of this class of programs are applications ABBYY Lingvo, FineReader, etc. Thus, computer programs allow you to solve single application tasks only.

A set of software that defines the only problem domain and ensure its automation is called an information system (hereinafter – IS) [11]. Quite often in literary, IS is understood as a communication system for collecting, transmitting and processing information about an object that supplies employees with the data necessary to implement the management function [3]. Examples of such systems include 1C: Accounting and BOSS-HR, covering, respectively, the financial and human processes of the enterprise, which will be called automation standards later. The classical approach to the implementation of projects for such class of software implies execution of tasks in the following areas [3]:

- processes;
- application;
- data;
- technology,

which in [17] are called implementation levels. At the process level, the organizational structure and business processes of the enterprise are designed in AS-IS and TO-BE models, which allows a more balanced approach to identify requirements and bottlenecks in the company's work. The determined business processes are implemented in the IS at the application level, which requires software development for particular problem area. The processes reflected in the IS are continuously connected with the creation, modification and deletion of data, for which corresponding level is allocated. The data level within the IS is about tasks of data control and normalization, which requires the deployment of database management systems (hereinafter – DBMS). And, finally, the technical level at which the developed IS and implemented DBMS «live» and successfully functioning. Using levels of software implementation, it is possible to formulate following statement characterizing the IS.

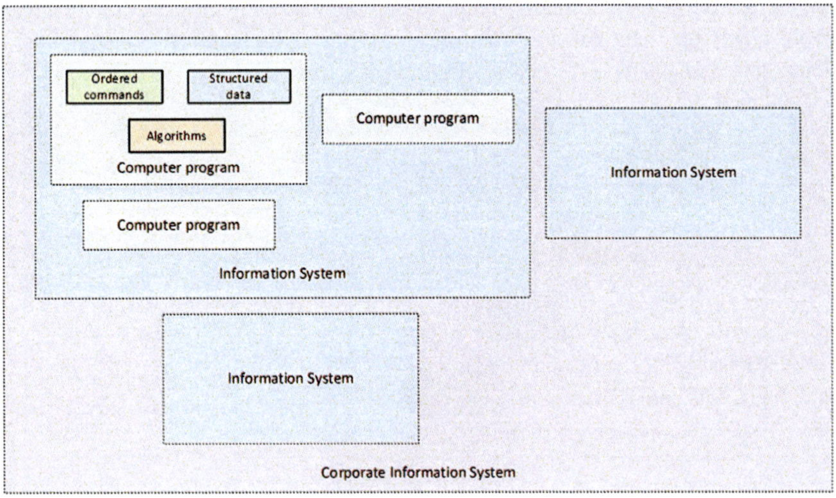

Fig. 1. Differences between computer programs, information systems and corporate information systems.

Statement 1. In information systems implementation projects, much attention is paid both to the development of software that automates a certain subject area of the company, and to processes, data and technology, without which successful deployment of the solution is impossible. Thus, increasing the number of programs to be developed leads increasing additional project tasks indirectly related to software.

The development of information systems provides automation of the only subject area of the company. And what happens if we try to process several problem areas at once? In this case, we are talking about corporate information systems (hereinafter – CIS).

In [18], the corporate information system is defined as a set of methods and solutions for creating a unified information space for management and ensuring the company's activities. In short, the CIS is a set of IS combined together [17]. Examples of such systems are software solutions SAP ERP, Oracle EBS, 1C: ERP, etc. Then it is natural that corporate systems have the same implementation levels that are used in IS projects, is not it? This is true, however with a few comments: an increase in the number of IS creates the need for better management of project activities on the part of both the supplier and the customer, moreover data priorities are also changing – more attention is paid not to control, but to migration [19].

Statement 2. The implementation levels of corporate systems in comparison with information systems are complemented by the tasks of project management and changes, moreover, the data level is used mainly to the issues of migrating information from legacy to the target software system.

The change level makes it possible to ensure a smooth transition from the previous software solution to the one being developed, to prepare the company and its employees for the use of a new software, as well as to consolidate both technical and organizational changes in order to obtain significant advantages over competitors [20]. At the project level, planning, execution, control and adjustment of activities of all other levels are carried out, which is mainly regulated by the international project management body of knowledge (hereinafter – PMBoK) [15]. Figure 2 clearly demonstrates all six implementation levels of corporate information systems.

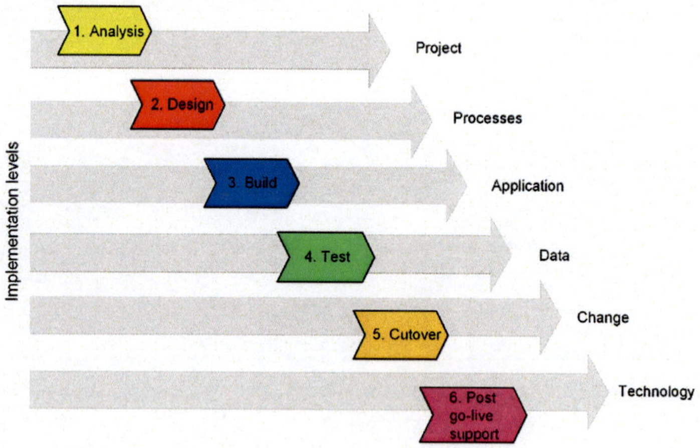

Fig. 2. Implementation levels of corporate information systems.

3 The Theory of Corporate Information Systems

Unlike user programs and IS, CIS implementation projects are characterized by a long duration and significant efforts due to the following distinctive features:

- mandatory adherence to the approved methodology for the implementation of software products, which sets the order and composition of tasks;
- planning of all works, as well as their further control and correction;
- the relationship of almost all activities performed and the need for their regular comparison and alignment;
- big volume of the tasks being implemented and the mass involvement of technical and business specialists;
- formalization of all performed tasks, as well as documentation of the results,

where the first four points correspond to the project level, and the last one is relevant to all other implementation levels.

Let's consider the development and implementation of CIS, starting from the project level. The fundamental standard regulating the list of tasks at this level is PMBoK [15], which identifies ten control parameters shown in Fig. 3. Since PMBoK is used for all kinds of projects, not limited only by implementation of software, we will clarify it for the purposes of CIS, then:

- integration will mean the charter of the project, which sets, among other things, the implementation methodology for CIS;
- resources and cost will determine the resource plan and the project cost plan calculated on its basis;
- timeline will set the project schedule aligned with the resource plan.

The meaning of the parameters of stakeholders, communications, risks, quality and supplies stay unchanged. The remaining parameter characterizing the content requires more in-depth study to adapt it to the CIS deployment projects.

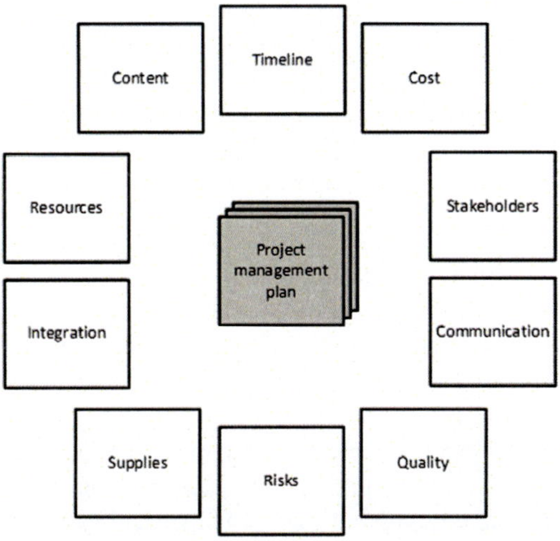

Fig. 3. Parameters of the project level.

Despite the fact that statement 2 introduces additional groups of tasks that distinguish CIS implementation projects from IS, their description remains unclear. Having analyzed the typical difficulties that technical specialists face in practice when implementing ERP-systems [17, 21], it is possible to identify extra activities inherent in CIS projects: requirements analysis, roles and authorization, business cutover and post go-live support. We have already noted earlier that CIS implementation projects are characterized by the formalization of all tasks performed. In practice, this means that any activity executed is a planned work related to a specific strategy for delivering CIS project. Now we can define the project parameter responsible for the content with the following statement.

Statement 3. The content of the CIS implementation project is set by the strategies of analysis, design, roles and authorization, changes, build, technical cutover, migration, training, testing, business cutover and post go-live support related to the implementation levels of processes, applications, data, technology and changes (Fig. 4).

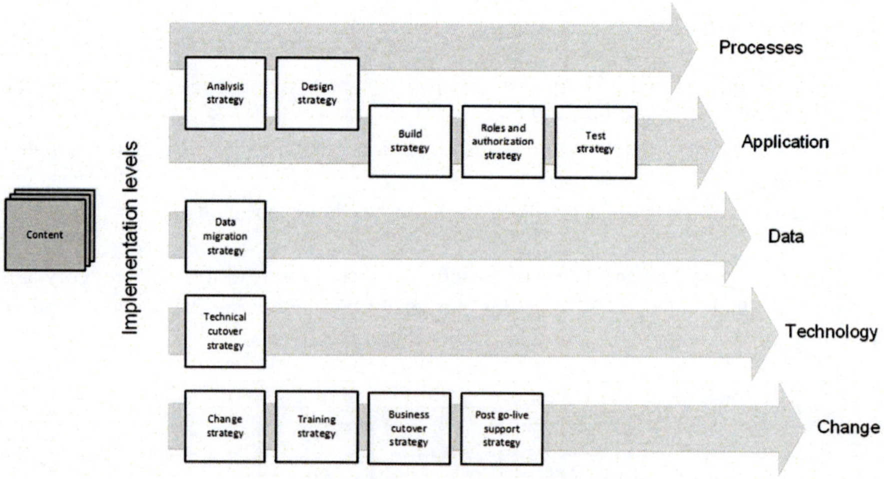

Fig. 4. Strategies per implementation levels.

Statement 3 allows us to clarify the content parameter related to the project level during the implementation of CIS (Fig. 3). Thus, we have all the information to summarize the distinctive features of the implementation of corporate information systems.

Definition 1. *The theory of corporate information systems* is an interdisciplinary field of knowledge describing deployment of corporate information systems based on implementation levels presented by processes, applications, data, technology, as well as project and changes, which in turn are characterized by strategies for analyzing requirements, designing processes and organizational structures, maintaining roles and authorizations, building programs, technical and business cutover, data migration, user training, software testing and post go-live support, as well as change management.

4 Application of the Theory in Real ERP Projects

Let's apply the theory of corporate information systems to real projects for the implementation of ERP-systems. The initial result of applying the theory is prepared strategies for managing and delivering the project content. The generated strategy documents are further detailed to specific steps, indicating the start and finish dates, as well as a set of documents to be prepared. Table 1 contains a list of key issues addressed in each of the strategies. The most typical solutions to the raised questions in practice are:

- a common analysis strategy is one in which the requirements are identified during the presenting of the demo version of the software system, and the estimator mechanism is used to calculate the efforts of development (allows for a pair of «type of development / configuration – complexity» to select the labor costs of preparing project documents and developing a program) [22];
- business process design involves creation of a business process map up to the 3rd level of detail, which makes it easier to identify requirements. ARIS VACD is often used as graphical notation for business processes at the top levels, SLD-based notation (Swim Lane Diagram) is used at the low levels, while detailing is carried out up to 4–5 levels [23];
- following the strategy of roles and authorization, usually a lot of business roles can be assigned to the end user, since in this case the creation and configuration of technical roles in the software system is greatly simplified [24];
- within technical cutover, the sandbox system is configured as an independent software subsystem, which provides the possibility of modifying critical functions and eliminates the impact on the CIS. Usually, separate copies of the quality control environment of CIS are created for all types of testing and migrations. The number of test technical cutovers mainly correlates with the number of tests and is equal to three [25];
- the change management strategy implies the choice of enterprise parameters to control their changes. When implementing CIS, usually only three parameters are considered: technologies, processes and people, they are enough to provide training for employees, updating job descriptions and work regulations [20];
- the naming convention provides a unified approach to the creation and naming of new program entities in CIS. Quality control of software products, in particular, the use of constant variables, mandatory indication of organizational levels at the program selection screens and verification of user credentials, allows you to pre-exclude typical algorithmic errors. The best practice for the implementation of corporate information systems implies the formation of such a build strategy, which confirms the use of the naming convention, as well as quality control procedures [26];
- a common practice is the allocation of a separate team responsible for the migration of master data, thereby indicating a functional organizational structure of the migration team. The number of test migrations is aligned with the number of tests: unit, integration and acceptance, thus three waves of migration. Each test migration requires loading of a certain amount of real data, the value of which is mainly close to 100%. Unlike transactional data, the master ones do not change so regularly, so they are often migrated long before go-live [19];

Table 1. Implementation strategies per level.

Level	Strategy	Parameter	Possible values
Processes, Application	Analysis	A way to identify requirements	Prototyping, demonstration of the system
		Requirements assessment method	Expert assessment, Estimator
Processes, Application	Design	The need to create a business process map	Yes, no
		Top-level design notation	ARIS VACD, IDEF0
		Low-level design notation	ARIS eEPC, BPMN SLD
		Depth of low-level description	3–5 levels
Application	Roles and authorization	The number of roles to be assigned to user	1 or more
Technology	Technical cutover	Type of sandbox system	Independent subsystem, dependent environment
		Number of copies of the quality control environment	1–3
		Number of test technical cutover	1–3
Change	Change	The need to assess of changes in technologies, processes and people	Yes, no
Application	Build	The need to use naming convention for technical objects	Yes, no
		Use of program quality control procedures	Yes, no
Data	Migration	Organizational structure of the migration team	Functional, matrix
		Number of test migrations	1–3
		%-data uploads for test migration waves	Meaning %-uploads
		The need for early migration of master data	Yes, no

(*continued*)

Table 1. *(continued)*

Level	Strategy	Parameter	Possible values
Change	Training	Type of training materials	Operational or process, technical or business
		Type of listeners	Key users, end users
		Training method	By the project team, by the key users
Application	Testing	Types of testing	Unit, integration, acceptance, stress, regression
		Criteria for successful completion of testing	%-passed test scenarios, the number of open critical defects
Change	Business cutover	The need for a company blackout rehearsal	Yes, no
Change	Post go-live support	The level of support at which the project team will work	1–3
		Criteria for the completion of productive support	Number of open critical defects

- the training strategy should ensure maximum reduction of labor costs. Therefore, not user instructions are being prepared, but training materials containing a technical description of the operations performed. The key users of the customer are trained so that they are the ones who will train the end users in the future [27];
- in the context of the testing strategy, unit, integration and end-to-end acceptance testing are most often carried out. The criteria for successful completion of tests are the number of open critical defects tending to zero or 1–5 according to the results of the acceptance test [27];

- in the process of working out a business cutover, the shutdown of the enterprise during blackout period is often emulated;
- post go-live support is organized as follows. All the questions that arise from end users go through key users. If key user cannot give an answer, then a defect is registered. Usually support service is organized the way that representatives of the customer's Help desk work on the 1st line. The project team is working only at the 2nd line of support, when defects have already been registered and allocated. The main criteria for the completion of support is the absence of critical defects during last 1–3 days.

As you can see from the table above, each strategy defines a group of tasks that are interconnected with all the others. That is why project schedule and resource plan are based on formed strategies, from which it is possible to understand the efforts. Thus, any project plan is a set of work plans for corresponding delivery strategies. The effect of the proposed theory does not end there, further, according to the Deming PDCA cycle, each of the tasks must be performed, it's result must be controlled, and also corrected in case of variances, which is done at the project level.

5 Conclusions

So, what prompted the author to write this article? After analyzing the literature, the following reasons can be distinguished:

- there are a large number of technical disciplines [3, 11–15] devoted to the design and development of information systems. Despite this, when you faced in real life with a project for the implementation of a corporate information system, it is not always clear and obvious how to use the knowledge gained;
- a detailed immersion in corporate systems implementation projects reveals a lack of knowledge, since the emphasis in them is on completely different tasks than described in textbooks and most articles. Moreover, the question of integration and alignment of all project activities remains open, because the implementation project has a single goal, regardless of its content;
- each project is unique due to the specificity of the customer's enterprise, which sets the focus and priorities of project tasks. The experience and knowledge gained on one project cannot simply be copied and applied to another. This means that there is no single universal formula for the implementation of corporate systems, some adaptation of the approach to implementation is always required.

Further, a more detailed analysis of the implementation levels of information systems was carried out [3], the results are as follows. If we look at the subject of databases in more detail, it is easy to find that the main focus is on creating balanced data tables, building the right relationships, as well as shaping SQL-queries [28]. But the situation is different in corporate systems implementation projects: the described tasks are typical and have already been implemented in most information systems. For the project, the activities of harmonization, extracting, transformation, loading and validation of data from the legacy to the target software system are much more important, which is practically not covered in the literature. A similar situation is observed with the design of business processes. Graphic notations are given in almost every textbook on information systems, which is explained by the need to identify bottle necks in the company's work [23]. But this is not so important in CIS implementation projects, moreover, the deployment of a software system already involves standardization and transparency of processes, which is called best practice. Therefore, CASE tools in corporate systems implementation projects can be considered more as a mechanism for visualizing and explaining processes than as a means of analyzing and improving them.

Things are better with software development and technical architecture. Many scientific and popular works are devoted to program coding [8, 9], but even here corporate systems contribute: more attention is paid to the unification and standardization of all developments than to the optimality of the program code. The technical architecture in the context of corporate information systems remains virtually unchanged, however, development, testing and productive environments are being introduced, since the requirements for software products may change during their use in real time. The problems of project management, as well as changes during the implementation of CIS, are practically not mentioned in the literature. The maximum that can be easily found is three classic software development models: cascading, iterative and spiral [4], as well as a description of the difficulties of implementing any changes in the company, which only indirectly highlights the issues mentioned. All this has created the need to offer a new discipline that combines the design of information systems, databases, programming, as well as project and change management, called the theory of corporate information systems.

Unlike subjects devoted only to the design of information systems and technologies, the theory of corporate systems pays attention to the formalization, planning and consistency of all project tasks performed. In addition, great emphasis is placed on the scale of operations that require user involvement. Project tasks are grouped by implementation levels, represented by processes, data, application and technology, as well as project and changes. Each implementation level is led by a separate dedicated subteams and aligned with all others. The theory assumes formalization of strategies for analysis, design, roles and authorization, technical and business cutover, change management, build, data migration, training, testing and post go-live support, each of which is decomposed into subtasks, consistent with other activities, and then planned for execution. Planning plays a key role during the implementation of corporate systems, since most of the tasks involve receiving feedback from end users. Thus, the theory of corporate information systems combines the existing technical disciplines and expands their use to larger-scale IT projects.

References

1. Lang, V.: Digital Fluency. APress (2021)
2. Lorizio, L.: Digitalization: The New Normal of The Post-Pandemic World. Independently Published (2021)
3. Gvozdeva, T., Ballod, B.: Information Systems Design. Phoenix, Moscow (2009). (in Russian)
4. Samara, T.: ERP and Information Systems. Integration or Disintegration. John Wiley & Sons Limited (2018)
5. Gavrilov, D.: Production Management Based on the Standard MRPII. Piter, Spb. (2008). (in Russian)
6. Standards of corporate information systems: Corporate information systems, Journal (2022). http://corpinfosys.ru/knowledgebase/standards. Accessed 14 July 2022
7. Stepanov, D.: Shaping ERP3 standard to manage corporate information systems in the time of Industry 4.0. In: Abstracts of 3rd International Conference on Control Systems, Mathematical Modeling, Automation and Energy Efficiency, Russia, Lipetsk, 10–12 November 2021
8. Knuth, D.: The Art of Computer Programming, vol. 1, Fundamental Algorithms. Addison-Wesley Professional (1997)

9. Knuth, D.: The Art of Computer Programming, vol. 2, Seminumerical Algorithms. Addison-Wesley Professional (1997)
10. Knuth, D.: The Art of Computer Programming, vol. 3, Sorting and Searching. Addison-Wesley Professional (1998)
11. Ostroukh, A., Surkova, N.: Information System Design. Lan, Moscow (2019). (in Russian)
12. Laudon, J., Laudon, K.: Management Information Systems. Prentice Hall (2002)
13. Sudakov, V.: Corporate Information Systems. MAI, Moscow (2016). (in Russian)
14. Tsiteladze, D.: Project Management. Infra-M, Moscow (2022). (in Russian)
15. Shirenbek, H., Lister, M., Kirmse, S.: A guide to the project management body of knowledge, 6th edn. PMI (2017)
16. Pfleeger, S., Atlee, J.: Software Engineering: Theory and Practice. Pearson (2009)
17. Stepanov, D.: Analysis, design and development of corporate information systems: theory and practice. Russian Technol. J. **8**, 227–238 (2015). (in Russian)
18. Oleinik, P.: Corporate information systems. Piter, Spb. (2012). (in Russian)
19. Ya, E.: Features of data migration to SAP ERP. Corporate Information Systems, 22–28 (2019). (in Russian)
20. Kozhevina, O.: Change Management. Infra-M, Moscow (2012). (in Russian)
21. Habadi, A.: An introduction to ERP-systems: architecture, implementation and impacts. Int. J. Comput. Appl. **167**, 1–4 (2017)
22. Yu, M.: Analysis of Requirements for Automated Information Systems. Binom, Moscow (2008). (in Russian)
23. Kovalev, S., Kovalev, V.: Secrets of successful enterprises: business processes and organizational structure. Bitek, Moscow (2012). (in Russian)
24. Petrov, S.: Strategy of roles and authorizations in ERP projects. Corporate Information Systems, 53–58 (2018). (in Russian)
25. Pryadilnikov, E.: Strategy of technical cutover in ERP projects. Corporate Information Systems, 44–49 (2018). (in Russian)
26. Pryadilnikov, E.: Strategy of development in ERP-systems implementation projects. Corporate Information Systems, 10–16 (2018). (in Russian)
27. Stepanov, D.: Problems of implementation of corporate information systems: application level. Management Today, 180–191 (2015). (in Russian)
28. Scott, L.: Relational database and transact SQL, 2nd edn. Independently published (2021)

Modeling the Dynamics of User's Mood Based on the Fokker-Planck Equation and Changes in the Parameters of Network Graphs of Their Comments

J. P. Perova(✉), D. O. Zhukov, and V. N. Kalinin

MIREA-Russian Technological University, 78, Vernadsky Avenue, Moscow 119454, Russia
jul-np@yandex.ru

Abstract. The paper studies the dynamics of the moods of Internet media users based on the Fokker-Planck equation and changes in the parameters of the graphs of networks of their comments. The article introduces the concept of the base state vector of the comment network graph, the elements of which are: the average value of the mediation coefficient, the average value of the clustering coefficient, the proportion of users in a particular state. The time dependence of the distance between the base vector and the current state vector forms a time series whose x-values can be thought of as a "wander point". This time series is considered on the segment [Lmin, Lmax], where Lmin и Lmax – given boundaries of the interval of possible values of the network state. The current state of the comment graph can be determined using network analysis tools and text analysis methods. The solution of the Fokker-Planck equation made it possible to obtain an analytical time dependence of the probability density of detecting the value of the state of the network in one or another value x. Based on this approach, an algorithm can be created for predicting the time to reach a given state of the user comments network graph (the allowable distance between the base and observed vectors) for a given probability level (this allows time to be predicted). Analysis of the model showed its adequacy, not inconsistency.

Keywords: Social networks · Network graph · Social network graph characteristics · Nonlinear dynamics · Fokker-Planck equation · Social process modeling · Monitoring · Management

1 Relevance of the Study

Changes in the opinions and moods of social network users are a good indicator of the real processes that are taking place in society. This leads to a change in trends and in social processes this is latent. Moreover, it begins to occur in social networks much earlier than it becomes noticeable in non-network structures. To detect such phenomena, it is possible to analyze large volumes of textual information in real time, which is generated daily by users of social networks. This is a complex scientific and technical problem.

A. A. Radionov and V. R. Gasiyarov (Eds.): RusAutoCon 2022, LNEE 986, pp. 35–45, 2023.
https://doi.org/10.1007/978-3-031-22311-2_4

Social processes can be investigated using macrostatistics and sociological surveys. However, this approach, on the one hand, is very laborious (preparing special questionnaires with a large number of questions, forming a representative sample, conducting the survey itself, processing data, etc., which significantly reduces the reliability of the data obtained.

One of the possibilities for solving the problems of monitoring and predicting the development of public sentiment can be the study of the activity of users who leave comments on blogs and news resources on various socially significant topics.

Using text analytics tools that allow you to cluster texts on selected topics, and tools for collecting open data from social networks and news network resources, you can determine the mood of users and build graphs of their connections within selected thematic groups. Each such graph will have its own set of characteristics (network density, average intermediation coefficient, average clustering coefficient, elasticity coefficient, and others), which can change daily over time, thereby forming a multidimensional time series.

2 Setting a Study Objective

In the presented article, we propose a vector representation for describing the network of comments. The elements of the vectors are the admissible parameters of the network values (density, the average value of the intermediation coefficient, the average value of the clustering coefficient, elasticity, and others), as well as such characteristics as the share of users, which can be attributed to one of four groups based on text analysis:

- loyalist (certainly supports the actions of the government and authorities).
- oppositionist.
- troll (a user using a resource to make a scandal, anger some and enjoy it).
- undecided or neutral user.

Achievement or realization of desired or not desired states of the entire social network as a whole can be given on the basis of basic vectors (we will discuss this in the article below).

The time variation of the distance between the base vector and the current state vector can be considered as a "wander point" on the interval [Lmin, Lmax] or as a random (or almost random) time series. And some given value of this distance (the state in which management decisions should be made) can be considered as a trap or a point of an acceptable threshold for implementation, where a "wandering point" can eventually fall. This allows you to build probabilistic sociodynamic models to predict the dynamics of public sentiment.

In the traditional description of the behavior of a "wandering point", as a rule, the diffusion model is used. However, in this case it cannot be considered reliable. As a rule, time series describing processes in complex systems (for example, financial indicators of stock and commodity exchanges) are not stationary, which is due to various reasons, including the presence of a human factor. Their selective distribution functions have a time-dependent mathematical expectation, which contradicts the simple diffusion model and shows a non-stationary time series.

In this regard, we are supposed to consider more complex models of behavior of the "wandering point", for example, based on the Fokker-Planck equations.

3 Data Collection and Processing

As one of the examples of a study on the analysis of the structure of comment graphs, one of the news on the Echo-Moscow portal was chosen and all user comments on it and their available data were collected. After that, they (633 comments) were processed and, based on the analysis of texts, marked as belonging to one of four types: loyalist; oppositionist; troll; undefined.

The "undefined" group was singled out because, due to the small amount of text component of the comment, it was almost certainly impossible to say anything about the user's affiliation with one of the other three groups. Some comments (77 pieces) at the time of data collection were deleted by the site moderators for rule violations, but due to the fact that the child comment retained the value of the parent comment, it became possible to restore information about their existence (but without texts). Therefore, the total number of nodes in the graph is 710. The stateful statistics of users are as follows: 10.28% loyalists (73 nodes); oppositionists 43.8% (311 knots); trolls 30.42% (216 knots); unspecified 4.65% (33 knots); 10.85% (77 nodes) deleted.

Figure 1 graphically shows us the structure of the graph obtained when processing comments on news. In color visualization of the obtained data, the nodes of the graph, depending on their state (assignments to one of four types), can be marked with different colors.

The links in Fig. 1 show the mutual commenting of users to each other. Thus, by the "color" of the nodes, one can judge their state, and by the edges of the graph about the interaction.

Figure 1 shows that there are many unrelated single vertices in this structure. However, you can also notice the presence of a related component of the graph, which is separately shown in Fig. 2. Closed oval lines in Fig. 2 show users commenting on themselves.

Let's consider the elements of the network state vector that we will use in our model:

- The proportion of nodes that have a certain condition (for example, those who are negative towards any event in public life.
- Clustering coefficient is a measure of the density of the connections of a given vertex with neighboring ones. The ratio of the real number of links that connect the nearest neighbors of a given node i to the maximum possible (such that all the nearest neighbors of a given node would be connected directly to each other) is called the node clustering coefficient, its value lies on the segment [0, 1]. The larger its value, the more significant this node is in the exchange of information.
- Coefficient of mediation - shows the ratio of the number of shortest paths between all pairs of network nodes passing through this node to the total number of all shortest paths in the network, its value lies on the segment [0, 1]. The larger its value, the more significant the role of this vertex in the exchange of information.

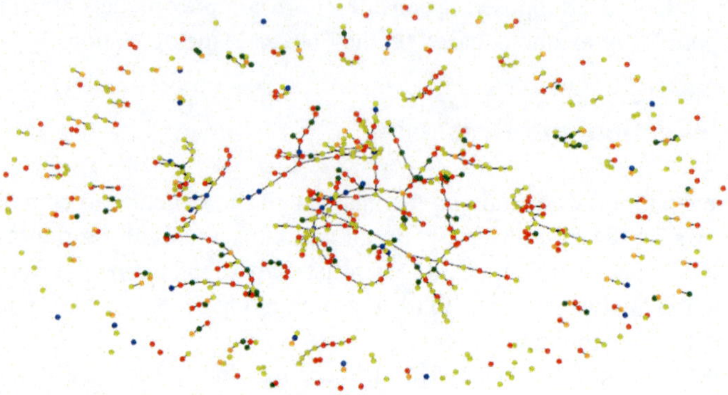

Fig. 1. The structure of the column of comments on the news in question.

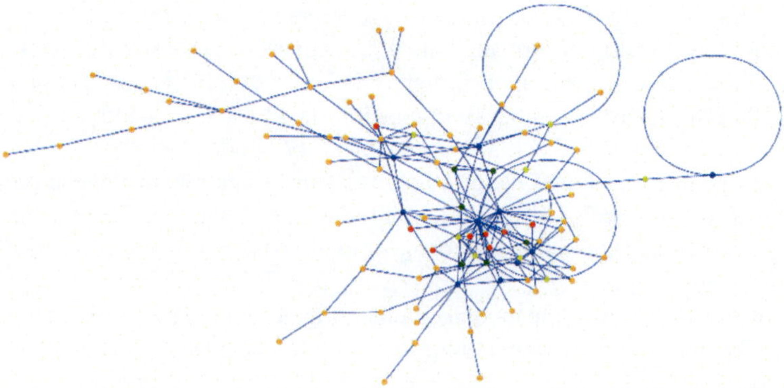

Fig. 2. Graph of user relationships by comments.

Determine the value of the elements of the basic network state vector (let's designate it as θ). They set thresholds, the transition of which is undesirable from the point of view of state management. Given that in any community there is always a 0.10 to 0.15 share of participants always disagreeing on any issue, we will accept the share of those who are negative towards the event in question equal to 0.12.

The desired average value (for all nodes) of the clustering coefficient of such a network is also assumed to be small, for example, equal to 0.05; and the average degree of mediation of nodes in such a network is also equal to 0.05. Thus, the base vector will be: $q = (0.12; 0.05; 0.05)$.

Note that the number of parameters with which you can describe the state of the network can be greater, and we have chosen only those that in our opinion are the most significant. In addition, the selected parameters are normalized (lie on the line [0, 1], therefore, they equally affect the calculation of the distance metric.

In our proposed approach, various columns of commenting on news on a selected resource on certain topics during the day can be combined into a single structure through connections between nodes that belong to users. Thus, we can highlight a large graph that will describe the activity of users of this network information resource during the day. Next, you can define the elements of the current state vector that describes its characteristics.

Changes in the components of this vector for each day for a certain time will form a multidimensional time series.

4 Theoretical Part

The resulting multidimensional time series can be used to describe the dynamics of the change in public sentiments of users of the intern network.

In the course of the study, articles were studied, on the basis of which it can be concluded that social network analysis methods are a useful tool for creating a complete picture of public sentiment at a time when events of a certain nature occur in a country or in a country. World. We can consider a number of works close to the topic of our study on the description of processes in complex social network structures.

Since finding similarities between nodes in a network is a time-consuming process as the network grows, researchers have used swarm algorithms to optimize the process of solving link prediction and community discovery problems [1]. Swarm-based optimization techniques used in social network analysis are compared in this article with community and link analysis based on traditionally used approaches.

In works [2–4] proposed the KroMFac technique, which conducts community detection using regulated non-negative matrix factorization (NMF) based on the Kronecker graph model. KroMFac combines network analysis and community discovery techniques in a single, unified framework. This technique links four areas of research, namely the detection of communities on graphs, the detection of overlapping communities, the detection of communities in incomplete networks with missing edges and complete networks.

In the work [5] proposed a new weighted summary measure for detecting influential users in social networks. This method combines the influence of several structural features of the network, as well as local and global information to obtain an estimate of weighted total centrality.

The authors of [6] propose a new index for analyzing the distribution of messages in social networks, based on the topological nature of networks and the strength of messages' influence. This indicator characterizes the strength of each node as a means of launching a message, dividing nodes into starters and non-starters.

Works [7–9] on the analysis of random networks presents physically justified models and effective algorithms for determining hierarchical ranks of nodes in directed networks.

The dynamics of changes in the public mood of Internet users can be attributed to stochastic processes. The presence of the human factor (many people with different opinions, preferences and behaviors) on the one hand creates a randomness of changes (due to the wide variety of user behavioral models), and on the other hand introduces elements of purposefulness into the dynamics of changes. A detailed description of the use of stochastic methods for modeling the dynamics of social processes can be found in [10].

The most promising in our opinion for creating models of the dynamics of change in public mood are models that can be created from the Fokker-Planck equation, which takes into account both ordered and random changes.

The Fokker-Planck equation is widely used to analyze and model the behavior of time series when describing processes in complex systems [11–14].

It should be noted that in addition to the Fokker-Planck equation, other approaches are used for modeling based on differential equations, for example, the Liouville equations [14, 15], the diffusion equations [13, 16] and several others.

To simulate social processes, not only models based on partial differential equations are used, but, for example, models based on game theoretic approaches and methods for making managerial decisions based on them [17].

The Fokker-Planck equation is widely used to analyze and model transients observed in various complex systems and provides good agreement with predicted behavior and observed data. Therefore, as a hypothesis, we will assume that the Fokker-Planck equation can be used to analyze and model the appearance of comments on news and blogs. The Fokker-Planck equation has the form:

$$\frac{\partial\rho(x,t)}{\partial t} = -\frac{\partial}{\partial x}[\mu(x)\cdot\rho(x,t)] + \frac{1}{2}\frac{\partial^2}{\partial x^2}[D(x)\cdot\rho(x,t)] \tag{1}$$

where $\rho(x, t)$ - time-dependent t probability density of state distribution x (in our case, state x is the number of comments observed at time t), $D(x)$ – state-dependent x factor determining random state change x, $\mu(x)$ – a state-dependent x coefficient defining a targeted state change x.

Applicable to our model $D(x)$ can be interpreted as user actions caused by a spontaneous impulse that arose when reading the news or other users' comments on it, when the event described in the news or blog is not significantly important, but the user is ready to spend time commenting or responding to another commentator (the user had a spontaneous desire to respond to this news). And $\mu(x)$ can be interpreted as targeted actions caused by the desire to respond to a significant news or blog for the user, as well as comment on the comment of another user if he touched on a topic important from the point of view of this user (the user is constantly interested in this topic).

Next, when you need to build a model, you need to make assumptions about the dependence $D(x)$ and $\mu(x)$ from state x and consider two conditions. First, we take into account the dimension of the terms included in Eq. (1), and secondly, we can make the assumption that with an increase in the state of x (an increase in the number of possible comments (the significance of the news or blog) of magnitude $D(x)$ and $\mu(x)$ should also increase).

Logic dictates that all terms of Eq. (1) must have the same dimension, which has $\rho(x)$. Both the first and second condition will be met if the dependencies $D(x)$ and $\mu(x)$ from state x will have the form: $\mu(x) = \mu_0 \cdot x$ and $D(x) = D_0 \cdot x^2$. In this form, on the one hand, growth is ensured $D(x)$ and $\mu(x)$ increasing the state of x, and on the other hand, the condition of maintaining the dimension will be met.

Solving the stationary Fokker-Planck equation:

$$-\frac{d}{dx}[\mu(x)\cdot\rho(x)] + \frac{1}{2}\frac{d^2}{dx^2}[D(x)\cdot\rho(x)] = 0 \tag{2}$$

Under the assumptions made has the form:

$$\rho(x) = [\gamma - 1]x^{-\gamma} \tag{3}$$

This is the power law of distribution of commentators by the number of comments observed in practice. Thus, this suggests that the Fokker-Planck equation can be used in practice to describe social processes.

To describe the change in the value of the distance between the magnitude of the current state vector and the given base vector over time, consider the solution of the unsteady Fokker-Planck equation, which may allow the construction of probabilistic sociodynamic models to predict the dynamics of public sentiment.

Let us formulate a boundary value problem, the solution of which will describe the process of changing the value of the distance between the value of the current state vector of the comment network graph and the given base vector in time.

The first boundary condition:

When selecting the first boundary condition, we will proceed from the following considerations: $x = L_{min}$ (the left border of a segment of possible states) determines the state through which the transition must be avoided (the area located on the segment to the left of this state is undesirable for us). The probability of detecting such a state of the system may not be zero. And the probability density, which determines the flow in the state $x = L_{min}$, must be taken equal to 0, since the states should not go beyond this border (here the reflection condition is implemented). Thus:

$$\rho(x, t)|_{x=L_{min}} = 0 \tag{4}$$

Second boundary condition:

We limit the area of possible states on the right to some value $x = L_{max}$ (the metric used in the calculations cannot be greater than the magnitude of the vector whose elements have maximum values in the space of the selected coordinates). The probability of detecting such a state over time will be different from zero. However, the probability density determining the flow in the state $x = L_{max}$, must be set to zero (the distance between the current and base vector of states is limited by the maximum values of possible coordinates in the vector space used (the reflection condition from the boundary is realized)):

$$\rho(x, t)|_{x=L_{max}} = 0 \tag{5}$$

To formulate the boundary value problem, it is necessary to specify the initial condition. Since at a point in time $t = 0$ system state (the distance between the base vector and the current state vector can be equal to some value x_0, then the initial condition can be set as:

$$\rho(x, t = 0) = \begin{cases} \delta(x - x_0)dx = 1, \int^x = x_0 \\ 0, \quad x \neq x_0 \end{cases} \tag{6}$$

The presence of a delta function leads to the fact that the solution of Eq. (1) under given boundary conditions and the assumptions made about $D(x)$ and $\mu(x)$ for time-dependent probability density of system state detection in one or another value x will be:

At equation $L_{min} \leq x \leq x_0$:

$$\rho_1(x,t) = -\varphi(x,t)\sum_{n=1}^{M} \frac{\sin\left\{\pi n \frac{\ln\left(\frac{L_{\max}}{x_0}\right)}{\ln\left(\frac{L_{\max}}{L_{\min}}\right)}\right\} \sin\left\{\pi n \frac{\ln\left(\frac{x}{L_{\min}}\right)}{\ln\left(\frac{L_{\max}}{L_{\min}}\right)}\right\}}{\cos(\pi n)} e^{-\omega(n,t)} \quad (7)$$

At equation $x_0 \leq x \leq L_{max}$:

$$\rho_2(x,t) = \varphi(x,t)\sum_{n=1}^{M} \frac{\sin\left\{\pi n \frac{\ln\left(\frac{x_0}{L_{\min}}\right)}{\ln\left(\frac{L_{\max}}{L_{\min}}\right)}\right\} \sin\left\{\pi n \frac{\ln\left(\frac{x}{L_{\max}}\right)}{\ln\left(\frac{L_{\max}}{L_{\min}}\right)}\right\}}{\cos(\pi n)} e^{-\omega(n,t)} \quad (8)$$

where $\alpha = \frac{1}{2} - \frac{\mu_0}{D_0}, \varphi(x,t) = 2\frac{x_0^{\alpha} \cdot x^{-[1+\alpha]} \cdot e^{-\frac{D_0\alpha^2}{2}t}}{\ln\left(\frac{L_{max}}{L_{min}}\right)}, \omega(n,t) = \frac{\pi^2 n^2 D_0 t}{2\left[\ln\left(\frac{L_{max}}{L_{min}}\right)\right]^2}$

Probability that by the time t the state of the system will be within a period from L_{min} to L_{max}, that is threshold state (θ) will not be reached can be calculated as follows:

$$P(\theta,t) = \int_{L_{min}}^{x_0} \rho_2(x,t)dx + \int_{x_0}^{L_{max}} \rho_1(x,t)dx \quad (9)$$

Probability $Q(\theta, t)$ that the threshold state is θ by the moment of time t will be achieved or exceeded, calculated by the formula:

$$Q(\theta,t) = 1 - P(\theta,t) \quad (10)$$

Defining the line boundaries of possible states from L_{min} to L_{max} we will discuss in the analysis section of the resulting model.

5 Prediction Algorithm for Achieving a Given State of the Network Comment Graph of Masmedia News Users

Predicting the dynamics of the moods of Internet media users based on the Fokker-Planck equation and changing the parameters of their commentary networks can be carried out according to the following algorithm:

- You need to collect text comments and metadata of users on a specific topic from online news media resources with date and time binding.
- Then you need to process data using text analytics and sentimental analysis, get a graph of user comments on a certain topic and calculate its characteristics (network density, average mediation coefficient, average clustering coefficient, elasticity, share of users with one or another mood).
- Next, you need to set the values of the elements of the base vector, which will determine the achievement of the desired or not desired state (θ) and form, based on the processed data and the given vector, a time series of changes in the graph of user comments on a certain topic over time

- Then we set the duration of the step τ (hour, day, week, etc.) and by time series values in a few steps for a given τ we determine using numerical calculations using observed data and Eqs. (9) and (10) model parameters μ_0 and D_0.
- We assume the last average value of the distance metric between the base vector and the vector of the current state of the network as the initial state x_0 and using the obtained values μ_0 and D_0, as well as Eqs. (9) and (10) perform calculations, and obtain a dependence on the time of probability of reaching, desired or not desired state. Next, you can set the probability value (for example, 0.95) and estimate the time to reach a given probability level (make a forecast by time). Analysis of the obtained model

For the graph shown in Fig. 2, you can determine its characteristics and elements of the vector of the current state at a given time t (which is taken as $t = 0$):$X(t) = (0.44; 0.11; 0.15)$. Distance between the specified base vector of the desired state $\theta = (0.12; 0.05; 0.05)$ and the current state vector $X(t)$ at time $t = 0$ will be equal to $x_0 = 0.34$. By analyzing the dynamics of the time series of changes in the state of the network over the previous few days and using the equations of the model, you can solve the opposite problem and determine the values of the model parameters μ_0 and D_0. In our case $\mu_0 = 0.0003$. and $D_0 = 0.007$.

Right boundary of a line of possible states Lmax can be specified as the distance between the base vector (θ) and vector of maximum possible values of network parameters $X(t) = (1; 1; 1)$. in the case under consideration $L_{max} = 1.61$. The left boundary for insurance can be defined, for example, as half the length of a given base vector (in this case $|\theta| = 0.14$), thus L_{min} will be equal to 0.07.

As the results showed, if the network is not affected, then under current conditions the required state can be achieved with a probability of 0.8 for 375 days, and with a probability of 0.9 for 525 days. The result obtained is quite possible, but the question of assessing its accuracy remains, which requires additional research.

Figure 3 shows the results of the simulation as a function of the probability time of reaching a given threshold network state.

6 Conclusion

In conclusion, we note that the complex nature of process dynamics in complex social systems can be described not only on the basis of models created on the basis of the Fokker-Planck equation. For example, in [18–20], models are presented for describing the stochastic dynamics of changes in the state of complex social systems, taking into account the processes of self-organization and the presence of memory.

This allows one to take into account memory and describe not only Markov, but also non-Markov processes. In these studies, a non-linear differential equation of the second order was derived, which makes it possible to set and solve boundary value problems to determine the probability density function of the amplitude of deviations of parameters describing the observed processes of non-stationary time series, depending on the values of the time interval of its determination and the depth of memory, which significantly distinguishes it from the Fokker-Planck equation.

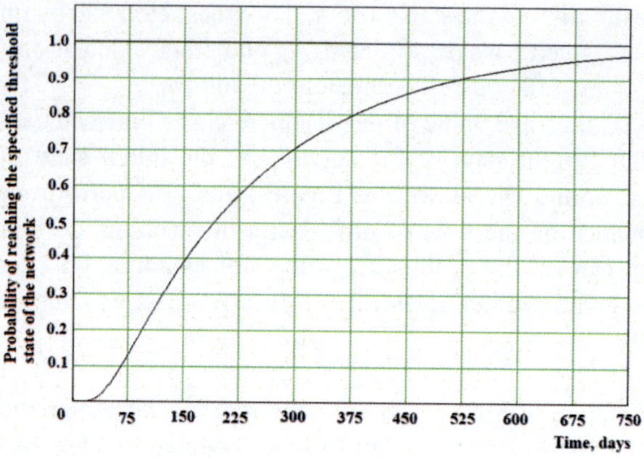

Fig. 3. The probability of reaching a given network threshold state versus time for the example dis-cussed.

Acknowledgment. The study was carried out with the support of the Russian Science Foundation (RSF), grant № 22-21-00109 "Development of models for predicting the dynamics of social sentiment based on the analysis of time series of text content of social networks using Fokker-Planck equations and nonlinear diffusion".

References

1. Pulipati, S., Somula, R., Parvathala, B.R.: Nature inspired link prediction and community detection algorithms for social networks: a survey. Int. J. Syst. Assur. Eng. Manag. 1–18 (2021). https://doi.org/10.1007/s13198-021-01125-8
2. Tran, C., Shin, W.-Y., Spitz, A.: Community detection in partially observable social networks. ACM Trans. Knowl. Discov. Data **16**(2), 1–24 (2022). https://doi.org/10.1145/3461339
3. Chen, Z., Li, L., Bruna, J.: Supervised community detection with line graph neural networks. In: Proceedings of the 7th International Conference on Learning Representations, New Orleans, United States (2019)
4. Hoffmann, T., Peel, L., Lambiotte, R., Jones, N.S.: Community detection in networks without observing edges. Sci. Adv. **6**(4), 1–11 (2020)
5. Jain, S., Sinha, A.: Discovering influential users in social network using weighted cumulative centrality. Concurrency and Computation: Practice and Experience **34**(1), e6521 (2022). https://doi.org/10.1002/cpe.6521
6. Martins, P., Martins, F.A.: Launcher nodes for detecting efficient influencers in social networks. Online Social Networks and Media **25**, 100157 (2021). https://doi.org/10.1016/j.osnem.2021.100157
7. Xue, L., Zhang, P., Zeng, A.: Maximizing spreading in complex networks with risk in node activation. Inf. Sci. **586**, 1–23 (2022). https://doi.org/10.1016/j.ins.2021.11.064
8. Arafeh, M., Ceravolo, P., Mourad, A., Damiani, E., Bellini, E.: Ontology based recommender system using social network data. Futur. Gener. Comput. Syst. **115**, 769–779 (2021). https://doi.org/10.1016/j.future.2020.09.030
9. De Bacco, C., Larremore, D.B., Moore, C.: A physical model for efficient ranking in networks. Sci. Adv. **4**(7), eaar8260 (2018). https://doi.org/10.1126/sciadv.aar8260

10. Gardiner, C.: Stochastic Methods: A Handbook for the Natural and Social Sciences. Springer-Verlag (2009)
11. Lux, T.: Inference for systems of stochastic differential equations from discretely sampled data: a numerical maximum likelihood approach. Ann. Finance **9**(2), 217–248 (2012). https://doi.org/10.1007/s10436-012-0219-9
12. Hurn, A., Jeisman, J., Lindsay, K.: Teaching an old dog new tricks: improved estimation of the parameters of stochastic differential equations by numerical solution of the Fokker-Planck equation. In: Gregoriou, G., Pascalau, R. (eds.) Financial Econometrics Handbook. Palgrave, London (2010)
13. Elliott, R.J., Siu, T.K., Chan, L.: A PDE approach for risk measures for derivatives with regime switching. Ann. Finance **4**(1), 55–74 (2007). https://doi.org/10.1007/s10436-006-0068-5
14. Orlov, Y., Fedorov, S.L.: Generation of non-stationary trajectories of a time series based on Fokker-Planck equation. MFTI Proceedings **8**(2), 126–133 (2016)
15. Chen, Y., Cosimano, T.F., Himonas, A.A., Kelly, P.: An analytic approach for stochastic differential utility for endowment and production economies. Comput. Econ. **44**(4), 397–443 (2013). https://doi.org/10.1007/s10614-013-9397-4
16. Savku, E., Weber, G.-W.: Stochastic differential games for optimal investment problems in a Markov regime-switching jump-diffusion market. Ann. Oper. Res. **312**, 1171–1196 (2020). https://doi.org/10.1007/s10479-020-03768-5
17. Krasnikov, KE.: Mathematical modeling of some social processes using game-theoretic approaches and making managerial decisions based on them. Russian Technol. J. **9**(5), 67–83 (2021). https://doi.org/10.32362/2500-316X-2021-9-5-67-83
18. Zhukov, D., Khvatova, T., Millar, C., Zaltcman, A.: Modelling the stochastic dynamics of transitions between states in social systems incorporating self-organization and memory. Technol. Forecast. Soc. Chang. **158**, 120134 (2020). https://doi.org/10.1016/j.techfore.2020.120134
19. Zhukov, D.O., Zaltcman, A.D., Khvatova, T.Y.: Forecasting changes in states in social networks and sentiment security using the principles of percolation theory and stochastic dynamics. In: Proceedings of the 2019 IEEE International Conference "Quality Management, Transport and Information Security, Information Technologies", IT and QM and IS 2019, 8928295, pp. 149–153 (2019)
20. Zhukov, D.O., Lesko, S.A.: Stochastic self-organization of poorly structured data and memory realization in an information domain when designing news events forecasting models. In: The 2nd IEEE International Conference on Big Data Intelligence and Computing, Auckland, New Zealand (2016). https://doi.org/10.1109/DASC-PICom-DataCom-CyberSciTec.2016.153

Algorithm for Selecting a Priority Task When Planning the Movement of an Autonomous Transport Platform

S. I. Timakov(✉), Yu. V. Frantzuzova, and A. V. Demidova

Tula State University (TulSU), 92, Lenina Prospekt, Tula 300600, Russia
timblok68348@gmail.com

Abstract. This article reveals the relevance of the area under study, determines the advantages and disadvantages of autonomous transport platforms, highlights the most vulnerable place of such platforms - the efficiency and safety of the algorithms under which they operate, regardless of the environment (static or dynamic), considers the creation of an alternative algorithm choosing a priority task when moving an autonomous transport platform. This algorithm should provide efficient system selection to optimize the performance of platform components and rationally allocate critical resources such as time and energy. The paper analyzes the existing concepts for the creation of such systems. The article presents a general scheme of the developed algorithm for selecting a priority task and a scheme for the interaction of platform subsystems. It also presents the basic configuration of an autonomous transport platform, a verbal algorithm for moving it to a given point, and describes the operation of a priority transition controller between subsystems.

Keywords: Autonomous · Transport · Platform · ATP · Algorithm · Subsystems · Tasks · Priority · Path · Planning

1 Introduction

Path planning is the most important task in the field of unmanned vehicle navigation. It mainly includes three aspects. First, the planned path must run from a given starting point to a given end point. Secondly, this path must ensure the movement of the object with avoidance of possible obstacles. Thirdly, the path must, among all possible paths that satisfy the first two requirements, be optimal in a certain sense [1].

A high degree of autonomy determines the need to adapt machines to a dynamically changing operating environment. They must make their own decisions in a complex and uncertain environment. Therefore, autonomous machines must have an intelligent control system and perform the tasks assigned to them [2].

ATP is an autonomous transport platform equipped with an automatic control system that can move without human intervention. Autopilot is a device or software and hardware complex that leads a vehicle along a certain trajectory given to it. Most often, autopilots are used to control aircraft (due to the fact that the flight most often takes place in a

A. A. Radionov and V. R. Gasiyarov (Eds.): RusAutoCon 2022, LNEE 986, pp. 46–56, 2023.
https://doi.org/10.1007/978-3-031-22311-2_5

space that does not contain a large number of obstacles), as well as to control vehicles moving along rail tracks. A modern autopilot allows you to automate all stages of flight or movement of another vehicle [3].

2 Relevance

The relevance of this topic is determined by continuous technological progress, covering all spheres of human activity. It encourages the creation of objects that do not require human control, which can perform not only everyday household tasks, but also tasks with increased risk, such as space and deep-sea research, maintenance of nuclear power plants, and elimination of the consequences of man-made accidents and disasters [1].

If projected onto the automotive industry, it is undergoing a significant transformation: the largest car manufacturers, together with IT specialists and developers, are moving towards the creation of vehicles with the possibility of fully autonomous driving. In the future, self-driving vehicles will become a mass phenomenon, but there are still many tasks to be solved on the way to the era of fully autonomous cars [3].

2.1 Advantages and Disadvantages

The advantages of autonomous transport platforms are:

- transportation of goods in hazardous areas, during natural and man-made disasters or military operations - safety;
- cost reduction of transporting goods and people due to savings on the wages of drivers;
- more economical consumption of energy, fuel and use of roads through centralized traffic management;
- saving time spent on transport platform management;
- minimization of emergencies, road traffic accidents with human casualties.

Also, autonomy has a number of shortcomings, which are gradually being eliminated, but at the moment they are still relevant:

- software reliability and efficiency of its algorithms;
- reliability of the technical component of the platform;
- suppression of unauthorized access to the platform;
- more expensive production of a transport platform than a classical platform under human control.

Some systems rely on infrastructure systems (such as those built into or near the road), but more advanced technologies allow the simulation of human presence at the level of steering and speed decisions through a suite of cameras, sensors, radar and satellite navigation systems [3].

3 Existing Concepts of the Navigation System of Autonomous Objects

Existing navigation systems for autonomous objects. The article by A. Guryev dated December 19, 2019 Strong Artificial Intelligence Model describes the concept of information representation and processing, which is called Action-Based Dynamic Semantic Network. It defines the process of selecting necessary actions to achieve goals based on needs.

During the life of the system, the algorithm put into action can change, improve, but it seems that it should have at least the following properties:

- the choice of the next action should not be random, but should be purposeful;
- the performed action should be assessed as success/failure. The failure of the action must be considered in the next selection;
- when choosing an action, prompting signals from the level of associations (connections of type 2 of the previous section) should be considered.

To implement the first property, the system must have information, firstly, about the goals, needs of the system at a given moment in time, secondly, about the expected results of each action known to the system, and, thirdly, about the conditions for successful completion of the action.

The big advantage of the proposed approach is that all three types of information (goals, expected results of the action and the necessary conditions for the action) can be represented by one entity - the concept-characteristic. For example, the characteristic "something near" may be present in the expected results of the action "Approach something", in the necessary conditions of the action "Take something", and also be a goal at some point in time.

The system has a list of goals, consisting of pointers to concepts-characteristics (the significance of this goal is also stored). With this list, as well as with the list of actions with active associative links ("hints"), the action "Select action" works. Also, each action known to the system includes two lists of pointers - one to the prerequisites for performing the action, and the other to the characteristics of the result. Based on this information, the Action Selection algorithm determines which action to choose, constructing a new composite action from separate known actions if necessary.

In general, it becomes clear how the concept of action, which originates from an elementary effector, is overgrown with auxiliary structures (receptors, concepts-characteristics). These structures form a kind of action model that allows you to evaluate the result of the action before the actual execution, which makes it possible to plan actions to achieve the desired result.

For example, if there is a goal "Hunger / Loss of charge", from the action model "Eat food / Replenish resources" (in the necessary conditions of which there is a concept-characteristic "Food nearby / Power stations nearby"), the goal "Food nearby / Power stations nearby" will be added, and for its implementation, the action "Approach food / Get to the station" was created [3].

Part of the concept of A. Guryev was used in the development of an algorithm for choosing a priority task when moving the ATP.

It is also known about older concepts underlying the adaptive behavior of autonomous objects, namely, about two closely related areas of research: "Artificial Life" and "Adaptive Behavior". "Adaptive behavior" - meant the study of the architecture and principles of functioning that allow animals or robots to live and act in a variable external environment, as well as the analysis of the evolution of animal cognitive abilities and the evolutionary origin of human intelligence [4–6].

Adaptive Behavior, as well as Artificial Life, mainly uses a phenomenological approach to the study of adaptive behavior control systems, i.e., it is assumed that there are formal rules for adaptive behavior, and these rules are not necessarily associated with specific microscopic neural or molecular structures, that are present in living organisms.

Research on adaptive behavior is ongoing at a number of universities and laboratories. Let us briefly characterize the work of one of the leading laboratories - the AnimatLab laboratory, which is headed by Jean-Arkady Meyer, one of the initiators of this direction.

The general approach of this laboratory can be summarized as follows. The animat shown in Fig. 1 exists in a real or simulated environment. It has sensors that receive information from the external and internal environment of the animat, and effectors through which it interacts with the environment, as well as a control system that coordinates the perception and actions of the animat. An animat's behavior is considered adaptive if the control system keeps the animat's vital variables (e.g. V1 and V2 in Fig. 1) within acceptable limits. ON picture (Fig. 1) the dashed arrow shows a possible trajectory that goes beyond the allowable area (gray background is an invalid variable area). The solid arrow shows the "corrected" trajectory, corrected with the help of a control system that ensures that the variables are maintained in the allowable (light) region.

The above scheme is of a general nature and does not reveal the internal structure of the self-government mechanism.

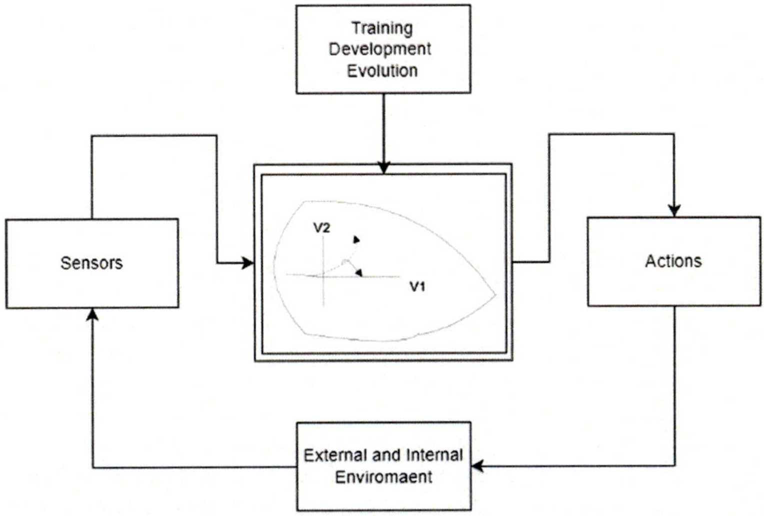

Fig. 1. The general scheme of the animat's behavior. The AnimaLab approach.

The concept of multi-agent systems in solving the problems of distributed artificial intelligence, which is focused on the intellectual sphere, has received rapid development [7–9, 11, 12].

A similar approach is presented in the work of Russell [13]. This concept considers a holistic "organism", called an agent. An agent is everything that can be considered as perceiving its environment with the help of sensors and acting on this environment with the help of actuators. The sequence of acts of perception of the agent is called the complete history of everything that was ever perceived by the agent. The agent's choice of actions at any particular point in time may depend on the entire sequence of percepts observed up to that point in time.

It is assumed that intelligent agents are able to maximize their performance indicators. The implementation of this property is simplified if the agent is able to assume the obligation to achieve the goal and strive for its satisfaction. Goals allow you to organize behavior by limiting the choice of intermediate steps that the agent tries to implement. The disadvantage of this approach is that the agent can limit the choice of intermediate stages, while in real life this is not always possible.

To solve this problem O. Yu. Yakovenko [14] decomposes the decision-making process into several stages: situation recognition; goal setting; building a plan to achieve the goal, and the set of consistent sequences of states that are guaranteed to transfer the object to the target state is the plan. When building a plan, a certain parameter is minimized (or maximized), which is called the cost of the plan. This is necessary in order to choose some of the most preferred subsets from the entire set of possible plans. The main disadvantage of this approach is its inability to plan actions in real time.

In the work of G.S. Osipov [15] on artificial intelligence, to solve modeling problems in planning behavior over discrete sets of actions, constructions equivalent to systems of rules are used. Among the rules, there are those that do not correspond to any actions, but only replenish the set of state factors, and rules that are some actions. Based on these rules, the decision is made. The disadvantage of this approach is not the ability to learn. To develop my algorithm, a part of the concept of G.S. Osipov was also used.

4 Work Goals and Objectives

The purpose of this work is to improve the efficiency of the algorithm for moving the ATP in dynamic space by developing new principles for its construction. To achieve this goal, the following priorities were identified:

1. to perform an analysis of existing trajectory planning algorithms and methods for finding a path in a dynamically changing space;
2. to develop an algorithm for choosing a priority task when planning the movement of ATP;
3. to choose the principles of organization and functional structure of the intellectual system for controlling the movement of ATP in a dynamically changing space;
4. to approve the algorithm for routing and controlling the movement of ATP in real time in a dynamically changing space;

5. to develop a simulation computer model and conduct experimental studies of the functioning of the ATP movement control system in order to test the operability of planning and control algorithms in dynamically changing environmental conditions.

In one of my articles, a patent search was implemented on existing ATP developments, and the methods and algorithms on which their behavior is based were also considered. This article has not yet been published. This article analyzes points 2 and 3 (on this material, another article was written for the Regional Master's Conference XVII, which also has not been published at the moment).

5 ATP Configuration

The object of research is autonomous transport platforms. When choosing the principle of organizing the internal structure, the main criterion was the possibility of implementing the tasks set for civilian platforms, such as delivery robots.

The main structural elements of the Autonomous Transport Platform were also identified:

- mechanical part (base, 4 wheels with motors);
- power cell (1 solar battery);
- sensors (2 infrared sensors to prevent collisions);
- navigation (GPS-module);
- controller (a system that provides interaction and control of nodes).

It is assumed that an autonomous object is self-managing and there are several subsystems in its structure, each of them is responsible for a specific job and all of them are interconnected. Moreover, the subsystems come into activity depending on the task, the goal towards which the movement occurs. Or, one might say, at a certain need of an autonomous object.

On picture (Fig. 2) shows a diagram of the interaction of subsystems. The subsystem of self-management includes the subsystem of self-preservation. The mechanical part is the effector subsystem, the remaining subsystems, respectively: the energy source is the battery, the sensors are the IR subsystem - elements, the GPS module.

Each of the presented subsystems is responsible for a specific work and they are all interconnected. Moreover, the transfer of control from one subsystem to another is carried out on the basis of an assessment of the priority needs of an autonomous object.

Subsystem of self-government. This subsystem is responsible for moving the object. For its work, the control subsystem uses signals from other subsystems, which are the values of the needs of these subsystems. The self-management subsystem evaluates these needs, as well as calculates its own need in time, and forms the ATP goal vector. In accordance with the chosen strategy, the self-management subsystem activates a certain ATP subsystem for a period of time. Acts as a controller.

Subsystem of self-preservation. The possible reaction of the self-preservation system is very diverse: an alarm signal, a change in the safe distance, a change in the parameters of movement, a force impact, and the neutralization of an external source of threat within

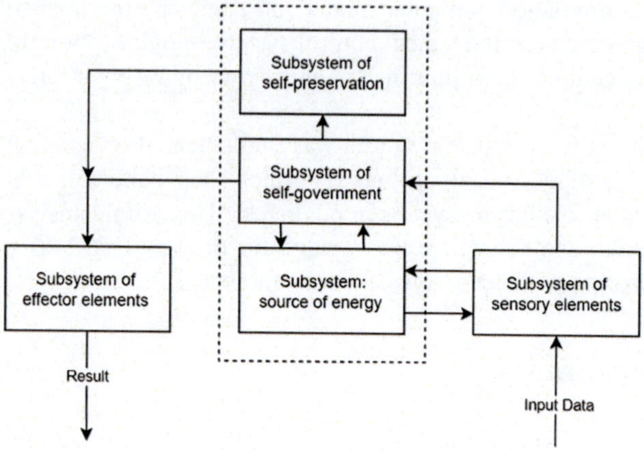

Fig. 2. Scheme of subsystems interaction.

the available resources. The developed autonomous object in the future will have the following features: a low battery alarm (switching to battery charging), changing motion parameters [16–18].

Subsystem of sensory elements. The alleged autonomous object has an infrared vision system, which allows avoiding collisions with objects and a GPS module for orienting on the ground and building a direction vector to the target.

Central source of energy. It is assumed that the energy source ensures the satisfaction of all needs for the implementation of the necessary functions for a long time. Moreover, this energy source in each case allows you to achieve the specified power. It is replenished with the help of solar energy when the movement of the object completely stops.

Subsystem of effector elements. Due to their presence, there is a force change in the motion parameters of an autonomous object. In our case, the object has the ability to move on the surface [19].

6 Movement of the Platform to a Given Point

Behavior planning, or AI planning, is the ability of an intelligent system to synthesize a sequence of actions to achieve the desired target state [10, 11].

The transport platform has initial coordinates, then the coordinates of the point you want to get to are transferred to it. Based on the results of the calculations, the vector of the direction of movement is constructed. The path is a straight line. There is an assessment of the energy resources expended and, if they are sufficient, the platform starts moving in a straight line towards the finish point.

If there are obstacles along the way, then the system is notified by infrared sensors and the priority is given to the self-preservation system, which decides to turn with a concomitant advance to the front to the side, until the obstacle disappears from the visibility of the sensors. As soon as the obstacle is gone, the priority is given to the self-management subsystem, in which the path is recalculated. Then the platform makes a turn towards the finish point until a direct motion vector is obtained [20].

In the process of reaching the end point, the power reserve is calculated and, if the power reserve is less than the distance to the target, then priority is given to the power source subsystem, which will immediately stop and charge up to the required amount of energy with a small margin to reach the finish line. The model of the general behavior of the platform and the choice of the priority system is shown in picture (Fig. 3) [21].

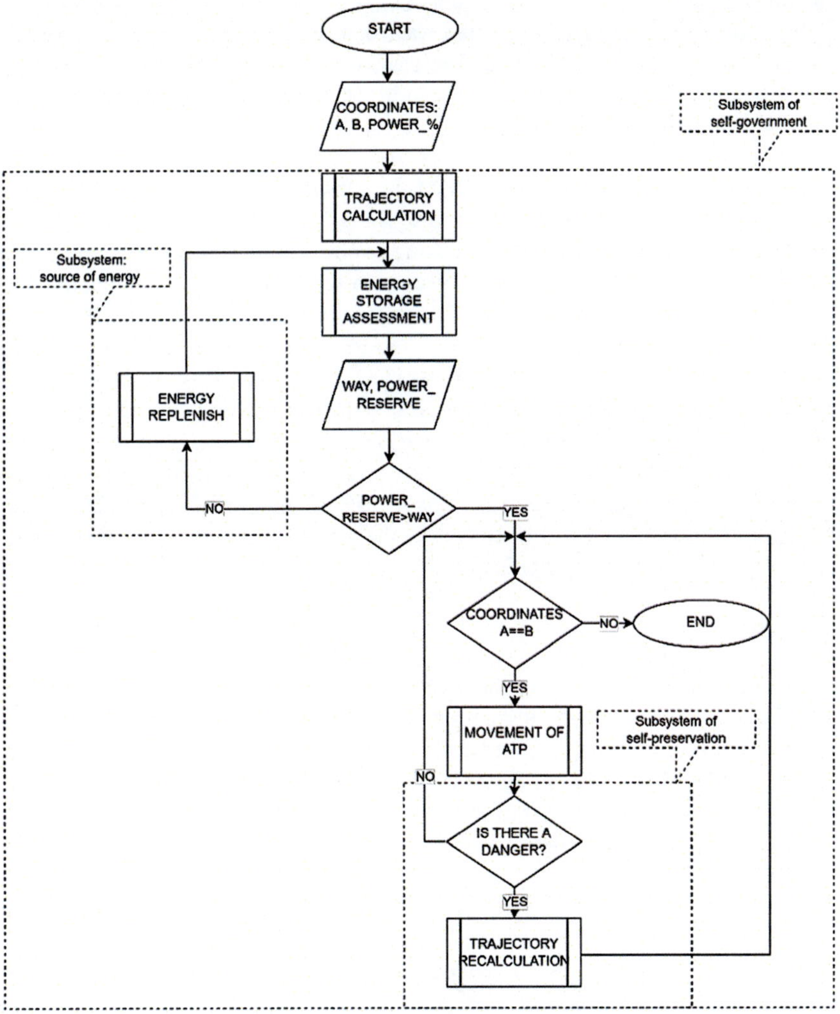

Fig. 3. Scheme of platform behavior.

Description of the algorithm:

1. 1All necessary data are submitted to the input, such as: ATP coordinates, finish line, battery level;

2. Calculation of the trajectory of movement, obtaining distances and estimating the energy reserve;
3. The next step is a comparison of the possible passage of the path on the remaining charge and the calculated path to the "finish";
4. If the charge is not enough, then we recharge the battery;
5. Otherwise, if there is enough energy, then we compare the current coordinates and the finish coordinates. If they are equal, then it's over. If not, then proceed to move;
6. If the sensor detects a danger on the way, then we recalculate the trajectory (the action continues until the obstacle disappears from the field of view of the sensors);
7. Then the algorithm goes through again, until the current coordinates are equal to the coordinates of the "finish" [20]

Priority transition controller. For each of the subsystems, it is assumed that there is a certain signal for the transition of priority to it.

There is a certain criterion for transferring control to the self-preservation system: this is the distance closer than which objects can cause a dangerous situation, cause damage to themselves and the platform. Depending on the speed at which the platform is moving, the safe distance is recalculated. Therefore, the faster the ATP movement, the less time the platform has to change the trajectory in front of the obstacle. The safety distance is dynamically recalculated according to the surrounding circumstances.

To transfer control to a self-sustaining subsystem, the criterion is the remaining range, which is also dynamically calculated from the energy consumption of the platform from external consumers. If this power reserve is $+\ 15\%$ (conditionally) less than required before the finish line, then the priority goes to the self-sustaining system, which in turn ensures the recharging of the batteries [22].

7 Conclusion

Based on the analysis of work on the creation of autopiloted autonomous intelligent objects, the main problems of their creation were identified. Despite the intensive development of such areas as robotics, artificial intelligence, etc. approaches to the creation of a universal algorithm have not been proposed. All previously created works, for the most part, were narrowly specialized for a specific situation or for a specific object, but most importantly, they all contain complex mathematical calculations that reduce the speed of the ATP [23].

As you know, speed is one of the most important characteristics for autonomous platforms, since the safety of others and the ATP itself depends on it.

In this paper, the first steps have been taken to solve this problem, thanks to the development of a simple algorithm that distributes the priority of subsystems, which facilitates the work of ATP, since only the necessary subsystems are involved at the right time.

References

1. Timakov, S.I.: Overview of autonomous transport platforms and their behavior in 3D space INTELLECT. In: Proceedings of the Conference. Intellectual and Information Systems, pp. 306–309. TulGU Publishing House (2021)

2. Evstigneev, D.V.: Software-algorithmic support of intelligent control systems for autonomous mobile robots. Dissertation, Moscow, p. 218 (2003)
3. “Autopilot. Unmanned vehicle.” TADVISER. State. Business. Technology. https://www.tadviser.ru/index.php/. Accessed 23 June 2022
4. Guryev, A.: Model of strong artificial intelligence. Habr. Artificial intelligence. https://habr.com/ru/post/481172/. Accessed 23 June 2022
5. Anokhin, K.V., Sudakov, K.V.: Systemic organization of behavior: novelty as a leading factor in the expression of early genes in the brain during learning. Advances in Physiological Sciences **24**(3), 53–70 (1993)
6. Anokhin, P.K.: Nodal Questions of the Theory of Functional Systems, p. 198. Nauka, Moscow (1980)
7. Brodsky, Y.I., Lebedev, V.Y., Ogaryshev, V.F., Pavlovsky, Y.N., Savin, G.I.: General problems of modeling complex organizational and technical systems. In: book. “Problems of Cybernetics”. AN USSR, Moscow, pp. 42–48 (1990)
8. Gorodetsky, V.I.: Multi-agent systems: basic properties and models of behavior coordination. Information Technologies and Computing Systems, no. 1 (1998)
9. Perevezentsev, A.A.: Multi-agent model of information processing. Radioelectronics, Electrical Engineering and Energy: Proceedings. Report Ninth International Scientific and Technical Conference of Students and Graduate Students. MPEI, Moscow, pp. 302–303 (2003)
10. Pospelov, D.A. (ed.): Directory. Artificial Intelligence. In 3 books. Book 2 Models and Methods, p. 287. Nauka, Moscow (1990)
11. Pospelov, D.A.: Multi-agent systems - present and future. Information Technologies and Computing Systems, no 1 (1998)
12. Tarasov, V.B.: From artificial intelligence to artificial life: new directions in the sciences of artificial. News of Artificial Intelligence **4**, 93–117 (1995)
13. Russell, S., Norvig, P.: Artificial Intelligence: A Modern Approach, 2nd edn. Per. from English. Williams Publishing House, Moscow, p. 1408 (2006)
14. Yakovenko, O., Shestakova, M.P., Averkina, A.N.: Simulation of Human Motion Control, p. 360. Collection of Scientific Papers, SportAcademPress, Moscow (2003)
15. Osipov, G.S.: Methods of Artificial Intelligence and the Problem of Control, pp. 161–169. Nauka, Moscow (1990)
16. Belokon, N.N., Galagan, M.V., Semenchev, E.A.: Functional organization of mechanisms of self-sustainment and self-preservation of active self-organizing objects of simulation systems. News of TulGU. Series. Computer Engineering. Information Technology. Control systems. Issue 4. Information Systems. Publishing house of TulGU, Tula, pp. 58–65 (2005)
17. Belokon, N.N.: Functional organization of the mechanism of self-preservation of active self-organizing objects of simulation systems. International Youth Scientific Conference “XXX Gagarin Readings”. MATI - RGTU im. K.E. Tsiolkovsky, Moscow (2006)
18. Belokon, N.N.: The principle of operation of the self-preservation mechanism of an active object of an intelligent modeling system. I Scientific and Practical Master’s Conference, TulGU, p. 173 (2006)
19. Demidova, A.V., Semenchev, E.A.: Simulation study of a universal algorithm for moving intelligent unmanned objects in a dynamically changing three-dimensional space. Bulletin of TulGU. Series. Computer Engineering. Information Technology. Control systems. Issue 6, Information systems. Publishing house of TulGU, pp. 77–86 (2012)
20. Demidova, A.V., Semenchev, E.A.: Research of a universal algorithm for moving intelligent unmanned objects in a dynamically changing space. Sat. scientific works of students, graduate students and young scientists “Scientific creativity of the XXI century”. Belgorod, November, pp. 82–86 (2012)
21. Makarov, I.M., Lokhin, V.M., Manko, S.V., Romanov, M.P.: Artificial Intelligence and Intelligent Control Systems. Nauka, Moscow (2006)

22. Beloglazov, D.A., Kosenko, E.Y., Kobersi, I.S., Soloviev, I.O.: Intelligent motion control of autonomous moving objects based on behavioral approach. Electronic scientific journal "Engineering Bulletin of the Don", no. 3 (2015). https://ivdon.ru/ru/magazine/archive/n3y2015/3077
23. Demidova, A.V., Semenchev, E.A.: Simulation model of movement of intelligent autonomous objects in space with obstacles based on a universal algorithm. Technical science. News of TulGU **12**(2), 284–291 (2009)

Using an Information System for Distributing Orders at an Industrial Enterprise for Efficient Capacity Utilization

S. N. Medvedev(✉), O. P. Aksyonova, and E. K. Aksyonova

First President of Russia B.N. Yeltsin, Ural Federal University, 19 Mira St, Ekaterinburg 620002, Russia
stefanmedvedev@mail.ru

Abstract. The article provides a comparison of enterprise management information systems. Their functionality is analyzed with the definition of the pros and cons when used in an industrial enterprise, and the presence of a simulation module in them is checked, which allows for preliminary modeling of production processes to determine the ability of the enterprise to fulfill its obligations under existing orders. The information system developed by the authors is described, which includes three different, but interconnected modules, namely, a module for determining the optimal order distribution plan, a simulation module for "running" the model and comparing data with the optimal result obtained in the module for determining the optimal plan, as well as an analysis module, accounting and moving parts. In this system, when modeling processes, the distribution of enterprise orders is carried out by modeling the process of manufacturing item items using a new approach of "order distribution". This approach is based on the Kanban method, which differs from the standard method in that it is extended by the use of intelligent agents that collect, analyze, distribute, and store enterprise orders.

Keywords: Information technology · Production planning

1 Introduction

The article compares corporate information systems used for production management at industrial enterprises, considers the main directions of production activities of enterprises and their automation through the use of the considered information systems. The availability of the simulation module is checked to determine whether it is possible to analyze the execution of production orders at existing production facilities with minimal labor costs. The developed information system is considered for the distribution of production orders with the functional ability to simulate the process of manufacturing products using the developed approach of "retailing" which is based on the Kanban method. This approach is complemented by intelligent agents that determine which production chain to produce products on and that are responsible for dividing orders, check available production capacity and the ability to fulfill an incoming order. The structure of the system is presented, as well as the algorithm of the action of intelligent agents used in the simulation module which is part of the system.

A. A. Radionov and V. R. Gasiyarov (Eds.): RusAutoCon 2022, LNEE 986, pp. 57–64, 2023.
https://doi.org/10.1007/978-3-031-22311-2_6

2 Formulation of the Problem

Choosing the optimal corporate information systems for solving the problem of enterprise management, including the subtask of obtaining an effective production plan for a calendar period, is not a simple task that includes an analysis of existing systems on the market by functionality according to the list of production areas characteristic of industrial enterprises [1–3]. Large industrial enterprises often [4, 5] have divisions that allow them to produce products of the same type. Here is an example of one of these cases.

Machine-building enterprise consist of two different productions: metallurgical and assembly equipment. For the release of goods, the procedure consists of a full production cycle, namely the supply of the necessary materials and the production of the final product. The output chain of metallurgical manufacture consists of four workshops that organize the production scheme for the items and transfer of goods to the assembly shop. The technological process of production.

We will conduct a comparative analysis of corporate information systems for the purpose of drawing up an effective production plan for loading the same type of structural divisions present at an industrial enterprise.

3 Theory

Various information systems, such as (OmegaProduction [6], SAP S/4 HANA Cloud ERP system [7], ERP—Galaxy [8], Microsoft Dynamics AX [9], SmartFactory [10]) are used in various enterprises and allow, depending on the implemented functional modules, to solve various tasks. Consider the systems.

3.1 OmegaProduction

The OmegaProduction information system is an item of the Byelorussian corporation OmegaSoftware. This software is developed to control the trade and resources of the enterprise. The software belongs to corporate information system and integration of automation in all management processes of the enterprise, being a kind of CALS system. The system is a 2-link client-server architecture.

Considering this system as a system for production planning at a machine-building enterprise, it should be noted that the calculation of production plans is made according to the most common MRP II and APS methodologies.

3.2 SAP S/4HANA Cloud ERP System

The SAP S/4HANA cloud ERP system is a product of the SAP company which is a multifunctional information system belonging to the class of enterprise systems. The use of several modules together, such as SapPLM, SapERP, SapCRM, forms a CALS system in industrial enterprises. The most common product of the SAP company is the Sap ERP corporate information system. The system is a three-tier architecture (client-application server—DB).

Due to the fact that the management of engineering data is carried out in another module in Sap ERP, there is only the maintenance of specifications, material consumption rates, and a description of operations for manufactured products. Production planning takes place according to the MRP II and APS planning methodology. If there is no accurate data in the system, the production plan will contain incorrect data.

3.3 ERP-Galaxy

The ERP-Galaxy information system is a Russian development and also belongs to the class of corporate systems. The system architecture can be two-tier or three-tier. The planning function of this software is based on the MRP II methodology, a planning standard that is used to prepare production plans in an enterprise.

3.4 Microsoft Dynamics AX

Information system Microsoft Dynamics AX belongs to the class of corporate systems and is a software product that allows you to manage the resources of the enterprise throughout the production process, putting under control the functions of the supply chain of materials, conducting trade and procurement procedures, personnel management, financing, production planning. Microsoft Dynamics AX supports a three-tier architecture.

The production management and planning function is also based on the MRP II and APS methodologies. The construction of the plan can be either “forward” or “backward”, taking into account the limited production capacity.

3.5 SmartFactory

The production planning system from SmartSolutions is a Russian development which was released relatively recently. The system is designed to work at the MES level. This system is based on a multi-agent approach as a basis for production planning which radically distinguishes it from the previously considered systems. The multi-agent approach is based on the method of networks of needs and opportunities (PV-networks) [11]. The idea of this method is to use separate intelligent programmed agents for each existing entity, for example, a machine, a production line, a process, a person, a site, a production shop. Agents exchange information among themselves, so-called negotiations, and come to mutually beneficial solutions to obtain the optimal result and the earliest possible fulfillment of an existing goal or goals. Thanks to the use of this approach, the system has a flexibility in comparison with other information systems, due to which there is no need to perform calculations of a previously created production task when a new order appears.

4 Results of the Comparison

After reviewing the system data by application and function, we will summarize all the data in a table and compare these systems. In Table 1 shows the functional features of information systems.

Table 1. Functional features of the systems

No	Parameters	OmegaProduction	SAP	ERP-Galaxy	Microsoft DynamicsAX	SmartFactory
1	Production dispatching	+	+	+	+	+
2	Production planning (including splitting the order between multiple routes)	+	+	+	+	+
3	Warehouse accounting	+	+	+	+	+
4	Production planning	+	+	+	+	+
5	Engineering data management	+	–	–	–	–
6	Simulation module	–	–	–	–	–
7	Bottleneck analysis	–	–	–	–	–

Basing on the data in the table, we can see that the systems considered allow us to solve similar tasks, for example, sales accounting, supply accounting, but there are also differences, for example, focusing on business analysis, accounting unit, design and technological preparation of production, while support for order separation is implemented in almost all systems which allows us to talk about the presence of the so-called "Capacity Balancing" function. This function works both in automatic and manual mode, which does not allow you to determine the most effective plan and also requires significant time resources to create it in other words, systems build a plan based on current constraints, without taking into account the fact that it is possible to offer solutions for improving/reengineering processes, which can have a positive effect on planning. The presence of a simulation module is absent in all the systems considered which means that it is impossible to create a model and play it back to determine the most effective plan.

The considered information systems are ready-made software products with the introduction of which there is a greater change in the business processes of the enterprise than the customization of systems to meet the requirements of the customer, this is especially clearly observed in foreign companies such as SAP.

5 Planning Methods

In the considered information systems, various methods are used to build a production plan: the method of PV-Networks, APS, MRP-II.

5.1 Method APS

The APS method is a concept of optimized (or synchronous) production planning. Main features: increased planning detail; reduction of production stocks; improvement of product supply activities; better use of fixed assets; increased rhythmic utilization of production resources; more flexible response to competitive demands.

5.2 Method MRP II

The method is a production planning strategy that provides operational and financial planning for production.

5.3 Method PV-Networks

The main idea of the approach is to use software agents for each process or operation, as well as for each worker, equipment of the enterprise or workshop. The APS method and MRP II are more planning methodologies that define planning criteria. In view of this, it is proposed to take the Kanban method as the closest in terms of criteria to the APS method for conducting a comparative analysis. Replace the MRP II method with the simplex method of linear programming, because in both cases only the available resources and their limitations are taken into account in planning.

The task of assembling a car assembly product was considered. In the divisions of the plant, 3 different parts are manufactured (part 1, part 2, part 3), each of which has a different manufacturing process and one assembly product consisting of these parts. The applicability of each part included in the assembly unit is equal to 1 pc. The participating units are shown in Fig. 1. Each part and assembly unit is assembled according to the technological process. On the basis of the Kanban and PV-networks methods, simulation models were constructed and the results were carried out, in the case of the simplex method, mathematical calculations were performed. The data is shown in Table 2.

Based on the data in the table, it can be seen that the optimal method is a simplex method, but this is a mathematical calculation that shows the maximum allowable number of products to be assembled. The closest result is shown by the new method proposed by the authors. This is the MPPR/Kanban that is used in the created system. Models built using intelligent agents allow you to get any level of abstraction: high, medium and low, as well as to combine them with each other, which allows you to get realism and flexibility in describing the system, the ability to simulate the most complex nonlinear feedbacks, use any necessary level detail. In the future, the agent-based approach will be referred to as a multi-agent resource conversion process (MPPR), which displays the production processes at the enterprise with the participation of agents - decision makers.

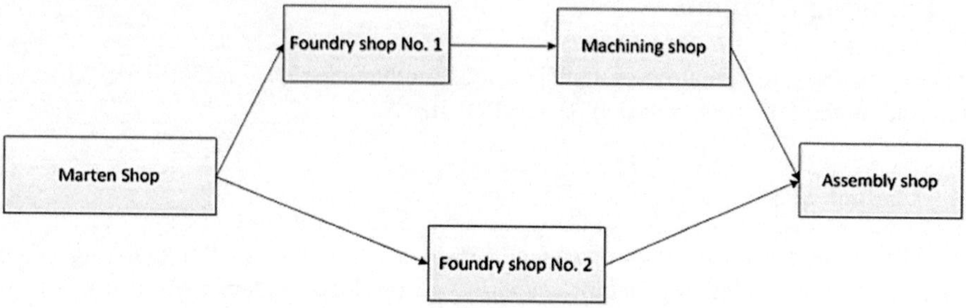

Fig. 1. Scheme of the manufacturing process.

Table 2. Model simulation results.

Parameters	Kanban	PV-Networks	MPPR/Kanban	Simplex method
Collected Products	62	70	71	87

6 New Information System

The developed information system is based on the objective function (1).

$$\sum_{j=1}^{z}\sum_{i=1}^{m}(c_{ij}) + c_v \rightarrow \min \tag{1}$$

where overhead costs $-c$, and c_v - shop costs with a decrease in production volumes. The developed order distribution system for the tasks of efficient order distribution at a machine-building enterprises is a system for obtaining an optimal plan for the distribution of orders along technological routes, simulation systems, taking into account resource constraints, and systems for accounting and movement of parts and assembly units that are integrated with each other. The relationship with other enterprise information systems occurs at the MSSQL Server database level.

The system includes:

1. The decision support system (DSS) BPsim [12]. DSS is designed to solve problems of technical and economic design and dynamic business modeling of organizational and technical systems, problems of dynamic intelligent modeling of multi-agent decision-making processes [13–15], problems of system analysis and development of models of production, business systems, bottleneck analysis, reengineering, optimization. Design of a conceptual model of the process of manufacturing components in the structural divisions of the enterprise. The problem was solved using the simplex method which was implemented by BPsim.DSS using the Transact SQL language.
2. The implementation of the method of "re-engineering" of parts and assembly units, in the distribution of production orders at a machine-building enterprise in various production planning approaches, is carried out in the system of dynamic situation modeling (SDMS) BPsim.MAS. BPsim.MAS is obtained as a result of the integration

of simulation, expert, situational and multi-agent modeling methods. The results obtained in the BPsim.DSS system are checked on the developed simulation model
3. The results obtained on the distribution of production tasks by structural units are transmitted to the shop MES system, where the production schedule is built for the planned period.

7 Discussion of Results

The application of the developed system in the conditions of the existing metallurgical production, which is a part of the machine-building enterprise, allowed to obtain an increase in the productivity of the divisions, in comparison with the original model, by 1.56 times. The use of the "re-engineering" method made it possible to evenly load the same type of structural divisions, i.e., to balance the capacities in the simulation model of metallurgical production.

The use of this system in large industrial enterprises will allow you to effectively load duplicate structural divisions by step-by-step solution of the problem:

1. obtaining the optimal result
2. testing it in a production simulation model, where there are various resource constraints
3. transmitting information to the MES system for distribution by site. In comparison with the ERP systems discussed above, carrying out these works involves minimal time and financial resources.

Acknowledgments. This work is supported by Act 211 Government of the Russian Federation, contract No. 02.A03.21.0006. The study was carried out with the financial support of the Russian Federal Property Fund in the framework of the scientific project No. 18–37-00183.

References

1. Medvedev, S.N., Aksyonov, K.A., Aksyonova, O.P.: Application of a decision support system in an industrial enterprise. In: Proceedings of the International Conference on Modern Trends in Manufacturing Technologies and Equipment (ICMTME), pp. 1–5 (2019). https://doi.org/10.1088/1757-899X/709/4/044026
2. Solovyeva, S.B., Ivanov, D.: Analysis of position optimization method applicability in supply chain management problem. Proceedings of the Int. Conf. "Stability and Control Processes" in Memory of V.I. Zubov (SCP), pp. 498–500 (2015)
3. Klebanov, B., Antropov, T., Riabkina, E.: Bases of imitation model of artificial society construction accounting of the agents' needs recursion. In: Proc. 16th Int. Multidisciplinary Scientific GeoConf. SGEM2016 (2016). https://doi.org/10.5593/SGEM2016/B21/S07.014
4. Borodin, A., Kiselev, Y., Mirvoda, S., Porshnev, S.: On design of domain-specific query language for the metallurgical industry. In: Proc. 11th Int. Conf. Beyond Databases, Architectures and Structures (BDAS), pp. 505–515 (2019)

5. Borodin, A., Mirvoda, S., Kulikov, I., Porshnev, S.: Optimization of memory operations in generalized search trees of PostgreSQL. In: Proc. 13th Int. Conf. Beyond Databases, Architectures and Structures (BDAS), pp. 224–232 (2017)
6. Overview of the system features—"Omega Production": (2021). https://www.eastsoft.su/omega.html
7. SAP S/4HANA Features: (2021). https://help.sap.com/doc/b870b6ebcd2e4b5890f16f4b06827064/2021.000/en-US/WN_OP2021_EN.pdf
8. Corporation Galaxy: Galaktika ERP system—Description of the system functionality (2017) https://www.sap.com/cis/products/s4hana-erp/features.html.ru/docs/ERP_about.pdf
9. Microsoft. Microsoft Dynamics AX 2012: Functional and technological capabilities (2012). https://www.rea.ru/ru/org/cathedries/isemkaf/Documents/AX2012---------.pdf
10. Smart Factory—Intelligent production management system. https://kg.ru/solutions/smart-factory/ (2021)
11. Wittich, V.A., Skobelev, P.O.: Multi-agent interaction models for the design of the nets of requirements and capabilities in open systems. Autom. Telem. **1**, 177–185 (2003)
12. Aksyonov, K., Aksyonova, O., Antonova, A., Aksyonova, E., Ziomkovskaya, P.: Development of Cloud-Based Microservices to Decision Support System. In: Ivanov, V., Kruglov, A., Masyagin, S., Sillitti, A., Succi, G. (eds.) OSS 2020. IAICT, vol. 582, pp. 87–97. Springer, Cham (2020). https://doi.org/10.1007/978-3-030-47240-5_9
13. Dreżewski, R.: A Model of Co-evolution in Multi-agent System. In: Mařík, V., Pěchouček, M., Müller, J. (eds.) CEEMAS 2003. LNCS (LNAI), vol. 2691, pp. 314–323. Springer, Heidelberg (2003). https://doi.org/10.1007/3-540-45023-8_30
14. Manuel Rodriguez, L.A., Manuel, M.: Good practices for the adoption of DataOps in the software industry. J. Phys: Conf. Ser. **1694**, 1–5 (2021)
15. Aksyonov, K., Aksyonova, O., Ayvazyan, H.: Identification of problems and limitations of the planning of fuel distribution through the gas station network and their solution. In: Proc. Int. Multi-Conf. on Engineering, Computer and Information Sciences (SIBIRCON), pp. 647–651 (2019). https://doi.org/10.1109/SIBIRCON48586.2019.8958191

Development of a Decision Support System for the Diagnosis of Clinical Diseases

A. A. Ukhanova(✉) and J. V. Frantsuzova

Tula State University, Ave. Lenina, 92, Tula 300012, Russia
ukhanova.anast@yandex.ru

Abstract. This article discusses the decision support system used in medical practice, which makes it possible to facilitate the process of making a diagnosis and prescribing treatment. The principle of operation of the decision support system is stated. A decision support system (DSS) is a computer automated system, the purpose of which is to help people who make decisions in difficult conditions for a complete and objective analysis of subject activity. The decision support system is designed to support multicriteria decisions in a complex information environment. At the same time, multi-criteria are understood as the fact that the results of decisions made are evaluated not by one, but by the totality of many indicators (criteria) considered simultaneously. Information complexity is determined by the need to take into account a large amount of data, the processing of which is practically impossible without the help of modern computer technology. Under these conditions, the number of possible solutions, as a rule, is very large, and the choice of the best of the solutions, without a comprehensive analysis, can lead to gross errors. Thus, in this paper, we consider the solution to the problem of a large information load on specialists and the design of the system being developed.

Keywords: Automation · System · DSS · Information · Diagnosis

1 Introduction

Currently, researchers pay special attention to the use of the Decision Support System (DSS) in complex analytical cases. It is known that when making any medical decisions, problems can be lack of knowledge, limited time resources, inability to attract a large number of competent experts, incomplete information about the patient's condition [1].

Making the right decisions in medicine is usually difficult due to the multiplicity of factors and signs of most diseases, their interactions. The general condition of the patient can be influenced by certain factors not only individually, but also in certain combinations of values.

2 Relevance

Decision-making in medicine is directly related to human health and life, so decisions obtained in medical DSS should be as reliable as possible. Processing a large amount

A. A. Radionov and V. R. Gasiyarov (Eds.): RusAutoCon 2022, LNEE 986, pp. 65–74, 2023.
https://doi.org/10.1007/978-3-031-22311-2_7

of information and identifying certain patterns is often beyond the power of a clinician. Modern computer technology with appropriate software can help to cope with this [2].

The problem of providing computer support for decision-making in medicine is relevant due to the increasing information load on the doctor and the inability to devote enough time to the initial survey and making a preliminary diagnosis.

This problem can be solved by an automated diagnostic method.

An automated diagnostic method uses DSS algorithms that provide assistance in solving a number of clinical cases and tasks that can cause certain difficulties for a specialist. The use of an expert system will also allow for preliminary diagnosis of nosological groups of diseases. It is allowed to use the system by patients independently for the primary collection of information about the disease. The proposed system will allow to work according to an algorithm that ensures the detection of pathology according to the symptomatic-syndromic principle [3].

3 Formulation of the Problem

At the moment, the greatest attention is paid to the problem of development and application of medical information systems, their integration in the direction of building a single information space. Therefore, it is possible to single out the tasks that can be achieved through the implementation of DSS:

- reducing the time of examination by reducing the number of tested symptoms and identifying only signs that are significant for diagnosis;
- reducing the percentage of medical errors in diagnosis and differential diagnosis by reducing the influence of the human factor on the decision-making process based on the available data on the pathological process.

4 Structure of the DSS

The DSS includes components (Fig. 1): data sources, data model, database models and a software subsystem, which consist of a basic data management system (DBMS), a basic model management system (BMS) and a user-computer interface management system.

DSS architecture is represented by different authors in different ways. For example, Marakas (1999) [4] proposed a generic architecture consisting of five different parts:

- the data management system (DBMS),
- the model management system (MBMS),
- the knowledge engine (KE),
- the user interface,
- the user(s) [5].

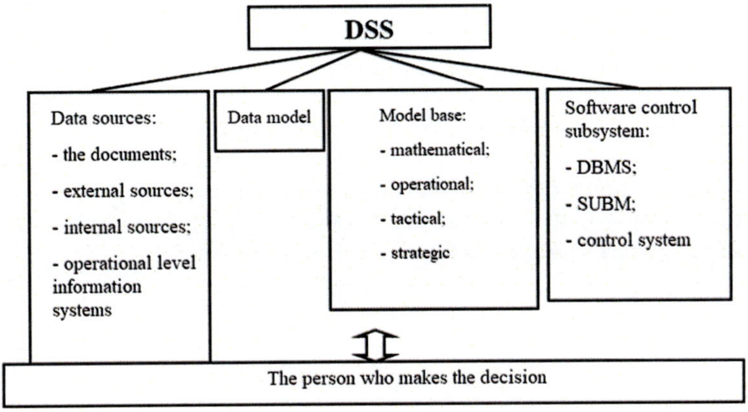

Fig. 1. Structure of the DSS.

The DSS information platform is the Data Warehouse [6].

The main functional set of DSS includes:

- formation of consolidated reporting (up to 200 preconfigured reports);
- multidimensional data analysis (OLAP);
- revealing hidden patterns (Data Mining);
- statistical analysis and forecasting of time series;
- formation of preconfigured requests (up to 500 – 600);
- intellectual search (by incomplete data and informal queries) [6].

DSS tool environment - integration systems based on open standards. These systems meet the requirements:

- information security;
- scalability;
- openness;
- multidimensional and multivariate presentation of data;
- intelligent interface;
- integrability with major platforms and business applications, integration of data from various sources, network integration (primarily web);
- provide a service for "cleaning" data when they are loaded into storages [7].

The technical support is connected with:

- data processing;
- secure data storage and integrity;
- archiving and data recovery;
- network and telecommunication support;
- cryptographic support;
- user access control;
- data loading, including using intelligent interface tools (pattern recognition: text, speech, images) [8].

5 The Principle of Operation of the DSS

The DSS provides information based on input data that helps people quickly and accurately assess the situation and make a decision.

From the functional diagram (Fig. 2), it can be seen that in order to receive a diagnosis and a prescription for treatment, it is necessary to enter the patient's testimony with his complaints into the DSS. Further, the system will automatically determine the preliminary diagnosis using the necessary medical documentation.

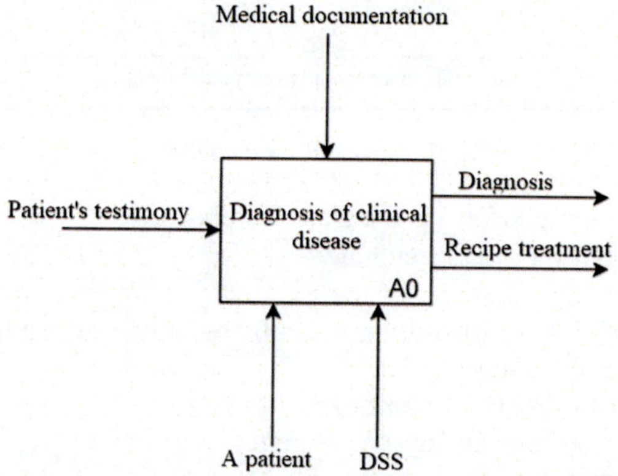

Fig. 2. Functional diagram of diagnostics using the application.

The decomposed process of establishing a diagnosis and prescription of treatment consists of three successive stages. After entering the patient's testimony, the DSS establishes a preliminary diagnosis, and based on it displays an approximate treatment prescription (Fig. 3).

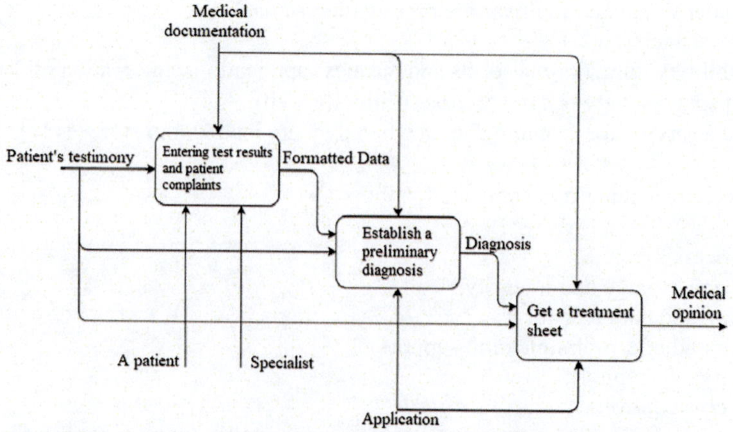

Fig. 3. Decomposition of the process of establishing a diagnosis.

The whole process is automated and as a result takes a minimum of time.

In general, DSS in clinical medicine perform the following functions:

- differential diagnosis and choice of treatment in a wide range of diseases;
- the effectiveness of solutions regardless of the severity of the clinical manifestations of the disease, which involves diagnosis in early forms of diseases and an erased clinical picture;
- taking into account the background conditions (comorbidities) of the patient, which is especially important in the selection of treatment;
- analysis of the dynamics of the pathological process with the forecast of potential adverse situations [9].

The process of making medical decisions can be represented as a cycle consisting of sequential procedures following one after another. The first three procedures implement the collection, processing and analysis of medical information. The fourth procedure is decision support, including conceptual or mathematical modeling, development of alternatives and selection of those that best meet the goals, which means choosing the most optimal treatment option for the patient. The fifth and sixth procedures include the selection of the set of the most effective medical measures and their implementation. After that, the cycle closes and the collection of information begins again, etc. [1].

The subsystems included in the DSS during operation exchange information based on the HTTP protocol using AJAX/JSON formats, using the information exchange modules included in the software.

The executive part of the application is responsible for most of the tasks related to processing requests, searching for solutions and working with data, interacting with the data warehouse, processing information during the operation of the web portal, as well as providing an interface for calling various operations on the server side, with the ability to send results performing these operations on the public part (Fig. 4).

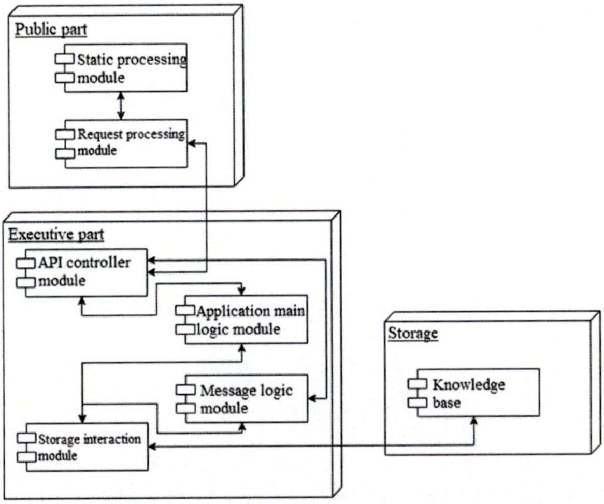

Fig. 4. Functional diagram of the system.

The public part performs request proxying and load balancing on the executive part, and is also responsible for processing static content and accessing the global cache.

6 The Method Used in DSS

The developed DSS will be based on the decision tree method. A decision tree can be called a visual instruction of what to do and how to act in various situations. A huge advantage of decision trees is that they are easily interpretable and human-readable.

The fundamental ideas that inspired the emergence and development of decision trees were laid down in the 1950s in the field of research on modeling human behavior using computer systems. Among them, it is worth highlighting the works of K. Hovland "Computer modeling of thinking" [10] and E. Hunt et al. "Experiments on induction" [11].

The main idea of any decision tree algorithm is as follows (Fig. 5):

- Select the best attributes using Attribute Selection Measures (ASM) to split the record.
- Make this attribute set a decision node and split the data set into smaller subsets.
- Start building the tree, repeating this process recursively for each child until one of the following conditions is met: (a) all tuples belong to the same attribute value; (b) no more remaining attributes; (c) no more instances [12].

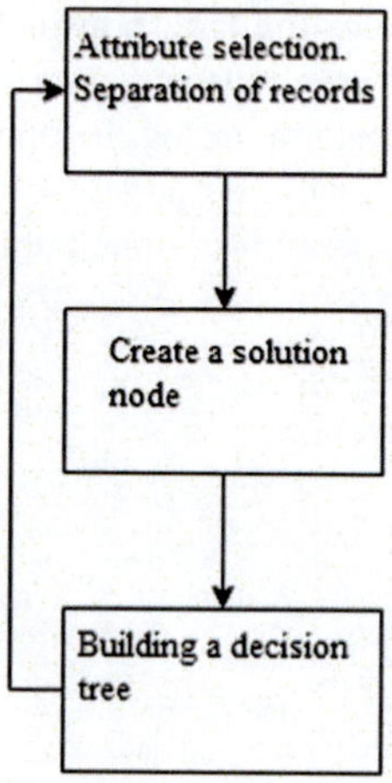

Fig. 5. Decision Tree Algorithm.

6.1 Description of the Learning Algorithm

Let a partitioning rule be applied to the training set, which uses the attribute A, which takes p values $a_1, a_2,\ldots,a_p$. [13]. As a result, p subsets $S_1, S_2,\ldots,S_p$ will be created, where examples will be distributed in which attribute A takes the corresponding value (Fig. 6).

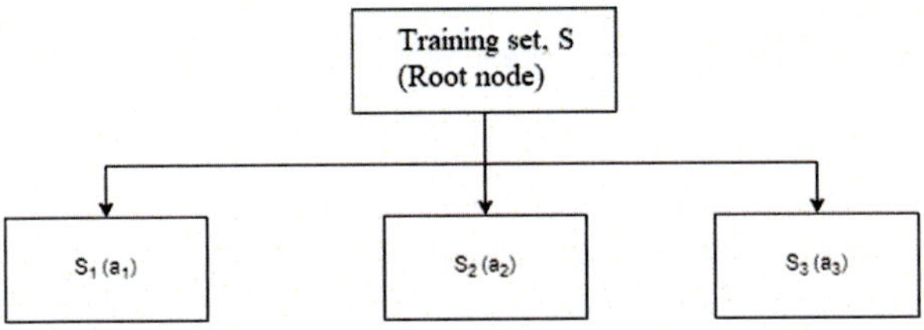

Fig. 6. Scheme of splitting into subsets by attribute A.

This raises the question: is the split by the selected attribute the best, or by choosing another attribute, we could get a better split? To answer this question, we use information about the number of examples of all classes in the training set and in each resulting subset [14].

The most popular selection criteria are information gain, information gain, and gain.

Information gain is a statistical property that measures how well a given attribute separates training examples according to their target classification [15].

To calculate the information gain, we have to introduce the term *entropy* of the dataset.

In physics and mathematics, entropy is called a randomness or an impurity in a system. In information theory, this refers to impurities in a group of examples. Getting information is a decrease in entropy.

6.2 Mathematical Support of the Decision Tree Method

The probability of class C_j in the set is presented in Eq. 1:

$$P = \frac{N(C_j S)}{N(S)} \tag{1}$$

where $N(S)$ is the total number of examples in the set S, $N(C_j S)$ is the number of examples of class C_j in the set S [16].

The Shannon entropy is defined for a system with N possible states by Eq. 2:

$$S = -\sum_{i=1}^{n} p_i \log_2 p_i \tag{2}$$

where p_i is the probability of the system being in the i-th state [17].

Formally, the increase in information (information gain, IG) when splitting the sample according to the attribute Q is determined by Eq. 3:

$$IG(Q) = S_0 - \sum_{i=1}^{n} \frac{N_i}{N} S_i \tag{3}$$

where n is the number of groups after splitting the sample, N is the number of sample elements for which feature Q has the i-th value [18].

The described procedure is applied to subsets Si and further, until the values of the criterion cease to increase significantly with new partitions or another stop condition is met [14].

7 Analysis of Existing Software Facilities

Currently, there are systems that automate the process of diagnosis. Research in the field of using artificial intelligence in medicine began in the late 60s and early 70s. XX century. The first developments were systems such as:

- AAHelp (University of Leeds), studying the search for the causes of severe pain and deciding on the need for surgical intervention. This system is highly specialized and does not make preliminary diagnoses [19].
- The INTERNIST system (University of Pittsburgh) solved the issue of assistance in making a diagnosis based on the observed symptoms. This system had a loaded and obscure interface. The consultation took from 30 to 90 min, which was too long for most clinics. The system was a universal computer diagnostic tool and worked only in offline mode [20].

From 2000 to the present, the following systems have been successfully developed:

- IndiGO (Archimedes) - provides processing of data of clinical, physiological nature and information about the management of the treatment process and forms individualized protocols for diagnosis and treatment. The system does not have an online mode of use and the possibility of data exchange between users.
- Auminence - a system of differential diagnosis, analyzes information about symptoms and other information, forms a diagnostic plan. The system is also intended only for the use of doctors without an online mode [21].

For clarity, we summarize the advantages and disadvantages of the considered approaches in Table 1.

Table 1. Advantages and disadvantages of existing analogues.

Analogues	Accomplishments	Flaws
AAHelp	Ability to determine the need for surgery	Narrow specialization, no diagnosis
INTERNIST	Ability to make preliminary diagnoses based on observed symptoms	Long consultation time. Loaded interface
IndiGO, Auminence	Systems make diagnoses and treatment plan	Intended for use by physicians. Does not have an online mode

The difference between the developed DSS and the existing ones is that this system will work directly with patients (without the participation of doctors, acting as a personal consultant), using the results of their analyzes and possible complaints and symptoms of the patient. If the system cannot determine the diagnosis, then it will be possible to send the entered data to the doctor for further consideration and conclusion.

8 Conclusion

Information and telecommunication technologies in the healthcare sector are developing at a rapid pace. The main effect that is planned to be achieved with the help of medical decision support systems is to provide better medical care to patients. The use of various methods of these systems should reduce the number of complications, deaths, reduce the frequency of using ineffective methods of treatment and prevention, and low-informative diagnostic methods.

The proposed DSS-based automation method makes it possible to use decision support systems in clinical practice for diagnosing diseases, reduce examination time by reducing the number of tested symptoms and identifying only signs that are significant for diagnosis, and reduce the percentage of medical errors by reducing the influence of the human factor on the decision-making process. Based on the available data on the pathological process.

References

1. Simankov, V.S., Khalafyan, A.A.: System analysis and modern information technologies in medical decision support systems. BinomPress, Moscow (2009)
2. Litvin, A.A., Litvin, V.A.: Decision support systems in surgery—News of surgery: UO GSMU Minsk **22**, 96–100 (2014)
3. Nora, S.A., Smirnova, I.V., Arkhipov, G.S.: Automated approach to the diagnosis of allergic diseases using decision support systems. Topical issues of fundamental, clinical medicine and pharmacy. NGOU im. Yaroslav the Wise, Veliky Novgorod, pp. 226–229 (2018)
4. Marakas, G.M.: Decision Support Systems in the Twenty-First Century. Upper Saddle River, Prentice Hall (1999)
5. Prokopenko NYu (2017) Decision support systems based on Deductor Studio Academic 5.3: Tutorial. N. Novgorod, p. 12
6. Barseghyan, A.A.: Data analysis technologies: Data Mining, Visual Mining, Text Mining, OLAP. BHV-Petersburg, St. Petersburg, p. 284 (2007).
7. Popov, A.L.: Decision Support Systems: Textbook. Yekaterinburg, p. 6 (2008)
8. Dichenko, S.A.: Control and ensuring the integrity of information in data storage systems. Science-intensive technologies in space research of the Earth **11**, 49–55 (2019)
9. Kobrinsky, B.A., Zarubina, T.V.: Medical informatics: textbook, p. 192. Academy, Moscow (2009)
10. Hovland, C.I.: Computer simulation of thinking. Am. Psyhol. **15**(11), 687–693 (1960)
11. Hunt Earl, B., Marin, J., Stone, P.J.: Experiments in Induction. Academic Press, New York (1966)
12. Nekrasov, N.V.: Automation of the "Decision Tree" Method: Actual Issues of Economic Sciences. RGRU, Ryazan, p. 3 (2013)
13. Paklin NB, Oreshkov VI (2013) Chapter 9. Business Analytics: From Data to Knowledge (+CD): Tutorial. 2nd ed. Peter, St. Petersburg, pp. 444–459
14. Savenkov, P.A.: Application of Data Analysis Algorithms in the Management Decision Support System. TulSU, Academy, Tula, p. 2 (2019)
15. Hastie, T., Tibshirani, R., Friedman, J.: Elements of Statistical Learning. Springer (2009)
16. Bolshakov, M.A., Molodkin, I.A., Pugachev, S.V.: Comparative analysis of machine learning methods to assess the quality of IT services. In: Models and Methods for Researching Information Systems in Transport. SPRT of Emperor Alexander I, Saint-Petersburg, pp. 142–148 (2020)

17. Kaziev, V.M.: Introduction to the analysis, synthesis and modeling of systems: Textbook, p. 243. BINOM, Moscow (2006)
18. Poletaeva, N.G.: Classification of machine learning systems, pp. 5–22. Bulletin of the Baltic Federal University. I. Kant, Kaliningrad (2020)
19. Polat, K., Akdemir, B., Gbne, S.: Computer aided diagnosis of ECG data on the least square support vector machine. Digit Signal Process **18**(1), 25–32 (2008)
20. Frolov, S.V., Kulikov, A.Y., Ostapenko, O.A., Strygina, E.V.: Decision support systems in medicine. Sci. J. TSTU, Tambov, p. 8 (2018)
21. Efimenko, I.V., Khoroshevsky, V.F.: Intelligent decision support systems in medicine: a retrospective review of the state of research and development and prospects. Materials of the conference "Open semantic technologies for designing intelligent systems." Minsk **7**, 251–260 (2017)

Check for updates

An Approach to Modeling the Interrelations Variability at Elements of Production Systems with a Variable Structure

J. V. Doronina and D. V. Moiseev(✉)

Sevastopol State University, Universitetskaya St, 33, Sevastopol 299053, Russia
DVMoiseev@Sevsu.ru

Abstract. The problems in modeling the interrelationships of production systems elements with a variable structure are considered, which generates a number of difficulties in description and research. The study of the functioning processes of complex production systems is fundamentally important due to the growing requirements for their efficiency and resource intensity. In enterprises where rapid changeover and/or rapid composition change o of production equipment is required, it is necessary to take into account the variability of the relationships between the elements. Changeover refers to the process of transition, for example, of a machine from the production of one product or part to the production of another by replacing molds, clamping joints, and so on.In most cases, modern production systems have a multicomponent composition, which imposes restrictions on the possibilities of modeling the relationships of their elements due to the dimension problem of the model. The component composition enlargement of production systems causes possible information losses in the description of each element associated with the analysis of the element significance and its relationships with other elements, taking into account the variable structure. An aggregated approach is proposed that takes into account the uniqueness degree of system components, changes in their significance by taking into account the component inclusion degree in the system, which will allow making informed decisions on CPS reconfiguration taking into account the variable structure.

Keywords: Production system · Variable structure · Component inclusion · Elements significance · Connections significance · Model enlargement

1 Introduction

The study of the functioning processes of complex production systems (CPS) is fundamentally important due to the growing requirements for their efficiency and resource intensity. In enterprises where rapid changeover and/or rapid composition change o of production equipment is required, it is necessary to take into account the variability of the relationships between the elements. Changeover refers to the process of transition, for example, of a machine from the production of one product or part to the production of another by replacing molds, clamping joints, and so on.

A. A. Radionov and V. R. Gasiyarov (Eds.): RusAutoCon 2022, LNEE 986, pp. 75–84, 2023.
https://doi.org/10.1007/978-3-031-22311-2_8

When using the methods of rapid changeover (Single-Minute Exchange of Dies (SMED), it is possible to reduce the time of equipment setup and changeover operations. It is known that there are two main types of SMED: operations that are performed after the equipment is stopped; operations that can be performed during the operation of the equipment [1, 2].

At the same time, a number of problems arise:

- The time required for changeover may exceed the critical level, according to the requirement of efficiency;
- Maintaining the components of the reconfigurable equipment in a “hot” reserve is problematic and expensive, however, in certain cases, the components of such equipment can be implemented on the principle of flexible production and transfer of functions to elements with similar or complex functions;
- Modeling of relationships between CPS elements should take into account different categories of relationships and the elements themselves;
- In addition, the significance of the links and the significance of the components themselves in the system may be incommensurable and non-stationary. Incommensurability is reflected either in the discrepancy of the measurement scales, or in a significant spread of values on a homogeneous scale. For example, the time to failure of CPS equipment can reach hundreds of hours, and the average time to implement one operation is several milliseconds. Non-stationarity is mainly associated with changes in indicators (for example, values or weights of components) over time.

A decrease in the number of states should not lead to a loss of accuracy in estimating the system output characteristics. When calculating approximate values of these characteristics, errors should be estimated, which complicates the mathematical model [3, 4]. The phase enlargement principles o have been developed, the essence of which is that the phase space of the initial system is split into a finite number of disjoint classes. The states of each class are “glued” into one state, and the probabilities of the initial states are summed up. An enlarged system is being built in a new, enlarged phase space [4]. This approach is based on asymptotic methods.

The synthesis of a complex system should take into account two aspects: the functional binding of functions (as a service) to the structural elements of the system and the uncertainty in the functional and structural components during the cycle of this system functioning. Functional binding means that there is a basic relationship between the elements of the system and the functions inherent in these elements. On the other hand, local changes of function bindings to structural elements are possible under certain conditions [3, 5]. The inconsistency of these two aspects can be solved by identifying stable (basic) structural elements with a clear binding of functions, as well as subsets of structural elements with a limited functional binding.

In [3], various methods of enlargement are considered and it is indicated that, unlike the existing ones, the author’s method consists in the fact that the states of the enlarged system are described by the same characteristics as the original one, that is, isomorphism of the original and enlarged system is ensured, including the enlarged system retains the markovity property. The idea of enlarging states by functional attribute, that is, by the functions performed by the system in each state, is considered.

The tasks of analyzing the elements significance in most cases are associated with the definition of methodological and computational aspects of technogenic risks in system functioning within the framework of applied subject areas [5], which, for the most part, concerns applied software.

The method presented in [6] for estimating the weighting coefficients of organizational and technical systems involves calculating the values of the objective function of the system without each element in turn. A tuple of the obtained values of the system objective function $F = < f1 \ldots fs \ldots fS >$, is constructed, where *fs* – the indicator value of the objective function with the excluded *s–th* element. The value of the redistributed weighted average coefficient, which is the same for all remaining elements of the system, is calculated as $w^* = 1/(S - 1)$. It is assumed that this approach will reduce the volume of expert assessments [7–11].

Thus, the existing studies [12–15], for the most part, do not focus on the non-stationarity of the weight functions that determine the significance of the component or its relationship with other CPS components [16, 17]. In addition, with the enlargement of multicomponent systems, the significance of boundary elements and their connections may change due to the formation of a new phase space of states of the system.

2 Problem Statement

The purpose of the study: modeling the variability of the interconnections of elements in the CPS by taking into account the inclusion degree of the component in the system and measures of components cross-inclusions in order to clarify solutions for managing the functioning of the CPS variable structure, for example, reducing the time for reconfiguration.

Let 's introduce a number of notations. Let be $f_i -$ Let be a function (service) that must be implemented by some component of CPS, $i = \overline{1, I}$, where I - is a finite number of CPS functions; F- is a finite set of IS functions; $S-$ a set of structural units, where $s_j -$ structural element of CPS, $j = \overline{1, J}$, J - s a finite number of structural elements of CPS.

Let's denote the significant subsets of the set $F = \{F_{if},\ F_f,\ F_s\}$, $F_{if} -$ (inflexible) a subset of functions that are rigidly bound to the structural units of the CPS, but can be implemented on other elements; $F_f -$ (flexible) a subset of flexible functions that are not tied to the structural units of the CPS, $F_s -$ stationary (fixed) functions that are distributed to certain components of the CPS. Examples F_{if} may be functions assigned to a specific section of the program code of a numerically controlled machine (CNC), which implies their binding to a specific structural unit of the CPS. The function of some intermediate type of product data control, on the contrary, can be implemented in various subsystems: a subsystem for data collection, a subsystem for their analysis or storage, which makes it possible to define it as F_{if}. The function of making a decision on the products rejection, for example, can only be in the corresponding subsystem and is generally defined as F_f.

In Fig. 1 shows possible combinations of subsets in the structure of the CPS, dotted lines A, B and C reflect possible connections of flexible functions and rigidly functionally oriented elements of the CPS.

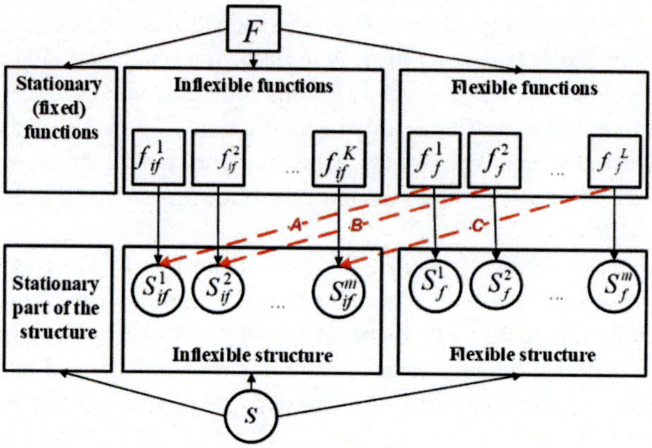

Fig. 1. The correspondence scheme of CPS structural elements and functions, taking into account their decomposition into stationary, flexible and inflexible.

The definition of these functions subsets that are rigidly tied to the structural units of CPS and flexible functions, as well as the construction of possible implementations of structures depending on various requirements for this system, are very problematic [16]. In order to simplify the description of the methods, a subset of stationary rigidly functionally oriented CPS structures is excluded from further consideration.

3 Methods and Results

Let us consider s_i and s_j – some components of the system S with weighting coefficients determining their significance in CPS: w_i and w_j respectively; l_{ij} and l_{ji}– he type of relationship between the components s_i,.. And their weighting coefficients w_i and w_j respectively. Then, in the general case, CPS can represent a tuple taking into account the variable structure (flexibility):

$$\begin{gathered} S_{if}\left\langle s_{if}^{i} | w_{if}^{i},\ s_{if}^{j} \middle| w_{if}^{j},\ l_{if}^{ij} \middle| w_{if}^{ij},\ l_{if}^{ji} \middle| w_{if}^{ji} \right\rangle, \\ S_{f}\left\langle s_{f}^{k} | w_{f}^{k},\ s_{f}^{m} \middle| w_{f}^{m},\ l_{f}^{km} \middle| w_{f}^{km},\ l_{f}^{mk} \middle| w_{f}^{mk} \right\rangle. \end{gathered} \tag{1}$$

With the nonstationarity of the weighting coefficients of the links for some $t_p \in T$ they can be represented by a weighting function of time:$w_i(t_p)$ and $w_j(t_p)$. Then expressions (1), taking into account their functionality, will take the form:

$$
S_{if}\left\langle \begin{array}{l} (s_{if}^{i}:f_{if}^{i},f_{f}^{i})|w_{if}^{i}(t_p), \\ (s_{if}^{j}:f_{if}^{j},f_{f}^{j})\Big|w_{if}^{j}(t_p), \\ l_{if}^{ij}\Big|w_{if}^{ij}(t_p),\ l_{if}^{ji}\Big|w_{if}^{ji}(t_p) \end{array} \right\rangle, \quad S_{f}\left\langle \begin{array}{l} (s_{f}^{k}:f_{f}^{k})|w_{f}^{k}(t_p), \\ (s_{f}^{m}:f_{f}^{m})\Big|w_{f}^{m}(t_p), \\ l_{f}^{km}\Big|w_{f}^{km}(t_p),\ l_{f}^{mk}\Big|w_{f}^{mk}(t_p) \end{array} \right\rangle. \tag{2}
$$

The entry $(s_{if}^{i}:f_{if}^{i},f_{f}^{i})|w_{if}^{i}(t_p)$ means, that the i-th element of the CPS structure s_{if}^{i} has an inflexible function f_{if}^{i} and can perform some additional function f_{f}^{i} provided the weight function $w_{if}^{i}(t_p)$ is entered.

Compliance with the structural element s_{if}^{i} of a certain weight function $w_i(t_p)$ is some assumption implemented by the decision maker (DM), since when transferring the function f_{f}^{i}, the value of the weight function may change.

Setting the weighting coefficients w_i is usually not difficult and the DM is implemented, while finding the weighting function $w_i(t_p)$ is problematic due to the presence of uncertainty and non-stationarity. It is proposed to use for these purposes the degree of inclusion of the element s_i in CPS. The expression for the degree of inclusion of the component in the system, taking into account (2), in general form can be written as

$$
u_{s_i}(t) = \frac{\sum_{i=1}^{N} w_i^{out}(t) \sum_{j=1}^{M} w_{ij}^{out}(t)}{\sum_{l=1}^{L} w_l^{in}(t) \sum_{h=1}^{H} w_{hl}^{in}(t)} \tag{3}
$$

where N, M, L, H – the cardinalities of the sets of external (out) and internal (in) components S, external and internal connections, respectively, relative to the S_i component.

The measure of component inclusion s_j in as a normalized number of intersections of their properties in is proposed to be evaluated as follows:

$$
W(s_i, s_j) = m(s_i \cap s_j)/m(s_i) \tag{4}
$$

In the case under consideration, the relation (4) can be interpreted as a normalized number of intersections l_{ij} and l_{ji} taking into account w_{ij} and w_{ji}. Then taking into account (4) for $t_p \in T$ the expression (3) will take the general form:

$$
u_{s_i}(t_p) = \frac{W(s_i, s_j)\left(\sum_{i=1}^{N} w_i^{out}(t_p) \sum_{j=1}^{M} w_{ij}^{out}(t_p)\right)}{W(s_j, s_i)\left(\sum_{l=1}^{L} w_l^{in}(t_p) \sum_{h=1}^{H} w_{hl}^{in}(t_p)\right)} \tag{5}
$$

To account for the impact of flexible functionality, it is advisable to use the expression (6):

$$u_{s_{if}^i, s_{if}^j}(t_p) = \frac{W\left[s_{if}^i; s_{if}^j\right]\left(\sum_{f_{if}^i f_f^i}\left(\sum_{i=1}^N w_{if}^{i,out}(t_p)\sum_{j=1}^M w_{if}^{ij,out}(t_p)\right)\right)}{W\left[s_{if}^j; s_{if}^i\right]\left(\sum_{f_f^i f_{if}^i}\left(\sum_{l=1}^L w_{if}^{l,in}(t_p)\sum_{h=1}^H w_{if}^{hl,in}(t_p)\right)\right)}, \quad (6)$$

where the designation $\sum_{f_{if}^i f_f^i}(\bullet)$ is an abbreviated description of the summation process by weight coefficients at a time $t_p \in T$ for various functions (flexible or inflexible) introduced into this element of the structure, which can significantly change $u_{s_{if}^i, s_{if}^j}(t_p)$. The estimation of this indicator may be complicated by the variability of the typification of connections l_{ij} and l_{ji}, n this study, this aspect is not taken into account, but will be considered in the future.

To model the variability of relationships between CPS elements, we introduce the assumption that in the simplest case, structural elements and functions coincide and can be represented by states [17–19]. For example, the operation of the software module controlling the changeover corresponds to the transition of the CPS to the changeover state [20–24]. Thus, under a number of assumptions, Fig. 2 presents a graph of states for the model under consideration in the following phase space:$E = \{E0, E1, E2, E3, E4\}$, where

E0—CPS operating state,
E1—CPS failure state,
E2—CPS recovery state after failure,
E3—CPS changeover state,
E4—failure state—is a state from which a transition can be made to a failure state, after which recovery is required, or to an initial state without recovery.

Thus, E4 is a state in which system recovery is not required. Transitions between states are indicated by arrows, there are stable states of the type in the model, reflecting the situation when at some control point in time the system did not switch to another state from the current one.

The transformation of the CPS block diagram into a state model is valid under the following assumptions:

Phase enlargement of the model with generalized states is applied in order to form a preliminary scheme for sequential reconfiguration of CPS changeover processes;

The assumption of one-sided functional and structural correspondences is introduced, Fig. 1.

When solving the problem of modeling the process of CPS functioning (without taking into account the flexible binding of functions during readjustment), it is obvious that at various stages of its functioning before the onset of degradation and after the development of degradation processes, the weights of the corresponding states, as a reaction to changes in the weight functions of elements or their functionality, may change: $w_1(t_1) > w_1(t_0)$, $w_4(t_1) < w_4(t_0)$, for $l_{0,4}$, $l_{0,1}$ accordingly:$w_{0,1}(t_0) < w_{0,4}(t_0)$. At the next stage $t_0 < t_1$ of the CPS operation, the values of the weight functions may change: $w_{0,1}(t_0) > w_{0,1}(t_1)$, $w_{0,4}(t_0) < w_{0,4}(t_1)$, for example, due to readjustment.

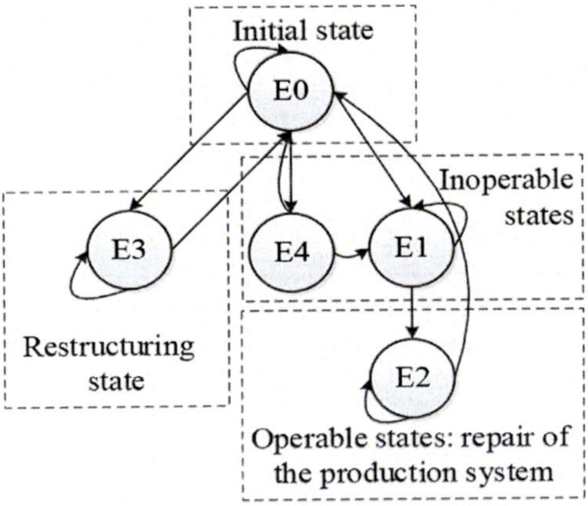

Fig. 2. The graph of states of the CPS.

Calculation procedures have shown that, for example, for some parameters:
$w_{1,2} = 0.41$, $w_{2,1} = 0.22$,
$w_1(t_1) = 0.41$, $w_1(t_0) = 0.79$,
$w_2(t_1) = 0.61$, $w_2(t_0) = 0.83$,
$w_{2,1}(t_0) = 0.38$, $w_{2,1}(t_1) = 0.87$,
$w_{1,2}(t_0) = 0.62$, $w_{1,2}(t_0) = 0.49$,
evaluations $u_{s_4}(t_p)$ equal:$u_{s_4}(t_0) \approx 0.068$, $u_{s_4}(t_1) \approx 0.057$.

This means that the inclusion degree of the CPS component (state E4) responsible for readjusting (for example, the corresponding algorithm of the control module) into the system, taking into account the normalized number of intersections, is reduced for $t_0 < t_1$, that is, at the beginning of degradation processes, the subsystem for registering short-term failures that do not require restoration is absorbed by the subsystem for registering all types of failures, which is associated with the enlargement these states.

Taking into account the flexible binding of functions during changeover with the same initial data, the state graph (Fig. 2) taking into account Fig. 1 is transformed into the graph shown in Fig. 3.

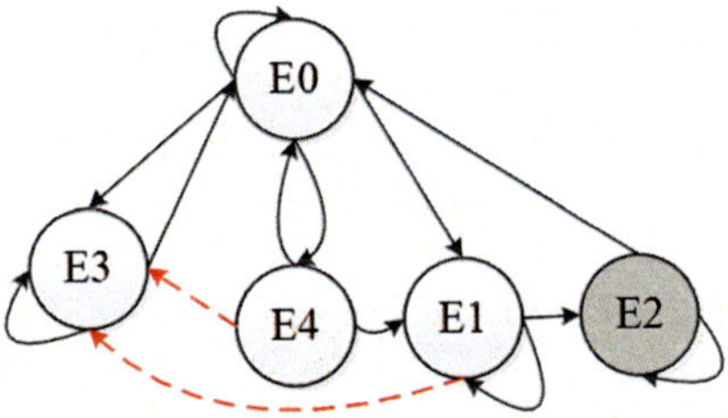

Fig. 3. The modified graph of states of the CPS Markov model with flexible binding of functions.

Assuming that

$$(s_2 : f_{if}^2, f_f^2) \to (s_3 : f_{if}^3, f_f^2); \tag{7}$$

$$(s_4 : f_{if}^4, f_f^4) \to (s_3 : f_{if}^3, f_f^4) \Rightarrow (s_3 : f_{if}^3, f_f^2, f_f^4). \tag{8}$$

This means that in the CPS recovery state after a failure (E2), part of the functions of the structural component responsible for recovery can be implemented on the structural components of the changeover.

When modeling the relationships of elements in these assumptions, estimates $u_{s_{if}^i, s_{if}^j}(t_p)$ are obtained:

$$u_{s4,\, s3}(t_0) \approx 0.066,\ u_{s4,\, s3}(t_1) \approx 0.049,\ u_{s4,\, s3}(t_2) \approx 0.044,$$
$$u_{s3}(t_0) \approx 0.051,\ u_{s3}(t_1) \approx 0.088,\ u_{s3}(t_2) \approx 0.089,$$
$$u_{s2\, s3}(t_0) \approx 0.094,\ u_{s2\, s3}(t_1) \approx 0.055,\ u_{s2\, s3}(t_1) \approx 0.043.$$

Therefore, for example, when transferring part of the functional from s2 to s3, the transition of the CPS to the E2 state can be replaced by the trajectory (7), and in case of a short-term failure, the trajectory (8) can be implemented with the transfer of the functional to the system element associated with the changeover. Thus, the involvement degree of the component responsible for restoring the system over time decreases from 0.094 at time t0 to 0.055 at time t1, which indicates the readjustment and absorption of the corresponding function and the importance of s3 increases from 0.051 to 0.088.

Figure 4 shows the results of comparing the obtained modeling results of relationships variability of production system elements with a variable structure. The given example of CPS reconfiguration with a variable structure is abstract and can be replaced by real CPS reconfiguration trajectories.

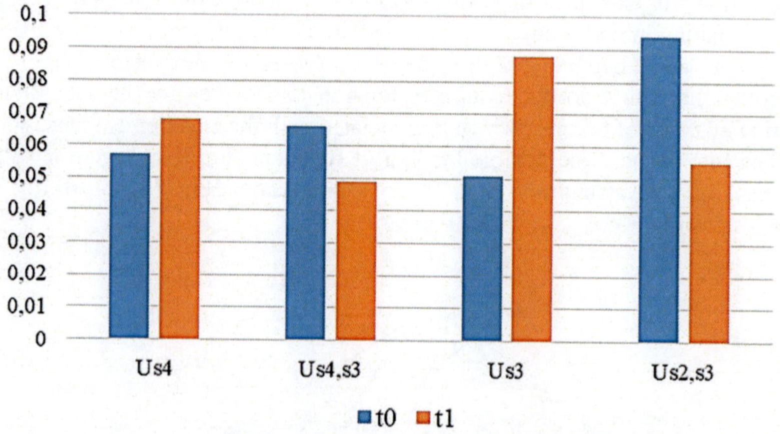

Fig. 4. Comparison results of estimates $u_{s4}(t_p)$ and $u_{s_{if}^i, s_{if}^j}(t_p)$.

The results of modeling estimates of the elements interrelationships in CPS indicate a noticeable spread of values in various situations requiring the use of the concept of flexible functions, which can be useful when making decisions about reducing the cost and resource intensity of processes in CPS, for example, when readjusting taking into account the degradation processes of the elements of production systems themselves.

4 Conclusion

The enlargement of the component composition of production systems causes possible loss of information in the description of each element associated with the analysis of the element significance and its relationships with other elements, taking into account the variable structure. The proposed aggregated approach takes into account the uniqueness degree of system components and changes in their significance by taking into account the inclusion degree of the component in the system. This makes it possible to refine solutions for managing the functioning of production systems, taking into account the variable structure. Using the example of the problem of modeling the degradation process of the production system as a whole and applying the proposed approach, it is possible to take into account the uniqueness degree of the system components and the change in the significance of the system components due to degradation processes.

The authors suggest further research in the direction of creating a decision support system for a preliminary assessment of the resource intensity and efficiency of the CPS reconfiguration process.

References

1. Chuvaev, A.V., Arzamastseva, M.A.: Implementation of rapid changeover tools (SMED) on the example of a CNC lathe. Act. Prob. Aviation Cosmon. **3**(13), 419–421 (2017)
2. Liker Jeffrey, K.: The Toyota Way: 14 Management Principles from the World's Greatest Manufacturer. McGraw-Hill, New York (2004)
3. Ryabchikov RV and Stepanov VM (2011) Reliability analysis of circuit diagrams of 110 kV switchgear substations of UNES: Proceedings of Tula State University. Tech. Scie. **6**:71–73
4. Serebrovsky, A.N.: Methodological and computational aspects of the significance of the risk of components of complex systems: Part I. Math. Mach. Syst. **1**(2), 145–154 (2012). (In Russian)
5. Boyko, A.A., Degtyarev, I.S.: A method for assessing the weight coefficients of elements of organizational and technical systems. Control Commun. Sec. Syst. **2**, 245–266 (2018). (In Russian)
6. Musaev AA, Skvorcov MS (2008) Metody parametricheskoj optimizacii nadezhnosti struktur no-slozhnyh tehnicheskih sistem (Methods of parametric optimization of reliability of structurally complex technical systems). Trudy SPIIRAN—SPIIRAS Proc. **1**(6):44–50
7. Habeeb, K., Verkhova, G.V., Akimov, S.V.: Modular technology in design of flexible complex systems. T-Comm-Telecommun. Transp. **11**(9), 86–90 (2017)
8. Pavlov, A.: Integrated modelling of the structural and functional reconfiguration of complex objects. SPIIRAS Proc. **5**(28), 143–168 (2013). https://doi.org/10.15622/sp.28.5. (InRussian)
9. Saravana Ram, R., Gopi Saminathan, A., Arun Prakash, S.: An Area Efficient and Low Power Consumption of Run Time Digital System Based on Dynamic Partial Reconfiguration. Int. J. Parallel Prog. **48**(3), 431–446 (2018). https://doi.org/10.1007/s10766-018-0578-6
10. Mabert, V., Watts, A., Charles, A.: A Enterprise Applications: Building Best-of-Breed Systems: Strategic ERP Extension and Use. Stanford University Press, Stanford (2005)
11. Capinski Vladimir Putin: Evolution of development and main features of mrp/mrp 2 class systems. Omsk Sci. Bull. **3**(28), 161–165 (2004)
12. Fornari, G., de Santiago Júnior, V.A.: Dynamically Reconfigurable Systems: A Systematic Literature Review. J. Intell. Rob. Syst. **95**(3–4), 829–849 (2018). https://doi.org/10.1007/s10846-018-0921-6

13. Rozewsky, P., Novikov, D., Zaikin, O., Bakhtadze, N.: (Eds.): New Frontiers in Information and Production Systems. Modeling and Analysis: Incentive Mechanisms, Competence Management, Knowledge-based Production: ISRL 98, Springer, Berlin (2016)
14. Volkov, V.F., Kovalev, A.P., Potryasaev, S.A., Salukhov, V.I.: Algorithm of sources management in deploying of information support systems for complicated technical complexes. SP **1**(44), 83–97 (2016)
15. Pavlov, A.N.: Integrated modelling of the structural and functional reconfiguration of complex objects. SPIIRAS Proc. **5**(28), 143–168 (2014)
16. Doronina, J.V.: The method of structural and functional synthesis of an information system taking into account the fuzzy binding of functions. Inf. Technol. **24**(9), 618–624 (2018). (In Russian)
17. Korolyuk, V.S., Turbin, A.F.: Phase consolidation of complex systems. Vishcha shkola, Kiev (1978). (In Russian)
18. Nam, H., Lysecky, R.: Latency, power, and security "optimization in distributed reconfigurable embedded systems. In: IEEE International Parallel and Distributed Processing Symposium Workshops, pp. 124–131 (2016). https://doi.org/10.1109/IPDPSW.2016.40
19. Doronina, J.V., Skatkov, A.V.: Cascade-hierarchical modeling in analyzing the dynamics of complex system resource Characteristics. Inf. Control Syst. **3**, 48–58 (2020). (In Russian)
20. Chowdhury, S., Maldonado, V., Tong, W., Messac, A.: New modular product-platform-planning approach to design macroscale reconfigurable unmanned aerial vehicles. J. Aircr. **53**(2), 309–322 (2016)
21. Saravana, R.R., Lordwin, P.M., Krishnan, S., Kamalraj, S., Dr, V.: Dynamic partial reconfiguration enchanced with security system for reduced area and low power consumption. Microprocess. Microsyst. **76**, 103088 (2020). https://doi.org/10.1016/j.micpro.2020.103088
22. Martins, L.E.G., Gorschek, T.: Requirements engineering for safety-critical systems: a systematic literature review. Inf. Softw. Technol. **75**, 71–89 (2016). https://doi.org/10.1016/j.infsof.2016.04.002
23. Savelov, N.S., Grechany, S.A., Lebedev, I.S.: New Approach to Modeling Dynamic Objects. International Conference on Industrial Engineering, Applications and Manufacturing (ICIEAM), Moscow, Russia, pp. 1–7 (2018). https://doi.org/10.1109/ICIEAM.2018.8728783
24. Kuzyakov, O.N., Lapteva, U.V., Sabitov, M.A., Makarova, L.N.: System For Determining Topology of Surface Based On Moire Method with Use of CBR Technology. In: International Conference on Industrial Engineering, Applications and Manufacturing (ICIEAM), Sochi, Russia, pp. 1–6. (2019) https://doi.org/10.1109/ICIEAM.2019.8742947

Monitoring of Parameters of Power Plants Based on Renewable Energy Sources

A. A. Berestinov(✉), D. V. Korobatov, and A. Z. Kulganatov

South Ural State University (National Research University), 76, Lenin Ave, Chelyabinsk 454080, Russia
berestinov.ru@mail.ru

Abstract. In recent decades, due to the depletion of minerals, renewable energy sources are gaining more and more popularity. In some countries, energy production from renewable sources already reaches more than 50% of total energy production. Monitoring the state of operation of renewable energy equipment is a critical task. Monitoring the parameters of power plants with renewable energy sources is an integral part of the entire operation of a power plant. This is an accounting of generated, consumed or lost electricity. And also, it is monitoring the technical condition of renewable energy equipment. In the article, the authors have developed a prototype device for monitoring the parameters of power plants with renewable energy sources. The block diagram of the device operation has been developed. A user interface has been developed where the energy indicators of a power plant with renewable energy sources will be displayed. The economic feasibility of a device for monitoring the parameters of power plants with renewable energy sources is considered. The principle of operation of the MQTT data transfer protocol is considered, the main advantages of the data transfer protocol, as well as the main disadvantages, are noted.

Keywords: MQTT · Renewable energy · ESP32 · HMI · IoT · Monitoring

1 Introduction

The operating and maintenance costs of the wind farms, as well as spare parts, are estimated at 10–15% of the wind farm's total revenue. Repairs and maintenance requiring the use of cranes result in additional costs due to the lack of cranes and the need for optimal weather conditions to operate. Maintenance optimization can be realized through condition monitoring and early fault diagnosis. 40% of induction generator failures are associated with bearings, 38% with the stator and 10% with the rotor [1]. Condition monitoring technologies such as vibration and surge analysis can detect some of these problems. There is growing interest in using generator current and power signals to detect generator faults.

The generator can meet the electricity needs of the modern era, and the types and characteristics of generators differ depending on its use, location and economy. Moreover, the role of the engine is inevitable in the real operations of any industry. In addition, it is

A. A. Radionov and V. R. Gasiyarov (Eds.): RusAutoCon 2022, LNEE 986, pp. 85–93, 2023.
https://doi.org/10.1007/978-3-031-22311-2_9

an integral part of processes such as power generation, oil and gas production, cement production, food processing and the fertilizer sector. Its use varies by industry, with motor ratings ranging from a few watts to megawatts depending on the nature of the job. In large industries, heavy duty generators and motors must operate at high loads [2]. Therefore, to manage the non-stop operation of engines and generators, we need reliable monitoring systems. Due to the continuous use of motors and generators, wear on parts is common and needs to be addressed. Predicting and preventing massive equipment failures is a vital need in today's industry. The question now arises: how can we track or anticipate malfunctions? The first basic rule is to take a proactive approach to maintenance. Preventive maintenance includes physical examination and periodic maintenance to identify any damaged or defective parts. Accordingly, non-compliant parts will be replaced before serious damage occurs. This is a little risky because periodic maintenance is scheduled according to the specified times, but over time the condition of the generator and engine continues to deteriorate.

Condition monitoring is progress in learning about the health, safety and performance of generators and engines where reliable, continuous and efficient operation is required. There are two types of monitoring systems: the first is offline monitoring, and the second is online monitoring [3]. In this monitoring method, various parameters are measured, transmitted and monitored at a control station where the operator is aware of the operating conditions of the generator and engine.

With the advancement of technology, smart systems are increasingly being used. These systems enable technicians, administrators and managers to monitor and control device performance from a safe distance. The monitoring system is very important when working with three-phase systems; some users and companies use smart monitoring software. These programs are installed on the user's smartphone or company computers so that employers can make decisions in the event of an error.

Thus, we are considering and offering promising solutions to some difficult issues for the renewable energy sector, namely monitoring of energy indicators and parameters.

2 Block Diagram

The Internet of Things refers to any object (or "thing") that can be connected to the network - from factory equipment and cars to mobile devices and smart watches. However, today the Internet of Things is usually understood as connected objects equipped with sensors, software and other technologies that allow them to exchange data with other "things". Traditionally, the connection was mainly via Wi-Fi, but today 5G and other types of network platforms are increasingly coping with large data sets, providing high speed and reliability [4].

In IoT and IIoT, sensors are used to collect information from different devices: smart bracelets, machine tools, cars. They can measure different physical parameters: from air temperature to the level of infrared radiation. To do this, sensors are equipped with sensitive elements, for example, photosensitive diodes or metal plates that change properties depending on the environment.

To transmit information to a regular server or cloud, where it will be processed and used further, the sensors are equipped with a transmitting module. In IoT, it is usually a wireless communication module, for example: Bluetooth, NFC, RF or Wi-Fi. Sometimes several sensors are connected to one transmitting module. Any sensors collect analog data. Such data is continuous it can be represented as a winding line, a continuous flow of information. It is not possible to transmit such data via cable or wireless communication first the signal must be converted into digital data. Digital data is a sequence of zeros and ones. To convert analog data to digital, a continuous analog line must be divided into several separate sections, and each section must be assigned a specific value.

Analog data can also be transmitted, for example, by radio. But computers only work with digital data, so they still have to be converted to digital. And it is better to do this before transmission in order to use more modern and faster communication channels. Simple analog sensors do not know how to convert a signal. To get information from them in digital form, we need a scale where digital values correspond to analog values.

2.1 Description of the Block Diagram

The developed block diagram of the device is shown in Fig. 1. The explanation is given below.

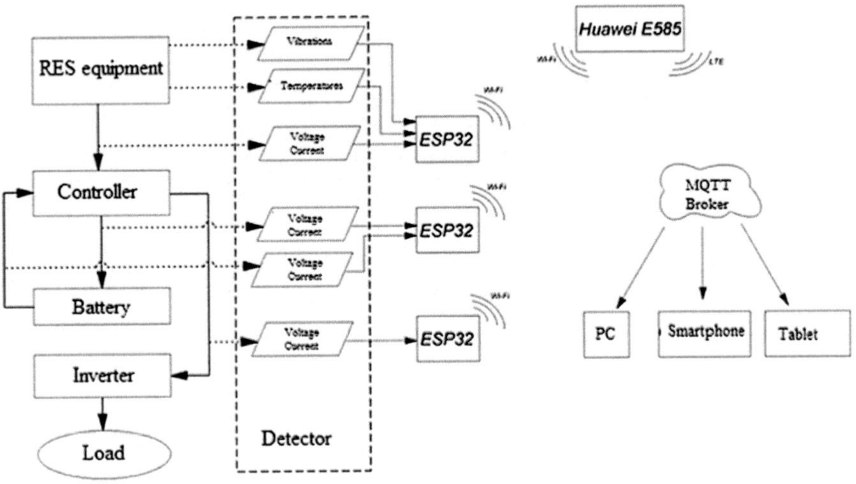

Fig. 1. Block diagram

"Renewable energy equipment" refers to wind turbines, solar panels, etc. The controller, battery and inverter are traditionally used equipment. We will measure voltage and current at each section. Then the received values from the sensors will be converted into a digital signal and sent to the MQTT Broker. A mini WiFi router was also providedin the absence of a stationary WiFi router next to the microcontrollers.

This block diagram shows sensors for monitoring parameters such as vibration, temperature of the RES installation itself, as well as voltage and current sensors in individual sections. Modernization and improvement of this scheme is possible due to the flexibility and mobility of sensors. For specific conditions of use of the monitoring device, it is possible to determine the optimal parameters of the device relative to renewable energy equipment. Today, the microelectronics market offers a large number of sensor options for various purposes.

In this chapter, a block diagram of the operation of a device for monitoring the parameters of renewable energy equipment was developed. Various modes of operation of the device are taken into account both with a fixed WiFi router and equipped with a mini WiFirouter.

3 Advantages of ESP32

ESP 32 is a device that is perfectly used in IoT technologies today, the company Expression Systems is engaged in production, the popularity of the microcontroller is growing due to its low cost, wide availability spector and a wide range of features. The core of the microcontroller system is a dual-core Extensa LX6 microprocessor with a frequency of 240 MHz, which provides significantly low energy consumption. The microcontroller is equipped with 520 kbytes of static random access memory and 448 kbytes of programmable permanent storage device. The board is equipped with both a Wi-Fi module and a Bluetooth module to meet a rich range of Internet of Things use cases [5].

To date, the Internet of Things (IoT), Internet of Energy (IoE) is formed from highly intelligent monitoring and management of electricity consumption in order to identify and optimize energy consumption by monitoring the consumption profile of users using sensor data, which allows the process to be more efficient and economical. And also, IoE is applicable to energy production technologies and, therefore, has various capabilities for the development and application of solutions for energy processes for the production of renewable energy sources (RES), it should also be noted that meteorological fluctuations affect sensors and microcontroller, but not significantly, which can become the basis for the sustainability of optimized electricity production/consumption [6].

The technical characteristics of the microcontroller are summarized in Table 1.

Table 1. Technical ESP32.

Parameter	Value
Processor	Xtensa 32 LX6
Number of cores	2
Processor frequency	160–240 MHz
RAM	520 Kb
ROM	448 Kb
Type of network	WiFi 802.11n
	Bluetooth
Programming language	Lua

(*continued*)

Table 1. *(continued)*

Parameter	Value
	Arduino IDE
Operating temperature	-40 … + 80
Flash memory capacity	4–16 Mb
Food (direct current)	2,2–3,6 V
Average current consumption	80 mA
Dimensions	18x19,2x3,2 mm

Foreign colleagues use the ESP 32 microcontroller in healthcare for remote monitoring of an electrocardiogram with the ability to determine the probability of a normal heart rate, abnormal heart rate and noise.

4 Equipment for Measuring Voltage and Current

Electricity is an extremely important thing in a person's life. Due to the increase in energy consumption, currently reducing energy consumption is an important task in this area. As a result, knowing the state of electrical energy consumption is the first thing to do to solve this problem. When we learn the electrical information of each device, it will help us to know about the consumption of electrical energy at a certain point in time. As a result, you can save up to 20% of energy. In order for this to happen, the first step is to select equipment that can correctly measure energy consumption. One of the most important things in this device is the current sensor, which is used to measure the current flowing through a certain phase.

A current sensor is an equipment that can measure the amount of current flowing and send an output signal in the form of a voltage level to any microcontroller for reverse calculation of the measured current value. To select a current sensor, you need to take into account many criteria, such as budget, measurement accuracy, installation complexity, etc. Today, inexpensive current sensors are on the market, which are very popular in many areas. And also, it is necessary to know the information and characteristics of modern inexpensive current sensors.

4.1 Power Meter Module PZEM-004T

The PZEM-004T module is widely used for monitoring current and voltage consumption. The PZEM-004T outputs the RMS value of voltage and current, and also calculates the active power and total energy consumption over time or total energy consumption. The PAM-004T module is equipped with a current transformer, which is used to convert the effect of the detected magnetic field on the electrical load system into an electrical signal that will be processed by the PZEM-004T module [7, 8].

The cost of this power meter module is 500–600 rubles. The technical characteristics of the PZEM-004T module are summarized in Table 2.

Table 2. Technical characteristics of the SCT 013 current sensor.

Parameter	Value
Operating voltage: - variable	80–260 V
Nominal current	0–100 A
Operating temperature	−25 to +70 °C
Mass	80 g

5 Visualization

To date, the development of new technologies, namely software and hardware devices of data collection and monitoring systems are constantly being upgraded. Data collection and monitoring systems have demonstrated trends in data exchange, configuration, unified interfaces and multifunctional auxiliary application software.

Currently, production processes are visualized and controlled using special components of the human-machine interface (HMI) [9]. These devices are usually located in fixed locations on the production floor. This creates significant limitations in terms of portability and extensibility [10]. Smartphones (tablets) represent an interesting additional HMI solution. Smartphones are a universal solution that enables engineers to access energy and technological data occurring in various processes of the devices (for example, sensor values) from anywhere in the field. And also, the software on smartphones can be easily expanded to provide additional functions, such as access to company services, Internet resources or interaction with Internet of Things devices installed in an industrial network [11].

5.1 Interface

Figure 2 show screenshots of the visualization. Extensive visualization is developed for specific renewable energy equipment.

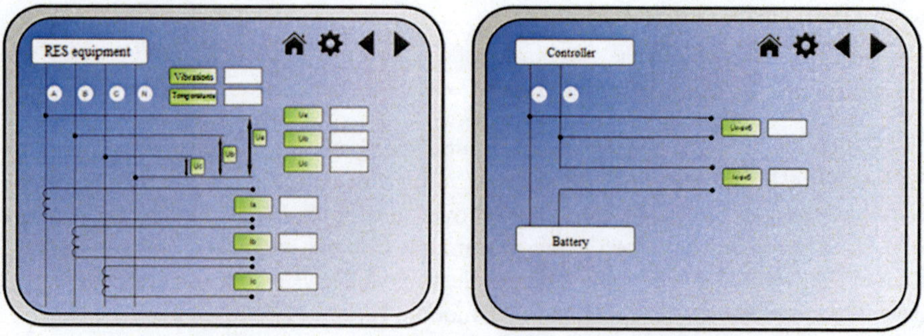

a) b)

Fig. 2. (a) Interface 1, (b)Interface 2.

In this section, a user interface has been developed for monitoring parameters in the renewable energy sector. The interface can be adjusted and adjusted according to the user's(operator's) wishes, can be supplemented with various schedules, etc. [12].

6 Data Transfer Protocol

For Internet of Things (IoT) devices, an Internet connection is a kind of requirement. An Internet connection allows devices to work with each other and with server services. The main network protocol of the Internet is TCP/IP. Built on top of the TCP/IP stack, MQTT (Message Queue Telemetry Transmission) has become the standard for IoT communication. MQTT can also run on SSL/TLS, which is a secure protocol built on TCP/IP to ensure that all data between devices is encrypted and protected [13]. MQTT (Message Queuing Telemetry Transport) a protocol for transmitting a sequence of messages with telemetry data, that is, information from temperature, humidity, light sensors, etc. The protocol that uses the publish/subscribe communication pattern is used for machine-to-machine communication (M2M). MQTT was proposed in 1999. Andy Standford-Clark as a protocol that would serve to transmit data on the state of the oil and gas pipeline in real time. The development was carried out by IBM for a new pipeline of the largest American oil company ConocoPhillips. As part of the creation of a dispatching control and data collection system (SCADA), it was necessary to ensure the guaranteed collection of a variety of information: the condition of pumps, bearing temperature, flow rate, valve condition, tank levels, etc. [14].

6.1 How Does MQTT Work?

The communication system built on MQTT consists of a publisher server, a broker server and one or more clients. The publisher does not require any settings for the number or location of subscribers receiving messages. In addition, subscribers do not need to configure for a specific publisher. There may be several brokers in the system distributing messages. MQTT provides a way to create a hierarchy of communication channels a kind of branch with leaves. Whenever the publisher has new data to distribute to customers, the message is accompanied by a delivery control note. Higher-level clients can receive each message, while lower-level clients can receive messages related to only one or two basic channels "branching off" at the bottom of the hierarchy. This facilitates the exchange of information ranging in size from two bytes to 256 megabytes [15]. Any data published or received by the MQTT broker will be encoded in binary format, since MQTT is a binary protocol. This means that to get the original content, you need to interpret the message. MQTT brokers can sometimes accumulate messages related to channels that do not have current subscribers. In this case, the messages will either be discarded or saved, depending on the instructions in the control message. This is useful in cases where new subscribers may need the most recently recorded data point, instead of waiting for the next dispatch. It is noteworthy that MQTT transmits security credentials in plain text, otherwise authentication or security features are not supported. This is where the SSL framework comes into play, helping to protect the transmitted information from interception or other forgery. In addition, in MQTT, you can use Ably authentication on tokens if you do not want to disclose your API key to the actual MQTT client at all (in the case of MQTT without SSL, tokens are required to prevent the transfer of API keys in plain text) [16, 17].

7 Conclusion

A block diagram of the operation of a device for monitoring the parameters of power plants with renewable energy sources has been developed. A user interface has been developed for displaying information on various devices, for example, on a smartphone, tablet, laptop or personal computer. The principle of operation of the MQTT data transfer protocol is considered, the main advantages of the data transfer protocol, as well as the main disadvantages, are noted. Based on the results of the work done, we can say that monitoring the parameters of power plants with renewable energy sources is an integral part of the entire process of the power plant operation. This is an accounting of generated, consumed or lost electricity. And also, it is monitoring the technical condition of renewable energy equipment.

References

1. Alkhasov, A.B., Popel, O.S.: (Proceedings of the conference Renewable energy: problems and prospects. Actual problems of the development of renewable energy resources. Makhachkala, p. 530 2020)
2. Fajardo Perez, M.G.: Twenty-second international scientific and technical conference "Radio electronics, electrical engineering and energy. Synchronous generator for an autonomous wind power plant, p. 27 (2016)
3. Komshin, A.S.: Standards and Quality. Equipment monitoring in the energy sector is a safety requirement, pp. 24–27 (2014)
4. Rolik, O., Telenyk, S., Zharikov, E .: IoT and cloud computing: the architecture of microcloud-based iot infrastructure management system. Emerging trends and applications of the internet of things, pp. 198–234 (2017)
5. Dokic, K., Radisic, B., Cobovic, M.: Micropython or arduino c for ESP32 - efficiency for neural network edge devices. Communications in computer and information science, pp. 33–43 (2020)
6. Maier, A., Sharp, A., Vagapov, Y.: Comparative Analysis and Practical Implementation of the ESP32 Microcontroller Module for the Internet of Things. 2017 Internet Technologies and Applications (ITA), pp. 1–6 (2017)
7. Tantidham, T., Ngamsuriyaros, S., Tungamnuayrith, N., Nildam, T., Banthao, K., Intakot, P.: Energy consumption collection design for smart building. In: 2018 Int. Conf. Embed. Syst. Intell. Technol. Int. Conf. Inf. Commun. Technol. Embed. Syst. ICESIT-ICICTES, Vol. 4, pp. 1–6 (2018)
8. Wasoontarajaroen, P.K., Chamnanphrai, V.: Development of an IoT device for monitoring electrical energy consumption, 2017 9th Int. Conf. Inf. Technol. Electr. Eng. ICITEE **2017**, 1–4 (2017)
9. Cacciabue Pietro, C.: Cognitive ergonomics: A key issue for human-machine systems, Le Travail Humain, pp. 359–364 (1991)
10. Kopytin, S., Markov, M.: The Way to Better HMI in the Consumer Market Components and Technologies, pp. 94–95 (2012)

11. Turygin, Y.V., Nistyuk, A.I.: Development of a Human-Machine Interface Based on the Operator's Workplace of a Robotic Complex. Bulletin of Izhevsk State Technical University named after M.T. Kalashnikov, pp. 94–95 (2012)
12. Shuxing, et al.: Study of method for computer aided ergonomics knowledge management and design Aiming at product design. In: Computer-Aided Industrial Design & Conceptual Design 2009. CAID & CD 2009. IEEE 10th International Conference on IEEE (2009)
13. MQTT Version 5.0": OASIS Committee Specification (May 2018)
14. Stanford-Clark A and Truong HL: MQTT For Sensor Networks (MQTT-SN). Protocol Specification Version 1.2 (2014)
15. Dikii, D.I., Artemeva, V.D.: MQTT data protocol in remote access control management model for Internet networks. Sci. Tech. J. Inf. Technol. Mech. Optics **19**(1), 109–117 (2019). https://doi.org/10.17586/2226-1494-2019-19-1-109-117. (in Russian)
16. Dan, Y., Xuesong, X., Jianjun, Y.: Research and implementation of IoT communication system based on MQTT protocol. Inf. Commun. **159**(3), 33–35 (2016)
17. Zhi Kong (2019) Design and implementation of message push cluster system based on MQTT. Xidian University
18. Pandey, T., Garg, D., Gore, M.M.: Publish/subscribe based information dissemination over VANET utilizing DHT. Front. Comp. Sci. **6**(6), 713–724 (2012)

Modern Methods of Searching for the Optimal Assessment of Product Quality

G. Pipiay(✉) and L. Chernenkaya

Peter the Great St.Petersburg Polytechnic University, 29, Polytechnic St, St. Petersburg 194021, Russia
gogpipiy@ya.ru

Abstract. The modern theory of product quality management, and in particular the subject of qualimetry, contains a large list of models and methods for assessing product quality, and over the past 5 years, this list has only expanded. However, the tools used in the processing of heterogeneous information, in the analysis of product quality and decision-making, are mainly focused on the opinions of experts or on mass production, which limits the functionality of the existing theory of product quality management. Authors proposes modern methods of two-level optimization for solving problems of product quality management. The authors give an example of applying the theory of bilevel optimization to find the optimal estimate of product quality. The example contains an algorithm for solving an analytical problem and the results obtained.

Keywords: Qualimetry · Bilevel optimization · Quality monitoring · Quality indicators

1 Introduction

Speaking about the current state of development of the theory and practice of product quality management, it is necessary first of all to mention the main reason for the development of this area, namely production systems and technologies for the production of modern products. Currently, modern production systems and technologies are combined in the concept of "Industry 4.0". The term "Industry 4.0" comprises the following technologies [1]:

1. Smart factory. Fully equipped production with sensors and autonomous systems.
2. Cyber-physical systems. Unification of physical and digital levels. By spanning the production level as well as the product level, systems arise whose physical and digital representation can no longer reasonably be differentiated.
3. Self-organization. Existing production systems are becoming increasingly decentralized. This is accompanied by the decomposition of the classical production hierarchy and the transition to decentralized self-organization.
4. New distribution and purchasing systems: Distribution and purchasing will be more individualized.

A. A. Radionov and V. R. Gasiyarov (Eds.): RusAutoCon 2022, LNEE 986, pp. 94–104, 2023.
https://doi.org/10.1007/978-3-031-22311-2_10

5. New systems in product and service development: product and service development will be more individualized.
6. Adaptation to human needs: new production systems should be designed to follow human needs, and not vice versa.
7. Corporate social responsibility at the heart of which is sustainability and resource efficiency.

In the theory of product quality management, the listed components are embedded in the concept of "Quality 4.0" [2]. Within the framework of this concept, big data technologies and data science are introduced and used, which allows for a comprehensive information accounting of indicators that characterize quality not only from a technical point of view, but also from an economic, technological and social one.

Based on the above descriptions of the concept of "Quality 4.0", it should be said that "Quality 4.0" is a set of modern methods and practices based on the digitalization of production in terms of quality management in the conditions of the fourth industrial revolution [3]. The concept of "Quality 4.0" is implemented through the following principles; Data, Analytics, Interaction, Collaboration, Applications, Scale, Management Systems, Correspondence, Culture, Leadership, Competences.

Today, there are many methods to achieve the requirements of the "Quality 4.0" concept. Among them are statistical methods, expert methods, fuzzy methods, analytical methods, expert-statistical methods, expert-analytical methods and expert-fuzzy methods. There are many articles devoted to these methods and at the moment they are well developed, which cannot be said about the methods of bilevel optimization and decision making.

The development of the theory of quality management based on bilevel optimization methods will make it possible to more effectively fulfill the requirements of the "Quality 4.0" concept and the "Industry 4.0" concept due to the ability to work with a large amount of heterogeneous information and make prompt and optimal decisions regarding the quality of products.

2 Description of the Problem Area

It is known that each company has strategic goals for building long-term plans. Strategic goals encompass all activities of the company, such as defining a goal in economics, goals in talent management, or goals in quality. These goals help companies navigate the process of product release and in the process of building business processes based on the results of decision-making.

The variety of existing factors influencing these goals and the availability of limited resources encourages companies to select priorities and proportionally prioritize their resources to improve the company's performance in a particular area. Hence, the problem of decision-making in the presence of several priorities arises. The need to consider heterogeneous information and decide based on the above factors requires the development of SMOPR for the instrument-making industry.

At present, the mechanisms of work of SMOPR are based on technologies such as "Business intelligence" (BI–systems), which include methods of processing and grouping data, methods of data analysis and analytics, methods of visualization and reporting.

Solving the problem of assessing and improving the quality of instrument-making products includes the following stages:

1. Determination of the scale for the assessment of single indicators of the quality of instrument-making products.
2. Development of a methodology for processing and differential assessment of single indicators of the quality of instrument-making products.
3. Formalization of target functions for assessing the quality of instrument-making products.
4. Approbation of the results obtained.

3 Overview of Bilevel Optimization Methods

Bilevel decision-making methods have been developed to address trade-offs between two interactive decision makers that are distributed throughout a hierarchical organization. In a bilevel decision-making process, decision makers (DMs) at the upper and lower levels are named leader and follower, respectively, and make their individual decisions sequentially with the goal of optimizing their respective goals. The original form of the bilevel process was proposed by Bracken and McGill [4]. A wide range of research in this area has been carried out using the notations of bilevel programming, bilevel optimization, and bilevel decision making. Bilevel decision making is basically a bilevel optimization problem and mainly uses bilevel programming techniques. It appears frequently in many decentralized control problems, and this has prompted a number of researchers to work on bilevel decision models [5], decision approaches [6], and applications [7].

Bilevel optimization problems are a tool for modeling decision-making processes in complex economic, environmental, and financial systems, which have a hierarchical structure. The analysis of such problems does not fit into the framework of the usual optimization theory and requires the development of new mathematical methods and approaches. Despite the external simplicity of the formulations, the solution to these problems is very difficult and a significant part of the research in bilevel programming is reduced to the selection of separate classes of problems that can be solved efficiently.

The development of research into bilevel decision problems can be traced from two sources. The first relates to the field of game theory: when studying oligopolistic market models, Stackelberg used the hierarchy of players to build descriptive decision-making models and establish game-theoretic equilibria. The second source refers to the field of mathematical programming, where bilevel optimization problems have arisen that contain nested internal optimization problems as constraints on the external optimization problem. Since that time, there has been a significant body of literature on bilevel optimization problems, which have been actively studied over the past few decades.

The general formulation of the bilevel optimization problem is formed as follows. We define two vectors of variables $x = [x_1, x_2, ..., x_n]$ and $y = [y_1, y_2, ..., y_n]$ where x- the leader, a y- the follower, wherein $x \in X, y \in Y$. For $x \in X \subset R^n$, $y \in Y \subset R^m$,

$F : X \times Y \to R$ и $f : X \times Y \to R$ linear programming is given as follows:

$$\begin{aligned} &\min_{x \in X} F(x, y) = c_1 x + d_1 y \\ &A_1 x + B_1 y \le b_1 \\ &\min_{y \in Y} f(x, y) = c_2 x + d_2 y \\ &A_2 x + B_2 y \le b_2 \end{aligned} \tag{1}$$

where $c_1, c_2 \in R^n$, $d_1, d_2 \in R^m$, $b_1 \in R^p$, $b_2 \in R^q$, $A_1 \in R^{p \times n}$, $B_1 \in R^{p \times m}$, $A_2 \in R^{q \times n}$, $B_2 \in R^{q \times m}$.

The problem of bilevel optimization is solved by finding the optimal point on the polyhedral set S, built from a set of constraints imposed on two levels of optimization. The admissible area for solving the problem of bilevel optimization is defined as follows:

$S = \{(x, y) | x \in X, y \in Y, A_1 x + B_1 y \le b_1, A_2 x + B_2 y \le b_2\}$.

As mentioned above, the follower makes decisions based on the leader's constraints, so the range of feasible decisions of the follower is limited by the set S(x):

$S(x) = \{y \in Y | B_2 y \le b_2 - A_2 x\}$.

The projection S on the decision space of the leader is given as follows:

$S(X) = \{x \in X | \exists y \in Y, A_1 x + B_1 y \le b_1, A_2 x + B_2 y \le b_2\}$.

Hence the rational reaction of the follower to optimize his objective function for $x \in S(X)$:

$P(x) = \{y \in Y | y \in \arg\min[f(x, \hat{y}) | \hat{y} \in S(x)]\}$,

where $\arg\min[f(x, \hat{y} | \hat{y} \in S(x)]$.

Permissible decision area for the leader after the follower's turn:

$IR = \{(x, y) | (x, y) \in S, y \in P(x)\}$.

Among the bilevel optimization methods, one can single out fuzzy bilevel optimization methods, bilevel optimization methods based on artificial intelligence, or meta-heuristic methods of bilevel optimization.

One of the problems of solving problems at two levels is that various uncertainties (including fuzziness and randomness) naturally appear in the decision-making process. Since the problem of uncertainties can arise both in the process of determining the parameters of the model and in the process of making a decision, the information associated with making decisions becomes very inaccurate and ambiguous, especially in this age of big data.

To solve bilevel decision-making problems under uncertainty, fuzzy sets and fuzzy systems were used in terms of both decision modeling and decision approaches formed by fuzzy bilevel decision-making methods.

The difference between the fuzzy method of bilevel optimization and the conventional method lies in the fuzzy definition of the parameters c_i, d_i, A_i, B_i и b_i. The fuzzy bilevel optimization model is given as follows:

$$
\begin{aligned}
&\min_{x \in X} F(x, y) = \tilde{c}_1 x + d_1 \tilde{y} \\
&\tilde{A}_1 x + \tilde{B}_1 y \leq \tilde{b}_1 \\
&\min_{y \in Y} f(x, y) = \tilde{c}_2 x + \tilde{d}_2 y \\
&\tilde{A}_2 x + \tilde{B}_2 y \leq \tilde{b}_2
\end{aligned}
\tag{2}
$$

In the category of meta-heuristics, Mathieu et al. [8] were the first to propose a bilevel programming algorithm based on a genetic algorithm (GABBA) for solving the DLP. Later Hejazi et al. [9] presented a method based on a genetic algorithm (GA). It is determined by comparing the results with the method proposed in [10]. Odugawa and Roy [11] developed a bilevel genetic algorithm, which is an elitist optimization algorithm designed to induce limited asymmetric interaction between two players to solve different types of DLPs in a single framework. Yin [12] has also proposed a GA-based approach to solve two PDLPs: road pricing and reserve capacity of a signal-controlled road network. Wang et al. [13] proposed an evolutionary algorithm for solving the non-linear bilevel programming problem. Recently, Wang et al. [14] developed a GA in which some methods were adopted to ensure that not only the original chromosomes are possible, but the chromosomes generated by the algorithm. Shahin and Kirik [15] presented an algorithm based on simulated annealing.

4 Decentralized Approach to Product Quality Management

Decentralized management refers to the process of distributing management functions to elements that are lower in the hierarchical structure. To avoid negative factors in the application of a decentralized approach, which undoubtedly exist, for example, loss of control over departments, increased time for making strategic decisions, the risk of violating the principles of total quality management in accordance with ISO 9000 and ISO 9001.

To avoid or reduce the risk of problem situations when application of decentralized management, it is necessary to determine the list of management functions or decision-making functions that will be transferred to lower levels of management. This problem is solved by solving the following tasks:

1. Determination of the dimension of the decision-making area, i.e. it is necessary to determine in respect of which elements of production and product quality the decision will be made.
2. Determination of the list of factors influencing product quality and sources of information for quantitative or qualitative registration of these factors.
3. Determination of structural divisions, whose information will be taken into account when making decisions and assessing product quality.
4. Delegation to certain structural divisions of management functions in relation to the collection of information, processing, coordination and intermediate assessment of product quality, for its part.

5 An Example of Applying a Bilevel Optimization Model to Find the Optimal Estimate of Product Quality

To demonstrate the possibility of using the bilevel optimization technique for product quality assessment tasks, we present a specific example on calculation.

Example. Product N consists of five blocks, production includes seven cost categories and requires three suppliers who supply three types of components. The input data for quality indicators are presented in Table 1.

Table 1. Input data.

No. of the block	Block quality indicators				
	x_1	y_2	y_3	z_2	z_3
E1	0.8	0.5	0.39	0.5	0.6
E2	0.6	0.42	0.39	0.48	0.7
E3	0.64	0.25	0.32	0.51	0.25
E4	0.72	0.36	0.15	0.7	0.39
Quality cost group No	Quality cost indicators				
	x_1	y_2	y_3	-	-
E1	0.1	0	0	-	-
E2	0	0.5	0	-	-
E3	0	1	0	-	-
E4	0	0	1	-	-
E5	0	0	0.9	-	-
E6	0.9	0.7	0	-	-
Supplier's No	Supplier's quality indicators				
	x_1	z_2	z_3	-	-
E1	0.3	0.1	0.6	-	-
E2	0.5	0.2	0.4	-	-
E3	1	1	1	-	-

Optimization of the bilevel model with two sequences will be carried out using the Kuhn-Tucker conditions and the simplex method.

The implementation of Kuhn-Tucker conditions eliminates the ambiguous determination problem of the main level (leader) function variables in the acceptable domain of the sublevel (follower), by replacing the problem of finding the optimal follower function with the Kuhn-Tucker conditions. The classical bilevel optimization task, after replacing the follower' task with Kuhn-Tucker conditions, is expressed by the model:

$$\min_{x \in X}\{Q(x, y) = c_1 x + d_1 y\}$$
$$A_1 x + B_1 y \leq b_1 \quad (3)$$

$$A_2x + B_2y \leq b_2 \tag{4}$$

$$uB_2 - v \leq -d_2 \tag{5}$$

$$u(b_2 - A_2x - B_2y) + vy = 0 \tag{6}$$

$$x \geq 0, y \geq 0, u \geq 0, v \geq 0. \tag{7}$$

The analytical model of the bilevel technical product quality assessment' optimization will take the following form:

$$\begin{aligned}
&\min_{x \in X}\{Q(x, y, z) = x_1 - y_2 - y_3 + z_2 + z_3 \\
&0.8x_1 - 0.5y_2 - 0.39y_3 + 0.5z_2 + 0.6z_3 \geq 0.22 \\
&0.6x_1 - 0.42y_2 - 0.39y_3 + 0.48z_2 + 0.7z_3 \geq 0.19 \\
&0.64x_1 - 0.25y_2 - 0.32y_3 + 0.51z_2 + 0.25z_3 \geq 0.17 \\
&0.72x_1 - 0.36y_2 - 0.15y_3 + 0.7z_2 + 0.39z_3 \geq 0.26 \\
&x_1 + y_2 + y_3 + z_2 + z_3 = 1 \\
&\min_{y_i \in Y}\{F_1(x, y) = (1 - x_1) + y_2 + y_3\} \\
&0.1x_1 \leq 0.1 \\
&0.5y_2 \leq 0.5 \\
&y_2 \leq 1 \\
&y_3 \leq 1 \\
&0.9y_3 \leq 0.9 \\
&0.9x_1 + 0.7y_2 \geq 0.8 \\
&\min_{z_i \in Z}\{F_2(x, z) = x_1 + z_2 + z_3\} \\
&0.3x_1 + 0.1z_2 + 0.6z_3 \geq 0.33 \\
&0.5x_1 + 0.2z_2 + 0.4z_3 \geq 0.37 \\
&x_1 + z_2 + z_3 \leq 1.
\end{aligned}$$

The algorithm for solving the bilevel optimization task consists of 13 steps:

1. Formulation of the task for the bilevel product quality optimization.
2. Transformation of linear constraints for the objective function of the sequence (4) to functions of the form $g(x, y) \geq 0$.
3. Setting Lagrange multipliers for the function and equality (5).
4. Setting the complementary non-rigidity (6).
5. The solution of the task by the simplex method without taking into account the condition (6), and in the case of finding a solution, fixed it as the initial solution $Q(x, y)_0$.

6. Checking the fulfillment of conditions (6) at the initial u_i (where i is the number of the founded Lagrange coefficients' vector). If the conditions are not met, go to step 7.
7. Changing values of previous coefficients u_i.
8. Checking the fulfillment of conditions (5)–(7) with new u_i.
9. If conditions (5)–(7) are not met, repeat steps 7 and 8. If the conditions (5)–(1.4) are met, go to step 10.
10. Changing the inequalities of linear constraints (4) to equality conditions under which coefficients u_i are nonzero.
11. Solving a new task with modified constraints by the simplex method. If $Q(x, y)_i \geq Q(x, y)_{i+1}$ and $Q(x, y)_{i+1} \geq F(x, y)_{i+1}$ (where i is the number of the objective function founded solution), repeat the iteration of steps 7–11, while preserving the values, consider the founded numerical value $Q(x, y)_{i+1}$ as optimal.
12. When $Q(x, y)_i \geq Q(x, y)_{i+1}$ and $Q(x, y)_{i+1} < F(x, y)_{i+1}$, repeat steps 7–11 until the conditions $Q(x, y)_i \geq Q(x, y)_{i+1}$, $Q(x, y)_{i+1} \geq F(x, y)_{i+1}$ are met.
13. When the values are found, repeat the iteration of steps 7–11, while saving the values, consider the founded numerical value $Q(x, y)_2$ as optimal.

Let's proceed to the definition of constraints for linear equations systems. If partial values of the i-th equation are differed from 1, then the average of these values is inserted in the right part with the sign $\geq$. If all partial values of the i-th equation are equal to 1, then the number 1 with the sign $\leq$ is inserted in the right side. If the left part of inequality consists of a single term, then numerical value of the left term is written in the right part the with the sign $\leq$.

The formulated task of product quality assessment is illustrated by Table 2.

Table 2. Example.

$Q(x, y, z)$	x_1	y_2	y_3	z_2	z_3	b
E1	0.8	0.5	0.39	0.5	0,6	0.8
E2	0.6	0.42	0.39	0.48	0.7	0.6
E3	0.64	0.25	0.32	0.51	0.25	0.64
E4	0.72	0.36	0.15	0.7	0.39	0.72
E5	0.8	0.5	0.39	0.5	0.6	0.8
$F_1(x, y)$	x_1	y_2	y_3		-	b'
E1	0.1	0	0	-	-	≤ 0.1
E2	0	0.5	0	-	-	≤ 0.5
E3	0	1	0	-	-	≤ 1.0
E4	0	0	1	-	-	≤ 1.0
E5	0	0	0.9	-	-	≤ 0.9
E6	0.9	0.7	0	-	-	≥ 0.8

(continued)

Table 2. (*continued*)

Q(x, y, z)	x_1	y_2	y_3	z_2	z_3	*b*
$F_2(x,z)$	x_1	-	-	z_2	z_3	$b^{//}$
E1	0.3	-	-	0.1	0.6	≥ 0.33
E2	0.5	-	-	0.2	0.4	≥ 0.37
E3	1	-	-	1	1	≤ 1

The numerical partial values of objective functions (quality indicators) are presented in Table 3, numerical values of all the functions are presented in Table 4.

Table 3. Output data.

No. of the block	Block quality indicators				
	x_1	$x_2(y_2)$	$x_2(y_3)$	$x_3(z_2)$	$x_3(z_3)$
E1	0.7048	0.0045	0	0	0.0552
E2	0.5286	0.00378	0	0	0.0644
E3	0.56384	0.00225	0	0	0.023
E4	0.63432	0.00324	0	0	0.03588
Quality cost group No	Quality cost indicators				
	$y_1(x_1)$	y_2	y_3		
E1	0.0881	0	0	-	-
E2	0	0.0045	0	-	-
E3	0	0.009	0	-	-
E4	0	0	0	-	-
E5	0	0	0	-	-
E6	0.7929	0.0063	0	-	-
Supplier's No	Supplier's quality indicators				
	$z_1(x_1)$	z_2	z_3		
E1	0.2643	0	0.0552	-	-
E2	0.4405	0	0.0368	-	-
E3	0.881	0	0.092	-	-

Numerical values of all the functions are presented in Table 4.

Table 4. Numerical values on objective functions.

Designation of objective functions	Numerical value of objective functions
$Q(x, y, z)$	0.881
$F_1(x, y)$	0.119
$F_2(x, z)$	0.776

6 Conclusion

The presented methods of bilevel optimization are a powerful tool for product quality management due to the possibility of their use when taking into account a large amount of heterogeneous information and searching for optimal ways to improve or ensure the quality of products.

Adapting these methods to the tasks of assessing product quality and making subsequent decisions will allow us to develop and expand the theory of product quality management.

The novelty of the improved methods and algorithms of bilevel optimization for the problems of product quality management is as follows:

1. Refinement of existing optimization methods and algorithms in terms of concepts and theory of quality management.
2. Designing the problem of bilevel optimization on the Pareto optimal area, taking into account the cooperation or relationship between followers.
3. Extension of the class of solvable applied problems of bilevel optimization, taking into account heterogeneous information from various sources.
4. Projection of the solution of the problem of bilevel optimization of product quality assessment with one leader and many followers on the Pareto area by using dialogue decision cards.

Acknowledgments. The reported study was funded by RFBR, project number 20-37-90012.

References

1. Lasi, H., Fettke, P., Kemper, H.-G., Feld, T., Hoffmann, M.: Industry 4.0. Bus. Inf. Syst. Eng. **6**(4), 239–242 (2014). https://doi.org/10.1007/s12599-014-0334-4
2. Arsovski S (2019). Social oriented quality: from quality 4.0 towards quailty 5.0. In 13th International Quality Conference 13:397–404
3. Sony M, Antony J, Douglas JA (2020) Essential ingredients for the implementation of Quality 4.0: A narrative review of literature and future directions for research. The TQM Journal 32(4):779–793. https://doi.org/10.1108/TQM-12-2019-0275
4. Bracken, J., McGill, J.T.: Mathematical programs with optimization problems in the constraints. Oper. Res. **21**(1), 37–44 (1973)

5. Lu, J., Shi, C., Zhang, G.: On bilevel multi-follower decision making: General framework and solutions. Inf. Sci. **176**(11), 1607–1627 (2006). https://doi.org/10.1016/j.ins.2005.04.010
6. Han, J., Lu, J., Hu, Y., Zhang, G.: Tri-level decision-making with multiple followers: Model, algorithm and case study. Inf. Sci. **311**, 182–204 (2015). https://doi.org/10.1016/j.ins.2015.03.043
7. Kalashnikov, V.V., Dempe, S., Pérez-Valdés, G.A., Kalashnykova, N.I., Camacho-Vallejo, J.F.: Bilevel programming and applications. Math. Probl. Eng. **2015**, 16 (2015). https://doi.org/10.1155/2015/310301
8. Mathieu, R., Pittard, L., Anandalingam, G.: Genetic algorithm based approach to bi-level linear programming. RAIRO-Operations Research-Recherche Opérationnelle **28**(1), 1–21 (1994)
9. Hejazi, S.R., Memariani, A., Jahanshahloo, G., Sepehri, M.M.: Linear bilevel programming solution by genetic algorithm. Comput. Oper. Res. **29**(13), 191–1925 (2002). https://doi.org/10.1016/S0305-0548(01)00066-1
10. Gendreau, M., Marcotte, P., Savard, G.: A hybrid tabu-ascent algorithm for the linear bilevel programming problem. J. Global Optim. **8**(3), 217–233 (1996). https://doi.org/10.1007/BF00121266
11. Oduguwa V, Roy R (2002) Bi-level optimisation using genetic algorithm. In Proceedings 2002 IEEE International Conference on Artificial Intelligence Systems (ICAIS 2002), pp. 322–327. IEEE. https://doi.org/10.1109/ICAIS.2002.1048121
12. Yin, Y.: Genetic-algorithms-based approach for bilevel programming models. J. Transp. Eng. **126**(2), 115–120 (2000). https://doi.org/10.1061/(ASCE)0733-947X(2000)126:2(115)
13. Wang Y, Jiao YC, Li H (2005) An evolutionary algorithm for solving nonlinear bilevel programming based on a new constraint-handling scheme. IEEE Transactions on Systems, Man, and Cybernetics, Part C (Applications and Reviews) 35(2):221–232. https://doi.org/10.1109/TSMCC.2004.841908
14. Guang-Min W, Zhong-Ping W, Xian-Jia W, Ya-Lin C (2005) Genetic algorithms for solving linear bilevel programming. In Sixth International Conference on Parallel and Distributed Computing Applications and Technologies (PDCAT'05), pp. 920–924. IEEE. https://doi.org/10.1109/PDCAT.2005.145
15. Sahin, K.H., Ciric, A.R.: A dual temperature simulated annealing approach for solving bilevel programming problems. Comput. Chem. Eng. **23**(1), 11–25 (1998). https://doi.org/10.1016/S0098-1354(98)00267-1

Modeling and Simulation

A Way of Exploring the Real World Using Virtuality with the Example of Creating Animation in the Unity Environment

L. A. Khaliullina(✉) and I. I. Bikmullina

Kazan National Research Technical University Named After A.N. Tupolev, 10, St. Karl Marx, Kazan 420111, Russia
liya.khaliullina@gmail.com

Abstract. This article describes the method of creating animation in the Unity environment as one of the possible ways to study the real world and simulate physical processes, which would give an optimal balance between the achieved result and the resources used. The result of the work done is a generated hybrid method for creating a cartoon based on combining animations performed in several techniques. The uniqueness and novelty of the proposed method lies in the synergy of various tools for creating products in the Unity environment that meet the criteria for effective animation. It is worth noting the simplicity and speed of implementation, the clarity and accessibility of the presentation of information, which make it possible to use this method for educational purposes when teaching students of any age to create animation and awaken interest in the production of cartoons.

Keywords: Unity · Animation · Image · Animation · Technology

1 Introduction

In connection with the development of modern technologies, the need for rapid organizational changes in the conditions of a huge flow of incoming information and large amounts of data, there is an acute issue of training people and automating this process.

The learning process can be viewed as a joint activity of participants in organizing the assimilation of an academic discipline, obtaining subjectively new or new subject and procedural knowledge, and exploring the real world [1]. Certain requirements are imposed on the learning process: quick adaptation to changes, flexible management taking into account various teaching methods, constant interaction of the participants in the learning process, as well as creating opportunities for creativity, reflection, for using the subjectivity of the participants in the process [2]. Automation of training allows you to improve, maintain and control the skill level of training participants.

It is advisable to consider the learning process on the example of educational work carried out with a child using animation and animation, because the fundamental foundations of knowledge are laid in a person in childhood.

A. A. Radionov and V. R. Gasiyarov (Eds.): RusAutoCon 2022, LNEE 986, pp. 107–118, 2023.
https://doi.org/10.1007/978-3-031-22311-2_11

2 Relevance

A person receives more than 70% of information from the world around him through his eyes, perceives it with the help of images, therefore, in the early stages of development, one of the ways to cognize the surrounding reality is to watch cartoons that reflect the model of reality [3].

Cartoons have characteristics that give reason to consider them as a learning tool. Due to its accessibility and clarity for understanding, animation takes a strong position in learning issues and allows you to make the educational process the most interesting, to visualize it. That is why the need to find optimal programs for creating cartoons and finding solutions to simplify their development is an urgent problem for improving the learning process with the possibility of its further automation.

From the point of view of modern development of science and technology, we are observing the evolution of figurative means. The use of the latest technologies in animation had a beneficial effect on it, accelerating the process of working on works, raising their quality level, which contributed to the emergence of new means of expression, new animation forms [4]. Technologies have opened up new ways of influencing the viewer's consciousness through spatial sound, stereoscopic images, and three-dimensional graphics.

At the same time, one of the main values of animation is the versatility of the language, which makes it possible to organize a comprehensive system of integrated developmental education for children.

3 Formulation of the Problem

To organize high-quality educational work with the possibility of further automating the learning process, it was decided to propose a method for creating animation in the Unity environment, which would give an optimal balance between the achieved result and the resources used. Subsequently, the resulting animations can be used to create a full-fledged training animation video (cartoon).

It is necessary to offer the final version of the cartoon, which consists of two parts, which makes it possible to implement all stages of development, including the installation of the finished project. In the first part of the cartoon, the movement of a boat on the surface of the water should be presented, taking into account the waves, in the second part of the cartoon, a scene of the abduction of livestock by an unidentified flying object will be proposed, taking into account sound and voice acting, creating a background and scenery. The result of the work done will be a generated hybrid method for creating a cartoon based on combining animations performed in several techniques.

4 Analysis of Existing Technologies for the Development of Cartoons in the Unity Environment

Animation is a type of multimedia technology.

The word animation comes from the Latin root Anima. Animation is the mobility of an image.

Animation is a technology that allows you to create the illusion of movement using inanimate objects. The most popular form is animation, which is a series of hand-drawn images.

Animation is a technique for creating the illusion of moving images using a sequence of still images (frames) replacing each other with a certain frequency. Animated films are a kind of feature films.

There are several types of animation, which, according to the method of creation, are subdivided into:

1) Plasticine animation - a type of animation, where the film is made by time-lapse shooting of plasticine objects, with their modification in the intervals between the frames taken.
2) Hand-drawn animation - animation technology based on time-lapse shooting of slightly different two-dimensional drawings.
3) Puppet animation - a method of volumetric animation. When creating, a scene-layout and doll-actors are used. The scene is photographed frame by frame, after each frame minimal changes are made to the scene (for example, the pose of the doll changes). When the resulting sequence of frames is played back, the illusion of movement of objects appears.
4) Computer animation - sequential display (slideshow) of previously prepared graphic files, as well as computer simulation of movement by changing (and redrawing) the shape of objects or showing sequential images with phases of movement.
5) Sand animation - an animation technology in which a light powder (usually cleaned and sifted sand, but also salt, coffee, or something similar) is applied in thin layers to glass and mixed, creating a moving picture (usually all actions are performed by hand, but brushes can also be used as accessories). The resulting image can be transmitted to the screen using an overhead projector or light board.

By duration

1. Feature-length cartoons: cartoons usually over 70 min in length.
2. Short cartoons (usually about 25–30 min).

Cartoons created using computer graphics are divided by type:

4.1 Animating a Rigid Body

Solid animation is used to create ready-made animation sequences that move or change the properties of objects, taking into account the fact that the objects are a single whole, as opposed to objects moving relative to the objects themselves in the part. Some examples of this type of animation are the movement of a racing car on the road, a hinged door opening, a spaceship flying through its trajectory space, and a piano falling from a building window. Despite the differences between the examples, they all have an important property. In particular, although changes are made to keyframes, they are made for a single and cohesive object. In other words, although the door hinges from closed to open, it still ends up as a door, with the same internal structure and the same structure as before. She does not transform into a tiger or a lion. It doesn't explode or jelly. It does not dissolve in raindrops. During the animation, the door retains its physical portability. It changes only in terms of its position, speed and scale. Thus, when animating a solid, changes to keyframes propagate only whole objects and their high-level properties. They are not related to subproperties and internal components. These kinds of animations can either be imported directly in the Unity Animation Editor, or in 3D animation programs (such as Maya, Max, or Blender) and then imported into Unity.

4.2 Animation with Sprites

Sprite animation is one of those things that, for all their primitiveness, have been successfully working and applied in computer graphics and games for more than a quarter of a century. Even in 3D games there are sprites - for example, explosion billboards. In many browser and flash games, sprite animation is used, since it is very simple and does not require high performance.

If one sprite is a still picture, then a series of these pictures quickly replacing each other makes up an animation called a sprite. This type of animation differs in that it is not the entire frame (frame) that changes on the screen, but only its small piece, where the sprites appear. Also, sprite animation is sometimes called software animation.

Figure 1 shows an example of the sprite animation.

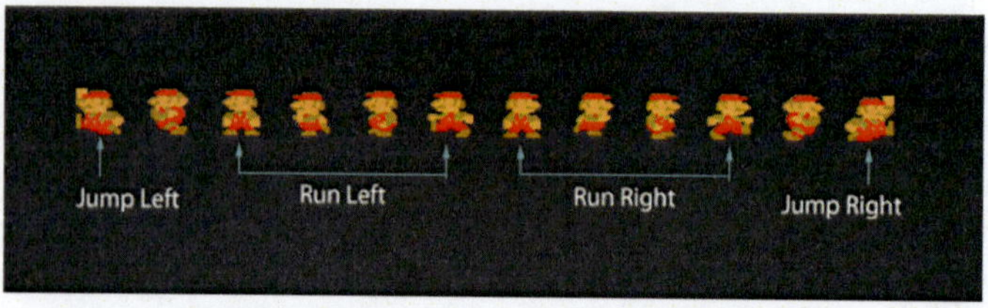

Fig. 1. Example of the sprite animation.

4.3 Animation Using Scripts

Scripting is an essential part of all games. Even the simplest games need scripts for player reactions and gameplay events. In addition, scripts can be used to create graphical effects, control the physical behavior of objects, or implement a custom AI system for game characters.

An example script is presented below:

```
using System.Collections;
using System.Collections.Generic;
using UnityEngine;
public class Wave_Controller: MonoBehaviour.
{ public float height;
public float time;
// Start is called before the first frame update.
```

```
void Start().
{
iTween.MoveBy(this.gameObject, iTween.Hash("y", height, "time", time, "loop-
type", "pingpong", "easet0ype", iTween.EaseType.easeInOutSine));
}.
// Update is called once per frame.
void Update().
{}}.
```

4.4 Rigged Animation, or Skeleton Based Animation

Skeletal animation is a way to animate 3D models in animation and computer games. It consists in the fact that the multiplier creates a skeleton, which is usually a tree-like structure of bones, in which each subsequent bone is "tied" to the previous one, that is, it repeats movements and turns after it, taking into account the hierarchy in the skeleton. Further, each vertex of the model is "snapped" to some bone of the skeleton. Thus, when a single bone moves, all vertices attached to it move as well. Thanks to this, the animator's task is greatly simplified, because there is no need to animate each vertex of the model separately, and it is enough to set the position and rotation of the bones of the skeleton.

Also, thanks to this method, the amount of information that is needed for animation is also reduced. It is enough to store information about the movement of bones, and the movements of the vertices are calculated based on them.

Figure 2 shows an example of a skeletal animation.

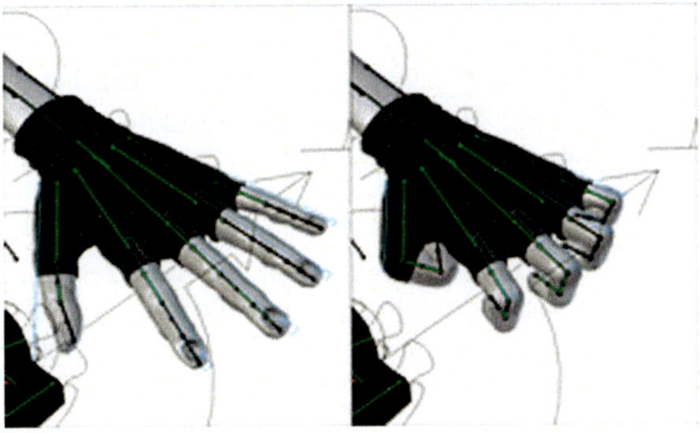

Fig. 2. Example of a skeletal animation.

Skeletal animation with weights of bones is a more advanced version of skeletal animation, in which each vertex of the model can be associated with not one, but several bones. In this case, for each bone, its weight is determined, that is, the magnitude of the influence of this bone on the displacement of the vertex. The greater the weight of the bone, the more the vertex moves under its influence. Thanks to the distribution of scales, you can animate smooth bends of surfaces, the movement of fabrics, flags.

Skeletal animation as a technology is also used in two-dimensional computer multiplication, for example, in the vector multiplication editor Anime Studio or the open Synfig Studio. Separate fragments of the model are "tied" to the bones, and then the multiplier sets the movement of the bones. The character model moves, following the movements of the skeleton.

5 Stages of Development of Cartoons

Usually, the development of a full-length cartoon includes several stages:

1) script writing;
2) determining the way to create a cartoon;
3) choosing a program for creating animation;
4) cartoon storyboard;
5) creating backgrounds and decorations;
6) soundtrack and voice acting;
7) installation of the finished project.

In order to understand the basic principles of cartoon development, we will consider the creation of computer animation using the example of the Unity game engine using various technologies.

Technology is a set and sequence of methods and processes for converting raw materials, which make it possible to obtain products with specified parameters [5].

Methodology is a field of knowledge that studies the means, prerequisites and principles of organizing cognitive and practical-transformative activities, the main tool is the method [6].

The term "educational animation" means any moving image that is used for educational purposes [7]. It can be just a moving arrow on a diagram or a full-fledged cartoon on your topic. The main reason many teachers create animations is to improve teaching efficiency.

Animation allows you to visually explain the content and meaning of complex concepts, abstract concepts. Information on a variety of topics can be presented in the form of animated images, and it is suitable for students of any age.

When adding animation effects, it is important to know some rules, because low-quality animation can alienate the viewer. Effective animation must have a number of characteristics:

- harmonious combination of text and images;
- not too much information;
- easy-to-read playback speed;
- emphasis on the most important elements;
- interactivity (the ability to pause, rewind, change the image scale, control the playback speed);
- voice guidance [8].

Difficulties may arise at each stage of the development of a full-length cartoon. In particular, the Unity3D game engine is suitable for the implementation of the cartoon. It has a number of advantages, such as cross-platform, good Community, which means that the various functions of the engine have a clear description with examples on the developer's website, which you can refer to at any time, and the Asset Store, where there are a huge number of different plugins and resources for creation of cartoons and games.

The Unity editor has ports for OS X and Windows, and was originally designed for OS X. Unity includes support for DirectX 11, which opens the way for your applications to the worlds of Windows 8 and Windows Phone 8. The Unity engine is especially valuable for its low barriers to entry for novice users, free indie version. The low barrier to entry is the result of well-designed application. Many things can be done using various editors without writing a single line of code (the code is written in JavaScript, C #, Boo). The C / C + + source code is closed, but this does not pose any obstacles due to the extended component structure of the Unity engine.

However, when analyzing technologies for the development of cartoons in the Unity environment, we faced the problem of the lack of clear algorithms for creating cartoons, there are only separate methods tied to specific types of animation. Therefore, in order to get the technology for creating cartoons, you need to develop your own algorithms and methods for the successful implementation of your cartoon.

6 Realization

To achieve the set goals in scientific work, it is necessary to use various methods, however, the problem of this study is that during the detailed analysis, it was revealed that there were no clearly formulated methods for developing cartoons in the Unity environment and ready-made practical solutions with detailed instructions. As a result, it became necessary to describe our own method as one of the possible ways to develop cartoons, which would give the optimal ratio between the achieved result and the resources used.

The described method is a synergy of various tools for creating products in the Unity environment that meet the criteria for effective animation.

Any development starts with creating an empty project. We place a Plane named "Water_Plane" and parameters in the Scale column x = 5, y = 5, z = 5. Then we add the Cloth component. It is a physical tissue modeling solution that works in conjunction with the Skinned Mesh Renderer technique for rendering bone animations [9].

You need to remove the Mesh Collider component that controls the main Unity graphics primitive (3D meshes) and select Plane in the Mesh column of the Skinned Mesh Renderer component.

You must select Edit cloth constraints in the Cloth component. The Stretching Stiffness value changes by 0.01.

In the next step, we hold down the Shift button and the left mouse button, select the Plane border, check the Max Distance checkbox and set the value to 0.2. These actions are repeated for all four boundaries (to check: the selected points should change their color).

It is necessary to uncheck the box opposite the Use Gravity column of the Cloth component. Then we download the iTween module package from the Asset Store and import it by unchecking the Sample box.

We add a script called "Wave_Controller" to our plane.

We set the values for height = 2.67 and time = 2.35 for our waves. It is necessary to change the value of the World Acceleration Scale of the Cloth component to 0.85, add a wave texture to the Assets and transfer it to Plane.

Change the values of the Tilling graph of the Wave Texture component to x = 5 and y = 5 and the Shader graph of the Wave Texture component to Legacy Shaders / Diffuse.

Download the Old Wooden Row Boat v2 unit package from the Asset Store and import it. Next, transfer the boat to Plane. Change the values for the boat to x = 1, y = 1, z = 1. Create an empty object named Boat and transfer Old_Boat_Prefab into it [10].

Next, we create the BoatFloater script. Change the name of Plane to Sea and add the Boat object to it. By independently changing the values of the Stretching Stiffness, World Acceleration Scale, height, time parameters, you can track how the waves change.

Next, install the Rock Package and import it. Add several stones to the Plane, changing their position and size.

This concludes the first part of the development of our cartoon. Figure 3 shows an example of the first computer animation using the Unity game engine.

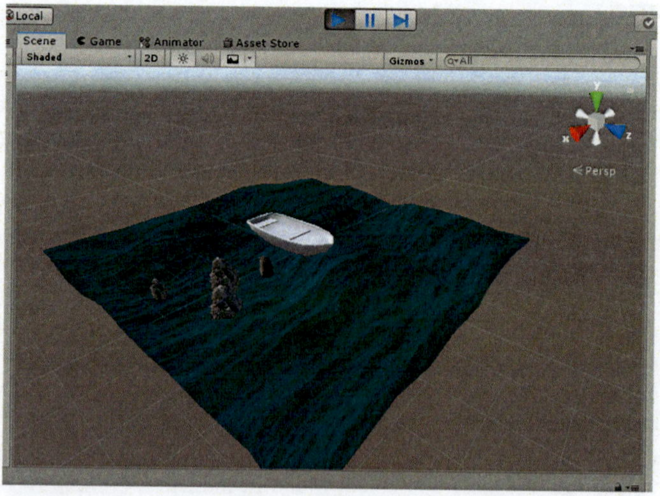

Fig. 3. Example of the first computer animation using the Unity game engine.

The second part of the development of our cartoon is the capture of cows using a teleportation beam before they reach a safe barn. We will create it as a 2D project, which at the installation stage can be connected to the first part [11].

We select the required platform in the Build Settings section of the File menu. Next, you need to select the size of the pictures that we will use. Before we start coding, we need to add the Asset to the Unity project. There are several ways to do this:

1. select Import New Asset from the Assets menu;
2. add details to the assets folder in the project;
3. drag the parts to the assets folder in the project.

After completing this step, we will see the assets of our project in the Assets folder in the Project panel [12].

Next, we create a scene for our animation by dragging and dropping objects into the Hierarchy or Scene panel. Then you need to transfer the background to the Hierarchy panel. We can select the Main camera in the Hierarchy panel and see what the camera will display. We will change the Size parameter for the Main Camera to 1.6 in the Inspector to see the whole scene.

The ship in our game is a static element and the player does not interact with it in any way. We'll position it in the center of the scene by dragging the aircraft from the Assets folder from the Project panel to the Scene tab. In the same way as with the UFO, we will transfer the barn for our cows to the stage.

To make sure that the barn reacts to a collision with a cow when she tries to go inside, you need to add one Box Collider 2D component [13].

Select the shed in the scene and in the Inspector tab, click on the Add Component button. From the list of components, select the Physics 2D section, and in it Box Collider 2D. Make sure there is a check mark next to the "Is Trigger" parameter.

When a cow hits the barn door, we want our trigger to fire. We need to make the collider smaller [14]. Going to the Inspector tab and changing the values of the collider of the Size and Center parameters, we achieve that the collider is located closer to the barn door.

The next step is to add an OnCollision script to make the application react to a collision when the cow enters the barn. This code snippet responds to a collision between a barn and an object named cow (Clone) This is one of the instances of a cow, the prefab of which we will create later. When a collision occurs, a sound is played and the cow object is destroyed.

To play a sound when a cow enters a barn, you first need to attach that sound to the barn. Click on the shed in the Hierarchy tab or Scene, click the Add Component button in the Inspector tab and select Audio Source in the Audio section. We have just added the Audio Source component. Now we need to associate our sound file with it. Click on the circle to the right of the Audio Clip item and select the sound barn. Uncheck the box next to Play on Awake. Drag the cosmic ray image from the Assets tab onto the stage and add a collider to it. This is necessary to detect the collision of the beam with cows. Make sure the "Is Trigger" switch is on.

Next, we'll add the Bullet script [15]. This simple code implies the following. First, we create an AudioClip instance called cowSound, which we will use to store the audio file. This is another option for playing sound if you don't want to bind two audio components to an object. We have declared the variable as public, so we can access it from the Inspector tab. Click on the little dot to the right of cowSound and select the cowSound audio file.

Then we made the ray invisible by turning off its rendering. We use the same object to save resources. This point is very important for optimization for low-power devices.

We detect screen touches, make the beam visible, and reproduce the sound of the beam. When our ray becomes visible, it begins to move downward to hit the cow. We also provided a check to see if our ray was out of the scene. If so, then we put it back in place and our UFO can fire again. The last part checks if the beam has touched the cow. If this happens, a sound is played and the cow disappears. Then the beam becomes invisible again, moves to its original position and the UFO is ready to fire again.

To add a soundtrack to our cosmic ray, you need to select our ray in the Hierarchy or Scene tab and click on the Add Component button in the Inspector tab [16]. Select Audio Source from the Audio section.

Uncheck Play on Awake and click the small dot to the right of the Audio Clip to select a sound file named bizzed.

Drag our cow from the Assets panel to the Scene tab. For the collision to be registered, we must associate the RigidBody2D component with at least one of the objects [17]. It is best to add the component to the cow because the cow can collide with both the barn and the beam. We also need to attach a collider to the cow to detect barn and beam collisions.

Let's create a script to move the cow. The MoveCow class animates the cow as it moves on the screen using the moveSpeed variable. The InvokeRepeating method changes the speed of the cow. This makes the animation more complex. Now that we have attached all the necessary components to the cow, it's time to convert it to a prefab.

A prefab is a reusable GameObject that is stored in the Project View [18]. The prefab can be inserted multiple times in any number of scenes. When you add a prefab to a scene, you create an instance of it. All instances of the prefab are associated with the original prefab and are its clones [19]. No matter how many copies you make for your project, any changes made to the original prefab will be applied to all instances of it.

Let's create a Spawner script that is responsible for displaying cows [20].

To convert a cow to a prefab, drag the cow from the Hierarchy tab to the Assets panel. As a result, the name in the Hierarchy will turn blue.

We use the InvokeRepeating method to clone the cows using the values set for spawnTime and spawnDelay. GameObject cow is set to public and created using the Inspector [21].

To create multiple instances of the cow prefab, use the cow graph that we added to the scene a few minutes ago [22]. Select it and remove its components. Then add the Spawner script.

Figure 4 shows an example of the second computer animation using the Unity game engine.

Fig. 4. Example of of the second computer animation using the Unity game engine

7 Conclusion

In this paper, a description of our own method of creating a cartoon in the Unity environment was given as one of the possible ways to develop cartoons, which would give an optimal ratio between the achieved result and the resources used.

There was a need to search for our own solutions, since during a detailed analysis it was revealed that there were no clearly formulated methods for developing cartoons in the Unity environment.

The uniqueness and novelty of the proposed method lies in the synergy of various tools for creating products in the Unity environment that meet the criteria for effective animation. It is worth noting the simplicity and speed of implementation, the clarity and accessibility of the presentation of information, which make it possible to use this method for educational purposes when teaching students of any age to create animation and awaken interest in the production of cartoons.

This method is ideal for studying the technologies of creating animation, for organizing high-quality educational work using animation at the stage of mastering a theoretical and practice-oriented course of academic disciplines.

References

1. Bikmullina, I.: Barkov Instrumentation and Control System for Test Bench. International Multi-Conference on Industrial Engineering and Modern Technologies (FarEastCon). EEE (2019) https://ieeexplore.ieee.org/author/37087040423

2. Laptev, O.S., Bikmullina, I.I.: Sightseeing Application Based on Location Marking and Convolutional Neural Network Building Recognition. International Russian Automation Conference (RusAutoCon). Conference Paper, Publisher: IEEE (2020) https://ieeexplore.ieee.org/search/searchresult.jsp?contentType=conferences&queryText=RusAutoCon&highlight=true&returnFacets=ALL&returnType=SEARCH&matchPubs=true&refinementName=Author&refinements=Author:Ilsiyar%20Bikmullina&refinements=Author:Ilsiyar%20I.%20Bikmullina
3. Bikmullina, I., Garaeva, E.: The Development of 3D Object Modeling Techniques for Use in the Unity Environmen. International Multi-Conference on Industrial Engineering and Modern Technologies (FarEastCon) (2020) https://ieeexplore.ieee.org/search/searchresult.jsp?contentType=conferences&queryText=FarEastCon&highlight=true&returnFacets=ALL&returnType=SEARCH&matchPubs=true&refinementName=Author&refinements=Author:I.%20Bikmullina&refinements=Author:Ilsiyar%20Bikmullina
4. Bikmullina, I., Ilnar, V.: Multi-Service Electronic Library of Scientific and Technical Articles. International Multi-Conference on Industrial Engineering and Modern Technologies (FarEastCon) (2020). https://ieeexplore.ieee.org/search/searchresult.jsp?contentType=conferences&queryText=FarEastCon&highlight=true&returnFacets=ALL&returnType=SEARCH&matchPubs=true&refinementName=Author&refinements=Author:I.%20Bikmullina&refinements=Author:Ilsiyar%20Bikmullina
5. Bikmullina, I., Kusyumov, N.: Orienteering Mobile App. Proceedings of the International Russian Automation Conference, RusAutoConf2020, Sochi, Russia (2020). https://doi.org/10.1007/978-3-030-71119-1_6
6. AYu, D., Kudinov, A.V.: Computer graphics. Tutorial. Tomsk Polytechnic University, Tomsk, p. 160 (2005)
7. Artem, A.: Surface modeling in SolidWorks. https://sapr.ru/article/16361
8. Vasiliev, V.E., Morozov, A.V.: Computer graphics: textbook. SZTU, SPb, p. 101 (2005)
9. Unity engine—features, advantages and disadvantages. https://cubiq.ru/dvizhok-unity/
10. Hawking, J.: Unity in action. Multi-platform development in C, Peter, SPb, p. 336 (2016)
11. Yazev, Y.: Review of the most popular game development engines. https://xakep.ru/2014/09/05/game-development-engines-review/
12. Blender 2.80 Manual. https://docs.blender.org
13. Blender. https://www.blender.org/about/license/
14. Cinema 4D. https://www.maxon.net/ru/produkty/cinema-4d/obzor/
15. GameMaker: Studio. https://la.by/game-engines/game-maker-studio
16. NURBS. http://www.ray-tracing.ru/articles249.html
17. Rendering: Theory. https://render.ru/ru/articles/post/10102
18. Unity Core Platform. https://unity.com/ru/products/core-platform
19. Unity. ProBilder. https://unity3-d.com/ru/unity/features/worldbuilding/probuilder
20. Unreal Engine. https://www.unrealengine.com/en-US/features
21. Vuforia Engine: developer portal. https://developer.vuforia.com/
22. Mytnikov, A.N., Mytnikova, E.A., Kuznetsova, L.N., Solin, S.: Mobile application development technologies. Theory and practice of modern science **4**(10), 504–507 (2016)

Model for Determining the Cloud Computing Services Vulnerability Level

S. V. Razumnikov(✉)

National Research Tomsk Polytechnic University, Lenina Str 30, Tomsk 634050, Russia
demolove7@inbox.ru

Abstract. In decision making when comparing alternatives, it is often necessary to work with different values and scores. Binary relations and graph theory proved to be efficient when conducting this type of analysis. Cloud computing services, like any other technical means, are exposed to vulnerability and prone to breakdowns. This article proposes a model for determining the vulnerability level of cloud computing sections based on the ELECTRE I method. This model allows identifying sections that are most at risk of breakdown or failure, as well as removing dominant alternatives that do not constitute a grave danger to the system. Based on identified preferences, it is possible to decide on discrepancies rectification in the selected sections of the cloud computing services.

Keywords: Computing services · Vulnerability · ELECTRE I method · Multi-criteria optimization · Non-dominated alternatives

1 Introduction

Cloud computing technologies have become an inherent element of the modern world. The number of businesses that consider transferring their IT infrastructure to the Cloud is increasing every day [1, 2]. The next step after choosing appropriate cloud computing services and their enterprise implementation is their maintenance and operational management [3–5]. Cloud computing services, like any other technical means, are exposed to vulnerability and prone to breakdowns. It is crucial to identify these malfunctions and remove them in time [6, 7].

When conducting analysis, it is often necessary to work with different values and scores. Binary relations and graph theory proved to be efficient evaluation instruments for this specific type of analysis [8–13]. One of the methods frequently used in the decision-making process is the ELECTRE I method. It offers a multi-criteria procedure for identifying the most preferred alternatives and consists of several phases.

Based on this method, we suggest a model aimed to determine the cloud computing services' vulnerability level. This model allows identifying elements exposed to breakdowns, and removing them.

A. A. Radionov and V. R. Gasiyarov (Eds.): RusAutoCon 2022, LNEE 986, pp. 119–126, 2023.
https://doi.org/10.1007/978-3-031-22311-2_12

2 Cloud Computing Services for Businesses

Cloud computing services are virtual IT infrastructures (servers, data-warehousing systems, network equipment, etc.) used by corporate programs and services. Examples of such services are accounting, databases, remote offices, CRM systems, etc.

The cloud computing services are stored in the provider's remote data centres. This infrastructure is accessible through system administration and management interface. The maintenance of the cloud computing services' physical components, i. e. servers, remains the provider's obligation.

The main difference between cloud computing services and other infrastructure solutions is configuration flexibility. For example, the web interface allows users to increase or reduce the amount of RAM, and activate or disable processor cores. At the same time, the cost of renting capacities varies accordingly: fewer resources – less payment.

Every business has unique priorities and demands. When one business aims to spare funds, another company is concerned about security issues, and the third one concentrates on optimizing its operative business. The use of cloud computing technologies is determined by business segment, its size, strategy, organizational structure, and even legal restrictions [5–7].

Consequently, maintaining the integrity and working capacity of cloud computing services proves to be crucial. Usually, providers are responsible for solving this issue; however, sometimes the problem can also be solved by the company's employees [14–20]. Now let us analyze using the ELECTRE I method for building a model for determining the cloud computing section vulnerability level.

3 Model for Determining the Cloud Computing Services Vulnerability Level

Based on this method, we propose a model that allows determining the vulnerability level of cloud computing services. Figure 1 represents a model (diagram) of using the ELECTRE I method for decision-making.

This model allows identifying sections that are most at risk of breakdown or failure, as well as removing dominant alternatives that do not constitute a grave danger to the system Based on identified preferences, it is possible to decide on discrepancies rectification in the selected sections of the cloud computing services.

4 An Example of a Cloud Computing Services Vulnerability Level Determining Problem

Let us analyze an example of using the ELECTRE I method for determining cloud computing services vulnerability level. The following 7 blocks will be analyzed: servers, virtual machines, storage, networks, OS, DBMS, and connection software. These sections are different according to these criteria: managing and customer support, productivity, reliability, and scaling. The criteria weights will be calculated through a pair wise comparison matrix.

The most important criterion when determining a section is reliability, the least important criterion is scaling (Fig. 2). Criteria weights are respectively equal to: $u = (0.19,\ 0.25,\ 0.47,\ 0.09)$.

The evaluation by the criteria will be conducted with the use of the qualitative scale (Table 1). In this scale, the gradations are encoded in numbers, i. e. if a section's score is equal to 4, that does not mean that it is 2 times higher than the sections with a score equal to 2.

Table 1. Criteria scale for determining cloud computing services vulnerability section.

Criteria	Gradations	Score (grade)
u_1- managing and customer support	Excellent Above average Average Below average Poor	5 4 3 2 1
u_2– productivity	Excellent Average Poor	3 2 1
u_3– reliability	High Average	2 1
u_4– scaling	Yes No	1 0

Let us compare 7 sections a1, …, a_7. Their scores u_{ij} by all the criteria are listed in Table 2. None of the rows in this table is dominant towards any other. Therefore all the variants to absolute dominance are non-comparable and constitute the Pareto set.

Then, using the ELECTRE I method algorithm, we can perform a pair wise comparison of the criteria indices group. First, let us analyze the (a_1, a_2) pairing. According to the 3 criterion, the first alternative is the best option for this pairing. According to the 1 criterion, it is a less suitable option. And according to the 2 and the 4 criteria, it is matched. Using formulas 1–4, we can calculate the values:

$$C(1,\ 2) = \{3\},\ D(1,\ 2) = \{1\},\ E(1,\ 2) = \{2,\ 4\}$$

$$c_{1,2} = 0.47, d_{1,2} = 0.19, e_{1,2} = 0.25 + 0.09 = 0.34$$

Using the formula (5), we can calculate the agreement index, considering that $\sum w_j = 1$.

$$s_{1,2} = 0.47 + 0.5 \cdot 0.34 = 0.64$$

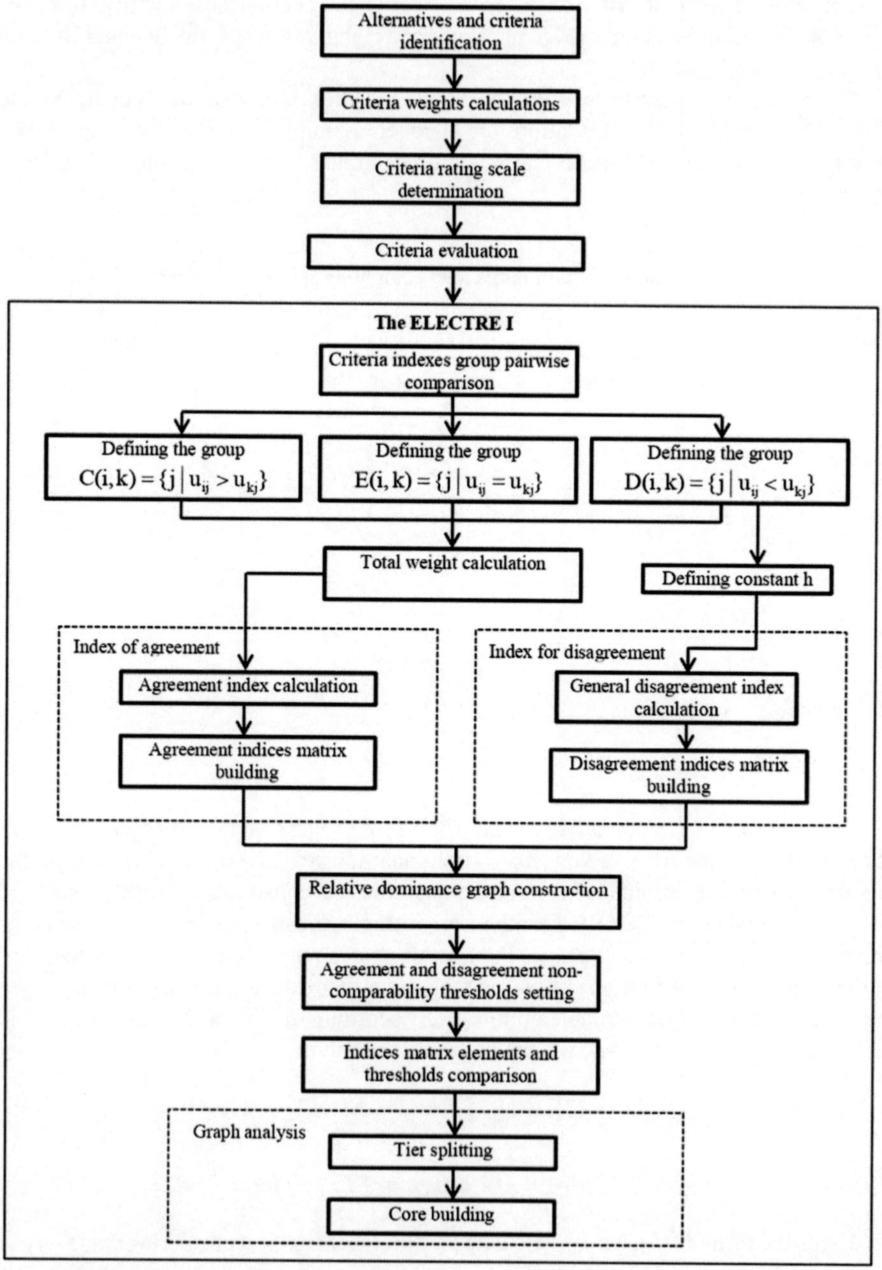

Fig. 1. Model for determining the vulnerability level of cloud computing sections using the ELECTRE I method.

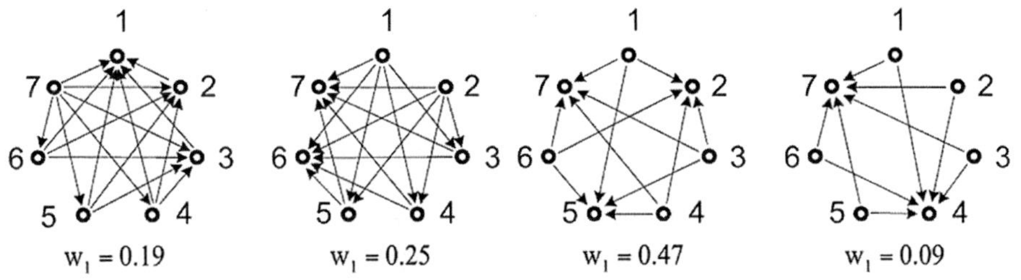

Fig. 2. Criteria preference graphs.

Table 2. Alternatives evaluation by criteria.

Alternatives a_i	u_{i1}	u_{i2}	u_{i3}	u_{i4}
a_1	1	3	2	1
a_2	2	3	1	1
a_3	2	2	2	1
a_4	3	2	2	0
a_5	3	2	1	1
a_6	3	1	2	1
a_7	5	1	1	0

Other indices can be calculated in the same way. Hence we get an S agreement indices matrix:

$$S = \begin{pmatrix} - & 0.64 & 0.53 & 0.58 & 0.76 & 0.53 & 0.81 \\ 0.36 & - & 0.39 & 0.34 & 0.53 & 0.29 & 0.58 \\ 0.47 & 0.61 & - & 0.45 & 0.64 & 0.53 & 0.81 \\ 0.42 & 0.66 & 0.55 & - & 0.69 & 0.58 & 0.76 \\ 0.24 & 0.47 & 0.36 & 0.31 & - & 0.39 & 0.58 \\ 0.47 & 0.70 & 0.47 & 0.42 & 0.61 & - & 0.68 \\ 0.19 & 0.42 & 0.19 & 0.24 & 0.42 & 0.32 & - \end{pmatrix}$$

Now we can calculate indices of disagreement. First, it is necessary to set constant values that indicate the protest. For example, $h_1 = 0.2$, $h_2 = 0.25$, $h_3 = 0.2$, $h_4 = 0.2$. That indicates that if we choose the alternative that loses by the managing and customer support criterion by one scale division, it provokes a conflict between the 1 criterion and the 0.2 intensity.

For example, let us analyze these calculations $r_{2,\,6}$: For a disagreement alternative (a_2, a_6) pairing (with an advantage) a_2 it is $D(2,\ 6) = \{1,\ 3\}$. The difference according to the first Δ_{26}^{1} and the third Δ_{26}^{3} criteria is that of one level. Consequently, the indices of disagreement are equal to $r_{26}^{1} = 1 \cdot 0.2$, $r_{26}^{3} = 1 \cdot 0.2$. Hence the general index is

$r_{26} = \max(0.2,\ 0.2) = 0.2$. Other indices can be calculated in the same way. The final R disagreement indices matrix:

$$R = \begin{pmatrix} - & 0.2 & 0.2 & 0.4 & 0.4 & 0.4 & 0.8 \\ 0.2 & - & 0.25 & 0.2 & 0.2 & 0.2 & 0.6 \\ 0.25 & 0.25 & - & 0.2 & 0.2 & 0.2 & 0.6 \\ 0.25 & 0.25 & 0.2 & - & 0.2 & 0.2 & 0.4 \\ 0.25 & 0.25 & 0.2 & 0.2 & - & 0.2 & 0.4 \\ 0.5 & 0.5 & 0.25 & 0.25 & 0.25 & - & 0.4 \\ 0.5 & 0.5 & 0.25 & 0.25 & 0.25 & 0.2 & - \end{pmatrix}$$

In the 3 phase, it is required to construct an absolute dominance graph. To do this, we need to consider agreement $s* = 0.5$ and disagreement $r* = 0.2$ threshold values. We need to iterate over the agreement and disagreement matrices elements. For instance, pairing (a_1, a_2) is compliant to the relative dominance ratio, as $s_{1,\,2} = 0.64 > 0.5$, $r_{1,\,2} = 0.2 \leq 0.2$; although pairing (a_1, a_5) is not compliant to the same ratio because $s_{1,\,5} = 0.76 > 0.5$, $r_{1,\,5} = 0.4 \leq 0.2$ etc. Given these non-comparability threshold values, we have the following ratio graph (Fig. 3).

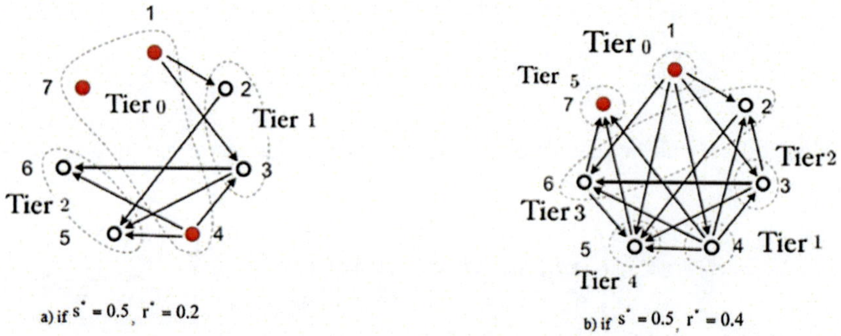

Fig. 3. Relative dominance graph with present non-comparability thresholds

The constructed ratio is not transitive, i. e. section a_1 is more preferred than a_3, a_3 is more preferred than a_6. Moreover there is no arc (a_1, a_6) in the graph. According to these algorithm phases, let us perform layering as demonstrated in Fig. 3a. The core is formed by vertices 1, 4, 7 colour-highlighted in the figure. As a result, the number of non-comparable elements has decreased from 7 to 3, i. e. it dominates all the other vertices.

To reduce the number of non-comparable vertices, it is necessary to alter non-comparability thresholds. If we extend the disagreement threshold r^*, new arcs will appear in the graph. Subsequently, there will be fewer vertices in the core. Figure 3b demonstrates the graph if $s^* = 0.5$, $r^* = 0.4$. Compared to Fig. 3a, in Fig. 3b there are 10 more arcs. Moreover, the core has 2 vertices −1 and 7. Arcs can be also added by the agreement threshold reduction s^*. However, in both cases, there is a risk of closed contours inception.

For example, let us perform the variation of the non-comparability thresholds. Figure 4 demonstrates a relative dominance graph if $s^* = 0.4$, $r^* = 0.2$. Here we can see one contour with two vertices −3 and 4. After pinching, we have a graph (Fig. 4b) that has a core with 4 vertices. This result is unacceptable.

Therefore, to have the optimum alternative, one should be attentive when calculating the non-comparability thresholds. Thus, vertices 1 and 7 are servers and connection software, i. e. 2 sections with the highest vulnerability level.

5 Conclusion

The ELECTRE I method proves to be efficient in problem-solving and decision-making. Based on this method, we propose a model that allows determining the vulnerability levels of cloud computing services. This model also allows identifying sections that are most at risk of breakdown or failure, as well as removing dominant alternatives that do not constitute a grave danger to the system. Based on identified preferences, it is possible to decide on discrepancies rectification in the selected sections of the cloud computing services.

The paper also provides an example of using the model for determining vulnerability levels in 7 cloud computing services sections. Two block with the greatest vulnerability level were identified.

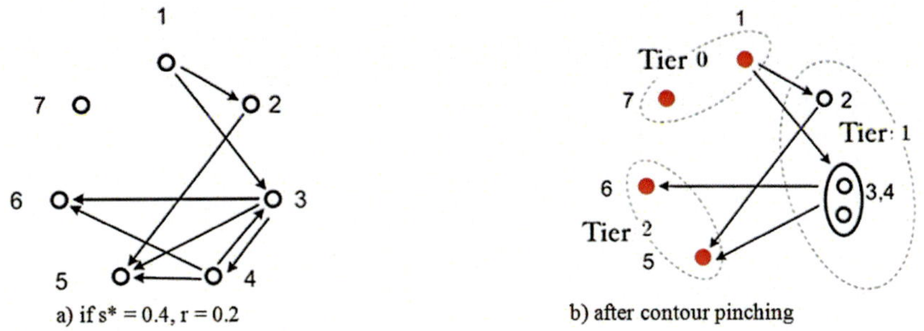

Fig. 4. Relative dominance graph on non-comparability thresholds variation

References

1. Lin, W., Wang, H., Zhang, Y., et al.: A cloud server energy consumption measurement system for heterogeneous cloud environments. Inf. Sci. **468**, 47–62 (2018)
2. Paul, P.K., Ghose, M.K.: Cloud Computing: possibilities, challenges and opportunities with special reference to its emerging need in the academic and working area of Information Science. International conference on modelling optimization and computing **38**, 2222–2227 (2012)
3. Reynolds, P., Yetton, P.: Aligning business and IT strategies in multi-business organization. J. Inf. Technol. **30**(2), 101–118 (2015)
4. Tsai, W.-L.: Constructing Assessment Indicators for Enterprises Employing Cloud IaaS. Asia Pac. Manag. Rev. **26**(1), 23–29 (2021)

5. Jones, S.: Cloud computing procurement and implementation: Lessons learnt from a United Kingdom case study. Int. J. Inf. Manage. **35**(6), 712–716 (2015)
6. Elamir, A.M., Jailani, N., Dakar, M.A.: Framework and architecture for programming education environment as cloud computing service. Procedia Technol. **11**, 1299–1308 (2013)
7. Sultan, N.: Knowledge management in the age of cloud computing and Web 2.0: Experiencing the power of disruptive innovations. Int. J. Inf. Manag. **33**(1), 160–165 (2013)
8. Breedveld, S., Craft, D., van Haveren, R., et al.: Multi-criteria optimization and decision-making in radiotherapy. Eur. J. Oper. Res. **277**(1), 1–19 (2019)
9. Singh, B.K., Roy, H., Mondal, B. et al.: Measurement of chip morphology and multi criteria optimization of turning parameters for machining of AISI 4340 steel using Y-ZTA cutting insert. Measur. J. Int. Measur. Conf. **142**, 181–194 (2019)
10. Wang, J.-Q., Li, S., Hedayati Dezfuli, F., et al.: Sensitivity analysis and multi-criteria optimization of SMA cable restrainers for longitudinal seismic protection of isolated simply supported highway bridges. Eng. Struct. **189**, 509–522 (2019)
11. Álvarez-Miranda, E., Garcia-Gonzalo, J., Pais, C., Weintraub, A.: A multicriteria stochastic optimization framework for sustainable forest decision making under uncertainty. Forest Policy Econ. **103**, 112–122 (2019)
12. Breedveld, S., Craft, D., van Haveren, R., Heijmen, B.: Multi-criteria optimization and decision-making in radiotherapy. Eur. J. Oper. Res. **277**(1), 1–19 (2019)
13. Razumnikov, S.V.: Building an Aggregate Rating of Popular SaaS Services Based on Organization of Customer Support Channels. Lecture Notes in Electrical Engineering **857**, 313–323 (2022)
14. Razumnikov, S.V., Kurmanbay, A.K.: Models of evaluating efficiency and risks on integration of cloud-base IT-services of the machine-building enterprise: a system approach. IOP Conference Series: Materials Science and Engineering **124**(1), 012089 (2016)
15. Razumnikov, S.V., Prankevich, D.A.: Integrated model to assess cloud deployment effectiveness when developing an IT-strategy. IOP Conference Series: Materials Science and Engineering **127**(1), 012018 (2016)
16. Micale, R., La Fata, C.M., La Scalia, G.: A combined interval-valued ELECTRE TRI and TOPSIS approach for solving the storage location assignment problem. Comput. Ind. Eng. **135**, 199–210 (2019)
17. Zhou, H., Wang, J.-Q., Zhang, H.-Y.: Stochastic multicriteria decision-making approach based on SMAA-ELECTRE with extended gray numbers. Int. Trans. Oper. Res. **26**(5), 2032–2052 (2019)
18. Liao, H., Jiang, L., Lev, B., Fujita, H.: Novel operations of PLTSs based on the disparity degrees of linguistic terms and their use in designing the probabilistic linguistic ELECTRE III method. Applied Soft Computing Journal **80**, 450–464 (2019)
19. Costa, A.S., Rui Figueira, J., Vieira, C.R., Vieira, I.V.: An application of the ELECTRE TRI-C method to characterize government performance in OECD countries. Int. Trans. Oper. Res. **26**(5), 1935–1955 (2019)
20. Akram, M., Waseem, N., Liu, P.: Novel Approach in Decision Making with m–Polar Fuzzy ELECTRE-I. Int. J. Fuzzy Syst. **21**(4), 1117–1129 (2019)

Simplex Method of Realization of Optimization Problem of Linear Programming

S. Oskin, A. Kuznecov(✉), and N. Koneva

Moscow Polytechnic University (Moscow Polytech), 38, Bolshaya Semyonovskaya Ul., Moscow 107023, Russia
sv-oskin@yandex.ru

Abstract. In a wide range of optimization tasks, it is necessary to search for an extremum. A feature of the simplex optimization method is the combination of studying the response surface with moving along it to the extremum. This is achieved by the fact that experiments are performed at the points of the factor space corresponding to the vertices of the simplexes. Thus, the simplex method has a number of positive qualities that allow it to be used in production conditions to solve optimization problems. The article presents the results of the implementation in the laboratory practical simplex method for the determination of the response function extremum at finding optimal experimental conditions.

Keywords: Linear programming · Simplex method · Design of experiments · Response function

1 Introduction

Originating in the 1930s, mathematical programming developed as a discipline dedicated to the theory and methods of solving control and planning problems. In the 1950s, mathematical programming became a separate section of the then emerging science of operations research. The new discipline was based on a set of methodological tools called system analysis.

From the 50s of the XX century. to substantiate solutions to complex problems in systems of a political, social, military, economic, scientific and technical nature, a set of methodological tools, called system analysis, began to be used. Systems analysis is used to solve such problems as the distribution of production and capacities for the production of various types of products, determining the future need for new equipment and labor of one skill or another, forecasting the demand for various types of products, as well as in solving problems related to the development and technical equipment of the armed forces, space exploration, etc.

Systems analysis is based on a number of applied mathematical disciplines and branches, in particular, operations research. When in the problem of system analysis there is one clearly defined goal, the degree of achievement of which can be assessed on the basis of one criterion, methods of mathematical programming are used.

A. A. Radionov and V. R. Gasiyarov (Eds.): RusAutoCon 2022, LNEE 986, pp. 127–135, 2023.
https://doi.org/10.1007/978-3-031-22311-2_13

The most developed section of mathematical programming is linear programming, which contains the theory and methods for solving extremal problems. The first publications on linear programming belong to the Soviet scientist L.V. Kantorovich, awarded in 1975, together with the American scientist T. Koopmans, the Nobel Prize for their contribution to the theory of optimization of resource allocation.

Mathematical programming methods were used already during the Second World War to plan military operations. In 1938, the United States coined the term "operations research" to characterize the type of activity of an unusual research group, created at the initiative of the Air Ministry Research Station and performing work on the analysis of military systems. As one of the tasks, the search for the optimal use of radar installations in the overall defense system of the country was considered. Subsequently, operations research was formed into a scientific direction.

Since the late 1940s, the field of application of operations research methods has steadily expanded, covering various aspects of human activity. Thus, operations research began to be used to solve both purely technical (especially technological) and technical and economic problems, as well as management problems at various levels.

The use of mathematics in economic research is inextricably linked with the name of the Nobel laureate L. V. Kantorovich. In the work "Mathematical methods of organization and planning of production" (1939), the main provisions of a new direction of mathematics, later called linear programming, were presented for the first time [1–3]. The practical implementation of the provisions of the theory made it possible to solve the problems of economic calculation of the best use of resources, as well as the effective use and development of transport.

When constructing optimal models of long-term planning, the simplex method has become widely used, the essence of which is determined by the algorithm for solving the optimization problem of linear programming by iterating over the vertices of a convex polyhedron in a multidimensional space.

The peculiarity of the simplex planning method developed in the early sixties is that the experimental points are set by the position of the correct simplex. It is known, for a regular simplex, all the distances between the vertices are equal. An example of a simplex of zero dimension is a point. A one-dimensional simplex is a straight line segment, a two-dimensional one is a triangle, and a three-dimensional one is a tetrahedron ("pyramid"). A regular k-dimensional simplex is a regular convex polygon with $k + 1$ vertices located in a k-dimensional space. A simplex in k–dimensional space allows us to represent a set of k independent factors acting on the object or process under study [4, 5].

2 Materials and Methods

A new simplex can be obtained from any simplex if one of the vertices is moved to a point that is mirror-symmetric with respect to the opposite face. Thus, the simplex ABC is transformed into the simplex CBA' by reflecting the vertex A in the "face" BC (Fig. 1).

The idea of applying the simplex method is based on the sequential transformation of the simplex when it "moves" to the extremum. This property is widely used in the optimization procedure and determines the special significance of the method for determining production conditions. The point at which the optimization process begins is

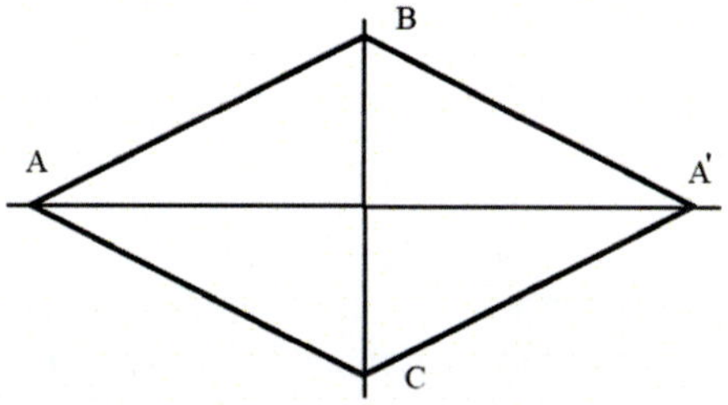

Fig. 1. An example of a 2nd-order simplex.

chosen from a priori considerations, for example, on the basis of existing regulatory provisions. In the case of solving applied problems in the field of economics, it is necessary to take into account the features of specific acting factors that ensure the achievement of the target function of the study. The selected point serves as the center of the initial position of the simplex.

Graphic constructions shown in Fig. 2 illustrate the nature of the advance to the extremum by mirroring the point with the worst results relative to the center of the opposite face of the simplex with the (k + 1)-th vertex.

In the common optimization methods [4, 5], it is possible to distinguish trial experiments designed to identify the direction of movement, and working steps that make progress to the extreme. A feature of the simplex optimization method is the combination of studying the response surface with moving along it to the extremum. This is achieved by the fact that the experiments are performed at the points of the factor space corresponding to the vertices of the simplices. To move to the extremum, it is necessary to move from the original simplex to the simplex located in the region of increased (if the maximum is sought) response values. This is achieved by comparing the response values at the vertices of the triangle.

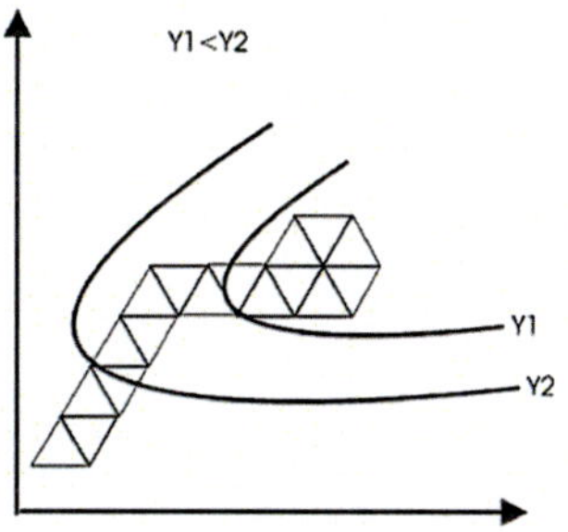

Fig. 2. Search for the extremum of the response function by the simplex method.

Initially, there is a vertex where the response is minimal. The next experiment is put at a point that is mirror-symmetric to this vertex. Thus, the simplex moves "away" from the unfavorable point. For the new position of the simplex, the values of the responses are compared again, the minimum is found, and the process of movement of the simplex is repeated. This happens until the simplex begins to rotate around a certain point, which should be taken as the optimal one – the one where the extremum is reached. The rotation of the simplex around a certain point indicates the localization of the extremum region. To clarify the coordinates of the latter, it is necessary to continue the experiments, having previously reduced the steps for independent variables.

The coordinates of the vertices of a regular k-dimensional simplex with the center located at the origin can be determined from the following matrix:

$$\begin{bmatrix} -r_1 & -r_2 & -r_3 & \ldots & -r_{k-1} & -r_k \\ R_1 & -r_2 & -r_3 & \ldots & -r_{k-1} & -r_k \\ 0 & R_2 & -r_3 & \ldots & -r_{k-1} & -r_k \\ \ldots & \ldots & \ldots & \ldots & \ldots & \ldots \\ 0 & 0 & 0 & \ldots & -r_{k-1} & -r_k \\ 0 & 0 & 0 & \ldots & 0 & R_3 \end{bmatrix} \quad (1)$$

where r_i and R_i are the radii of spheres, respectively inscribed and described near the i-dimensional simplex [4, 5].

If we take the edge length of an i-dimensional simplex equal to one, then the radii r_i and R_i are calculated using the formulas:

$$r_i = \frac{1}{\sqrt{2 \cdot i \cdot (i+1)}};\ R_i = \sqrt{\frac{i}{2 \cdot (i+1)}};\ (i = 1, 2, \ldots k) \quad (2)$$

In the matrix above, the coordinates of the vertices are given by the rows of the matrix. The first vertex corresponds to the first row, the second—the second, etc. The part of the matrix that is crossed out from the top left sets the coordinates of the three vertices of the triangle. We will first find:

$$r_1 = \frac{1}{2}, r_2 = \frac{1}{2\sqrt{3}}, R_1 = \frac{1}{2}, R_2 = \frac{1}{\sqrt{3}} \quad (3)$$

So, the first vertex has coordinates $(-\frac{1}{2}, -\frac{1}{2\sqrt{3}})$, the second $(\frac{1}{2}, -\frac{1}{2\sqrt{3}})$, the third $(0, \frac{1}{\sqrt{3}})$. The validity of the obtained results can be verified by examining the simplex shown in Fig. 3.

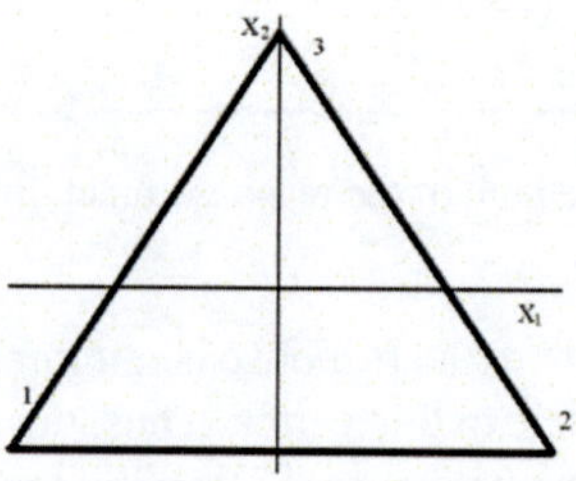

Fig. 3. Determination of the vertices of the simplex of the 2nd order.

We also give a formula that allows us to calculate the coordinates of a new vertex of the simplex, which is a mirror image of the previous vertex:

$$x_{i\,k+2} = \frac{2}{k}\sum_{n} x_{in} - x_i^* \tag{4}$$

where $\frac{1}{k}\sum_n x_{in}$—the average value of the coordinates of all points of the simplex, except for the previous one, discarded; x_i^*—the coordinate of the previous vertex.

Thus, the simplex method has a number of positive qualities that allow it to be used in production conditions to solve optimization problems. Simplicity, the ability to include additional factors at any stage of the study, automatic correction of gross errors are the most important properties of the method.

The conditions for conducting the experiment at the reflected point are determined by the expression:

$$X_{(k+2)i} = 2 \cdot X_{0i} \pm X_{hi}, \tag{5}$$

where $i = 1,2,\ldots, k$; X_{hi}—i-th coordinate of the point with the worst results; $X_{(k+2)i}$—the i-th coordinate of the new point obtained as a result of mirroring the point with the worst results.. X_{0i}—the i-th coordinate of the center of the opposite face, which is determined by the formula:

$$X_{0i} = \frac{\sum_{i=1}^{i=k+1} X_i}{k}, i \neq h \tag{6}$$

The numerator sums the coordinates of all points of the simplex with $(k + 1)$- th vertex, except for the coordinates of the point with the worst results ($i = h$).

A new k-dimensional simplex is obtained from the remaining face by adding a reflected point to it. It should be emphasized that this shift to the extreme occurs with each experiment.

The indicator of reaching the extremum area is the termination of the translational motion of the simplex and the beginning of its rotation around one of the vertices. In this case, the same point consistently occurs in more than $(k + 1)$ simplices. It should be noted that the direction of movement to the optimum, determined using the simplex, is generally steep, the trajectory of movement in this case is a broken line oscillating around the line of the steepest ascent.

3 Implementation of the Provisions of the Theory in a Laboratory Workshop

The implementation of steadily increasing requirements for the quality of the educational process is impossible without the broad involvement of students in solving actual engineering problems. The rapid development of information technologies based on the latest achievements of micro - and nanoelectronics clearly demonstrates the importance of introducing the provisions of theoretical science into practical activities. Such a combination of theory and practice in the framework of the educational process can be provided in the conditions of a laboratory workshop. The interactive nature of the methodological support of the educational process in the study of this topic suggests the possibility of a variable assignment of the initial data. This requirement is successfully implemented in a laboratory workshop developed with the involvement of linear programming methods for the course "Computer systems for processing experimental data".

The laboratory work "Simplex method for determining the extremum of the response function" is based on software developed during the implementation of the diploma design. In the practical part of the laboratory lesson, it is proposed to perform two tasks (I) and (II), each of which includes a computational part and an experimental one.

4 Practical Part

4.1 Conduct a Study of the Mathematical Response Function

$$Z = \sqrt{225 - (X_1 - 10)^2 - (X_2 - 10)^2} + 10, \tag{7}$$

where are the parameters X_1 и X_2 are set in the range from 1 to 23 dimensionless units.

The estimated part. The result of the mathematical analysis of the algebraic expression (6) is the theoretical value of the extremum.

The experimental part. For an experimental study of the function specified by expression (6), select the item "unknown" in the section "Study of response functions". At the same time, we need to set the initial parameters of the study and the steps for varying the parameters. The extremum (maximum or minimum) is found when the corresponding button is selected in the program window. The results of the calculations are given in the form of a table. To view the results in a graphical form, click the corresponding button in the program window. A simplex will be drawn in the program window that appears. To re-draw the simplex, double-click the left mouse button on the window where the simplex is drawn.

4.2 To Study the Function that Determines the Diameter D of the Electron Beam During Electron Beam Exposure

The estimated part. The theoretical dependence for the parameter d can be represented as:

$$d = (4 \cdot C)^{1/4} (\frac{I_n \cdot k \cdot T}{\pi \cdot J \cdot E})^{3/8}, \tag{8}$$

where C—the coefficient of spherical aberration (for various designs of electron-optical systems lies within 5–10^2); I_n—beam current, *A*; J—cathode emission current density, *A/cm*^{2}; T—absolute cathode temperature, *K*; E—electron energy, *eV*; k—Boltzmann constant 1,3807·10^{-23} J/K.

When studying the function of the diameter d of an electron beam in the case of electron beam exposure, two parameters change: electron beam current (I_n, X-axis) and the cathode emission current density (J, Y-axis). The electron beam current is set within

$1 \cdot 10^{-7} \div 1 \cdot 10^{-6}$ A with a step of variation within $0{,}5 \cdot 10^{-7} \div 0{,}5 \cdot 10^{-6}$. The parameter "cathode emission current density" is set within $0{,}5 \div 1$ A/cm^2 at the step 0,2 A/cm^2.

To study the function "electron beam diameter", you should set the initial values of the parameters of the study and the steps of varying the parameters: "electron beam current" set $10^{-6}A$, step of variation set $0{,}5 \cdot 10^{-6}$ and cathode emission current density set 0,5 A/cm^2 at the step 0,2 A/cm^2.

This function is investigated only for the minimum, therefore, by clicking the program window button, you should select "Search for the minimum". The results of the calculations are shown in the table. To view the results in a graphical form, click the button in the program window. A simplex will be drawn in the program window that appears. To re-draw the simplex, double-click the left mouse button on the window where the simplex was drawn.

Figure 4 shows a graphical representation of the user interface when performing one of the work options. The visual representation of the calculation results is provided by displaying the experimental points corresponding to the vertices of the simplex on the plane of sequential movement.

It should be noted that the simplex planning method allows you to localize only the region of the extremum (Fig. 4). Determining the immediate point of the extremum requires the use of other methods that are widely presented in the literature [6–12]. One of the possible ways to solve the problem is to continue research with a sequential decrease in the step of variable variation.

Figure 5 shows an example of an unsuccessful task implementation when constructing a simplex.

The possible discrepancy between theoretical estimates and practical results, even with a minimum step value, may be a consequence of errors in setting variables and measuring the response. The direct determination of the extreme conditions in the experiment makes it possible to find the right solution when using distorted initial data or with incomplete or inaccurate information about the object. This should be taken into account in production activities, where the cost of error can be especially high.

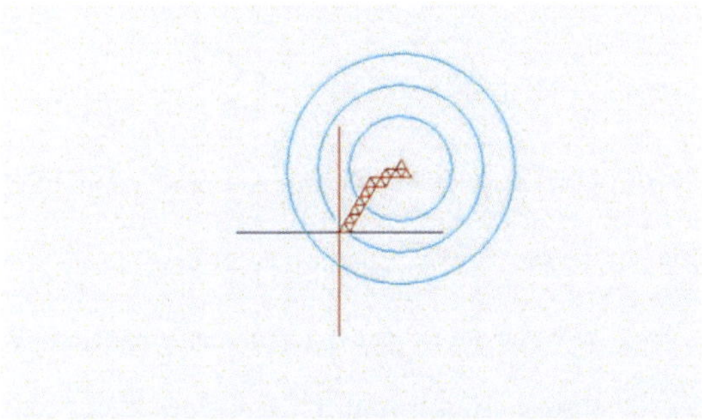

Fig. 4. An example of a successful building a simplex implementation.

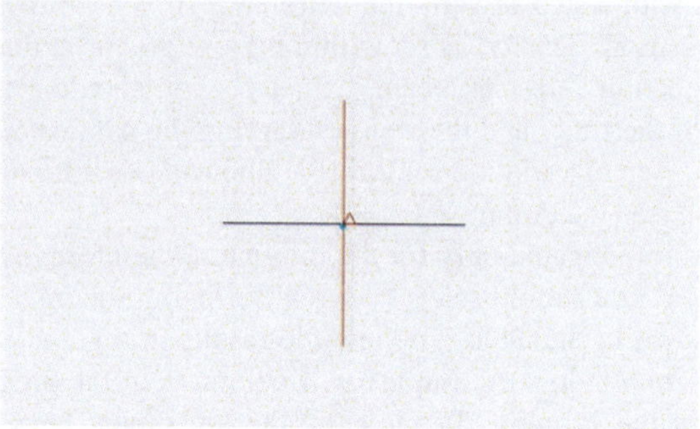

Fig. 5. An example of an unsuccessful implementation of a task when constructing a simplex.

5 Conclusion

The implementation of the simplex method in determining the extremum of the response function contributes to a deeper understanding of the theoretical foundations of linear programming in solving conditional extremal problems. The rapid development of high-speed computers makes it possible to reduce the solution of mathematical programming problems to a sequence of automatically performed operations specified by initial information. The involvement of a PC in solving specific tasks of operations research allows, within the framework of the educational process, to gain practical skills in solving an optimization problem of linear programming.

The implementation of the simplex method in determining the extremum of the response function contributes to a deeper understanding of the theory issues and allows you to gain practical skills in solving the optimization problem of linear programming within the educational process. The use during the experiment of the provisions developed by the founder of linear programming, the only Soviet Nobel laureate in economics, L.V. Kantorovich, opens up wide opportunities for the creative realization of young specialists and contributes to increasing the depth of scientific research.

References

1. Mathematical methods of organization and planning of production. LSU Publishing House, Leningrad (1939)
2. Economic calculation of the best use of resources. Publishing House of the USSR Academy of Sciences, Moscow (1959)
3. Optimal models of long-term planning. Application of mathematics in economic research. T. Z. M. (1965)
4. Berikashvili, V., Oskin, S.P.: Statistical data Processing, Experimental Design and Mathematical Description of Random Processes, p. 196. MGOU Publishing House, Moscow (2013)
5. Berikashvili, V., Oskin, S.P.: Statistical Data Processing, Experiment Planning and Random Processes: A Textbook for Undergraduate and Graduate Programs, p. 164. Bachelor and Master. Academic Course. Yurayt Publishing House, Moscow, Series (2019)

6. Oskin, S.P., Kuznecov, A., Grunenkov, N.: Physical modeling at experimental determination of a sensitivity area one of a passive infrared location instrument. Lecture Notes in Electrical Engineering **729**, 945–955 (2021)
7. Wenzel, E.S.: Introduction to Operations Research. Soviet Radio, Moscow (1964)
8. Montgomery, D.C., Peck, E.A.: Introduction to Linear Regression Analysis. John Wiley & Sons Inc, New York (1992)
9. Montgomery, D.C., Runger, G.C.: Applied Statistics and Probability for Engineers. John Wiley & Sons Inc, New York (1991)
10. Montgomery, D.C., Runger, G.C.: Gauge capability analysis and designed experiments. Part I: Basic methods. Quality Eng. **6**(1), 115–135 (1993)
11. Montgomery, D.C., Runger, G.C.: Gauge capability analysis and designed experiments. Part II: Experimental Design Models and Variance Component Estimation **6**(2), 289–305 (1993)
12. Myers, R.H., Montgomery, D.C.: Response Surface Methodology: Process and Product Optimization Using Designed Experiments, John Wiley & Sons, Inc., New York. Appendix H: References 369 (1995)

Mathematical Modeling of the Effect of Various Defects on the Parameters of Natural Vibrations of the Pipeline Under Operating Conditions

S. Gaponenko(✉)

Kazan State Power Engineering University, 51, Krasnoselskaya, Kazan 420066, Russia
sogaponenko@yandex.ru

Abstract. This paper presents mathematical modeling of the influence of various types of defects in pipelines on their oscillation frequencies. In the course of the work, pipelines with a defect were considered: 7 mm; 8 mm; 10 mm; cross-section; cross-section and longitudinal. The analysis of transient dynamic processes in the ANSYS software package made it possible to obtain a dynamic characteristic of pipelines with defects. Methods of mathematical modeling, numerical, logical-computational methods, and probabilistic-static methods of mathematical processing of measurement results were used in the performance of the work. As a result of studies conducted using modal analysis, the natural frequencies of pipelines without defects and with defects of various sizes were calculated. The obtained data can be used to determine the size of the defect by the presence of detected frequencies. The calculation results (modes) are presented in the table.

Keywords: Pipeline · Modal analysis · Oscillations · Defect

1 Introduction

The relevance of this study is to improve the quality of assessment of the technical condition and diagnostics of pipeline transport. The obtained information makes it possible to reliably assess the technical condition of pipelines, determine safe technological modes of equipment operation, establish the need and order of pipeline sections withdrawal for repair [1–6]. In recent years, the development of pipeline transport has been associated with the wide use of both metal and non-metal pipelines made of plastics and composites, which have many advantages [7, 8]. The main defects of pipelines are inhomogeneity of physical and mechanical properties and the presence of defects such as delamination, cracking (cracks), coarse grains of structure, non-glue, burnout, porosity, shrinkage shells, and various mechanical damages of the surface [9–11]. In this regard, the problem of non-destructive testing and diagnostics is especially relevant for these pipelines [12–14]. Defective pipelines with diameter d = 60 mm, wall thickness δ = 5 mm, length L = 1200 mm with defects of 7 mm were selected as the subject of the study; 8 mm; 10 mm; cross-sectional; cross-sectional and longitudinal.

A. A. Radionov and V. R. Gasiyarov (Eds.): RusAutoCon 2022, LNEE 986, pp. 136–146, 2023.
https://doi.org/10.1007/978-3-031-22311-2_14

2 Research of the Non-defective Pipeline

Figure 1 shows the imported geometry of the pipeline under investigation in the ANSYS environment [15–18], Fig. 2 shows a photograph of the non-defective pipeline under investigation.

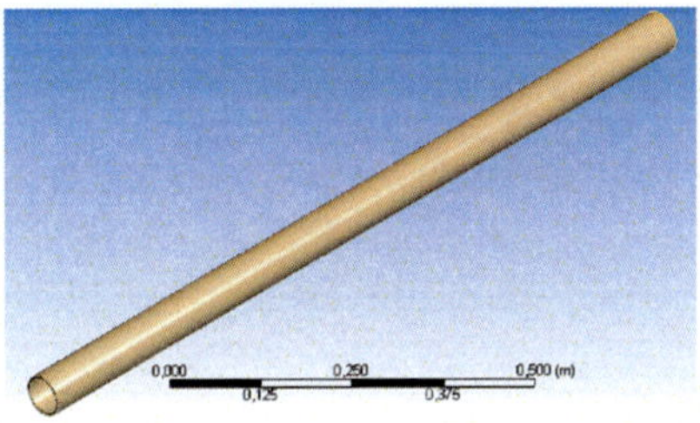

Fig. 1. Imported non-defective pipeline model.

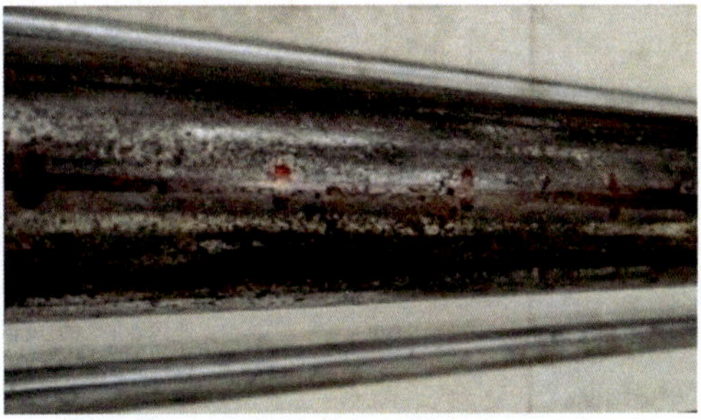

Fig. 2. Photo of non-defective pipeline.

As a result of the study, 48 modes of oscillation of a defective pipeline were determined [19–21]. The results are summarized in Table 1.

Table 1. Calculation of non-defective pipeline.

Vibration mode number	Frequency, Hz	Vibration mode number	Frequency, Hz	Vibration mode number	Frequency, Hz	Vibration mode number	Frequency, Hz
1	240.38	13	4413.9	25	6424.5	37	9131.7
2	645.43	14	4466.1	26	6521.7	38	9251.1
3	1220.2	15	4537.6	27	6858.7	39	9386.9

(continued)

Table 1. (*continued*)

Vibration mode number	Frequency, Hz	Vibration mode number	Frequency, Hz	Vibration mode number	Frequency, Hz	Vibration mode number	Frequency, Hz
4	1304.3	16	4571.7	28	7396.5	40	9578.1
5	1929.5	17	4645.5	29	7472.2	41	9939
6	2102.6	18	4749.5	30	7557.4	42	10105
7	2608.6	19	4956.7	31	7826.5	43	10438
8	2742	20	5217.2	32	7977	44	10577
9	3630.2	21	5595	33	8068.2	45	10658
10	3912.9	22	5929.2	34	8378.8	46	10860
11	4202.3	23	6295.6	35	8569.8	47	11408
12	4396.2	24	6368.1	36	8707.3	48	11561

The waveform of the non-defective pipeline at 3630.2 Hz is shown in Fig. 3.

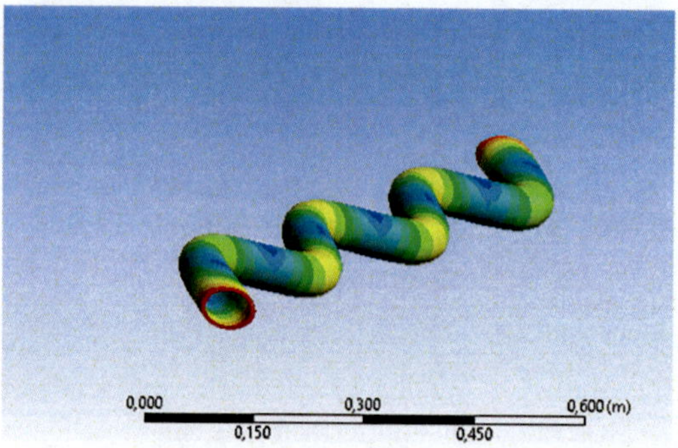

Fig. 3. The waveform of a non-defect pipeline at frequency 3630.2 Hz.

3 Inspection of 7 mm Defective Pipeline

Figure 4 shows a photograph of the inspected pipeline with a defect size of 7 mm.

As a result of the study, 48 first modes of oscillation of the calculated pipe were determined. The results are summarized in Table 2.

As a result of the conducted study, informative frequency ranges for control of a defective pipeline with a defect size of 7 mm were determined. The most characteristic frequencies are 5778.2; 5801.3; 6662.5; 9828.1; 10262 Hz.

The waveform of the tested pipeline with a defect size of 7 mm at a frequency of 5778.2 Hz is shown in Fig. 5.

Fig. 4. Photograph of investigated pipeline with defect size of 7 mm.

Table 2. Pipeline calculation with 7 mm defect.

Vibration mode number	Frequency, Hz	Vibration mode number	Frequency, Hz	Vibration mode number	Frequency, Hz	Vibration mode number	Frequency, Hz
1	241.85	13	3988.4	25	7006.5	37	8126.3
2	649.3	14	4202.9	26	7104.4	38	8347.5
3	651.37	15	4722.6	27	7146.3	39	8390.1
4	1231.6	16	5333.8	28	7182	40	8611.5
5	1233.4	17	5778.2	29	7264.3	41	8667.9
6	1322.9	18	5801.3	30	7471.2	42	8986.2
7	1952.5	19	6295.9	31	7640.2	43	9054.8
8	2099.9	20	6662.5	32	7703.5	44	9274.5
9	2652.9	21	6683.8	33	7869.4	45	9354.3
10	2788.7	22	6757.3	34	7913.2	46	9458.4
11	3713.3	23	6884.6	35	8046.9	47	9563.9
12	3717	24	6920.3	36	8111.8	48	9828.1

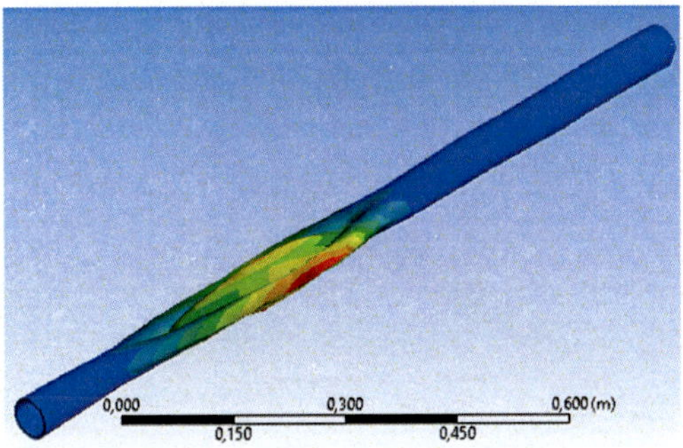

Fig. 5. The waveform of pipe with defect size 7 mm at frequency 5778.2 Hz.

4 Inspection of 8 mm Defective Pipeline

As a result of the study, 48 first modes of oscillation of the calculated pipe were determined. The results are summarized in Table 3.

Table 3. Pipeline calculation with 8 mm defect.

Vibration mode number	Frequency, Hz	Vibration mode number	Frequency, Hz	Vibration mode number	Frequency, Hz	Vibration mode number	Frequency, Hz
1	241.76	13	3976.8	25	6907.7	37	8081.7
2	242.54	14	4204	26	6991.3	38	8132.9
3	648.31	15	4702.8	27	7058.9	39	8355.8
4	652.39	16	5320.2	28	7241.4	40	8379
5	1230.1	17	5740.4	29	7292.9	41	8476.7
6	1326	18	5782.2	30	7391.7	42	8922.5
7	1950.5	19	6286.5	31	7474.1	43	8949.3
8	2110.6	20	6546.6	32	7575.8	44	9217.2
9	2651.5	21	6658.8	33	7731.6	45	9265
10	2782.9	22	6682.7	34	7879.5	46	9395.4
11	3701.9	23	6850.3	35	7996.5	47	9438.4
12	3715.7	24	6881.8	36	8044.2	48	9490.3

As a result of the conducted study, informative frequency ranges for control of a defective pipeline with a defect size of 8 mm were determined. The most characteristic frequencies are 5782.2; 6658.8; 6682, 7; 7058.9; 7241.4 Hz.

The waveform of the tested pipeline with a defect size of 8 mm at a frequency of 6658.8 Hz is shown in Fig. 6.

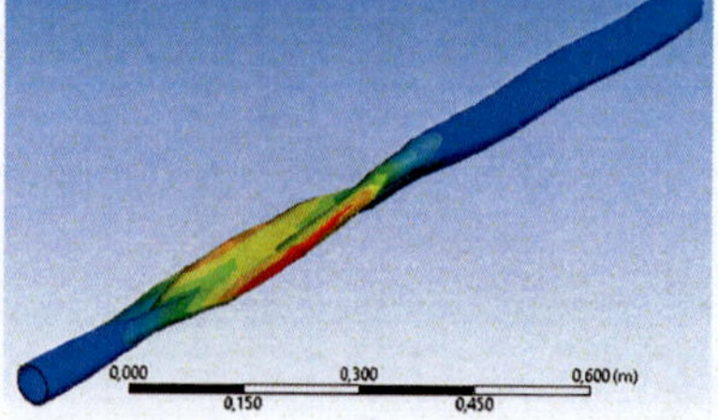

Fig. 6. The waveform of pipe with defect size 8 mm at frequency 6658.8 Hz

5 Inspection of 10 mm Defective Pipeline

As a result of the study, 48 first modes of oscillation of the calculated pipe were determined. The results are summarized in Table 4.

Table 4. Pipeline calculation with 10 mm defect.

Vibration mode number	Frequency, Hz	Vibration mode number	Frequency, Hz	Vibration mode number	Frequency, Hz	Vibration mode number	Frequency, Hz
1	240.23	13	3950.6	25	6837.9	37	7947
2	241.4	14	4219.7	26	6920	38	8013.7
3	647.77	15	4646.1	27	6957.7	39	8096.5
4	1223.7	16	5306.3	28	7058.6	40	8155.9
5	1230.9	17	5694.3	29	7184	41	8373.4
6	1315.5	18	5726.7	30	7220.1	42	8485.1
7	1946.1	19	6269.7	31	7394.7	43	8530.6
8	2095.6	20	6359.4	32	7403.7	44	9017.3
9	2655.1	21	6482.2	33	7467	45	9093.8
10	2769.3	22	6604.1	34	7650.1	46	9185.4
11	2787.6	23	6645.5	35	7764.1	47	9371.1
12	3685.2	24	6772.7	36	7901.2	48	9605

As a result of the conducted study, informative frequency ranges for control of a defective pipeline with a defect size of 10 mm were determined. The most characteristic frequencies are 5726.7; 6645.5; 7184; 7220.1; 9017.3 Hz.

The waveform of the tested pipeline with a defect size of 10 mm at a frequency of 7184 Hz is shown in Fig. 7.

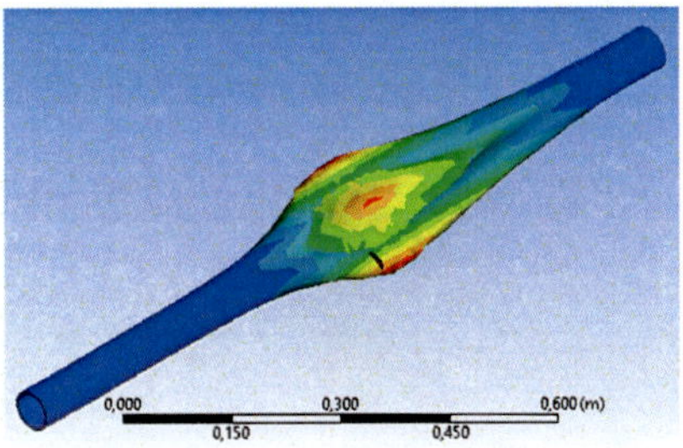

Fig. 7. The waveform of pipe with defect size 8 mm at frequency 7184 Hz.

6 Investigation of a Pipeline with a Cross Section

Figure 8 shows the imported cross sectional pipeline geometry, Fig. 9 is a photograph of the defective pipeline.

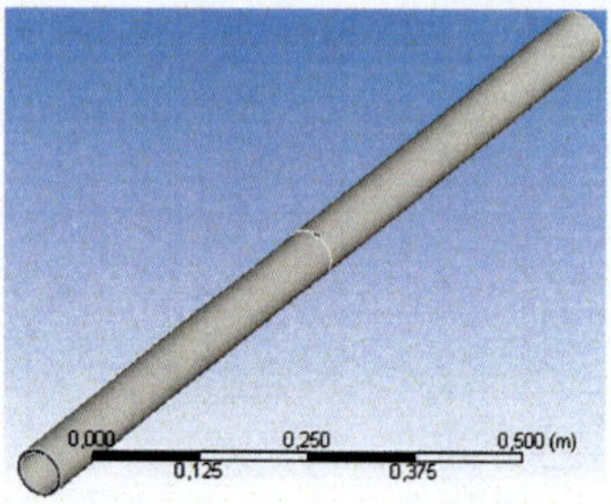

Fig. 8. Imported defective pipeline model.

Fig. 9. Photo of defective pipeline.

As a result of the study, 48 first modes of oscillation of the calculated pipe were determined. The results are summarized in Table 5.

As a result of the study, informative frequency ranges were determined to control the defective pipeline from the pipeline with a transverse section. The most characteristic frequencies are 6712.8; 7114.4; 7195.6; 8969.2; 9496.1 Hz.

The waveform of the defective cross-sectional pipeline at 6712.8 Hz is shown in Fig. 10.

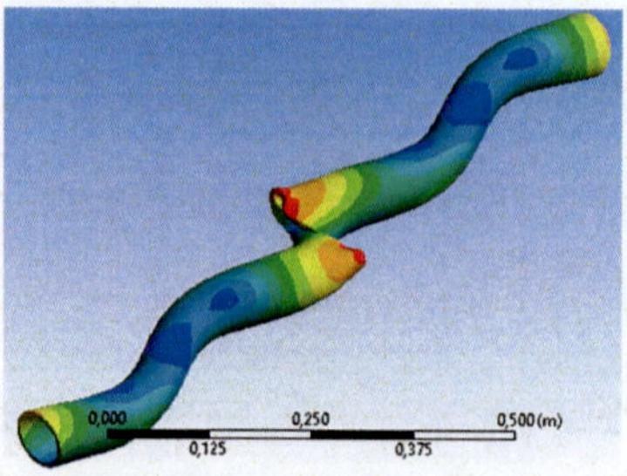

Fig. 10. The waveform of defective pipeline with a cross-section at frequency 6712.8 Hz.

Table 5. Pipeline calculation with a cross section.

Vibration mode number	Frequency, Hz	Vibration mode number	Frequency, Hz	Vibration mode number	Frequency, Hz	Vibration mode number	Frequency, Hz
1	181.1	13	3199.1	25	6492.5	37	7752
2	230.92	14	3660.6	26	6583.5	38	7945.3
3	637.94	15	3797.4	27	6627.9	39	8025.3
4	646.6	16	4202.6	28	6712.8	40	8381.8
5	1055	17	4349	29	6806.4	41	8434.6
6	1151.9	18	4593.8	30	6922.9	42	8629.2
7	1195.8	19	5125.2	31	7102.2	43	8761.5
8	1869.2	20	5305.5	32	7195.6	44	8948
9	1935	21	5646.9	33	7256.7	45	9101.2
10	2607.6	22	5751.7	34	7309.2	46	9496.1
11	2647.1	23	6129.6	35	7408.4	47	9707.4
12	2719.1	24	6369.8	36	7620.4	48	9830.8

As a result of the study, informative frequency ranges were determined to control the defective pipeline from the pipeline with a transverse section. The most characteristic frequencies are 6712.8; 7114.4; 7195.6; 8969.2; 9496.1 Hz.

The waveform of the defective cross-sectional pipeline at 6712.8 Hz is shown in Fig. 11.

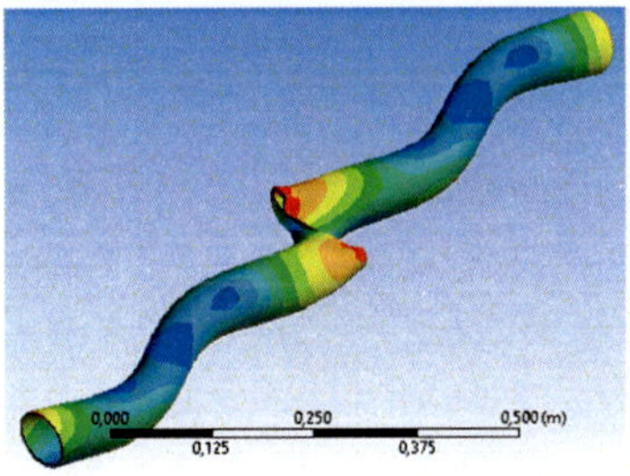

Fig. 11. The waveform of defective pipeline with a cross-section at frequency 6712.8 Hz.

7 Investigation of a Pipeline with a Transversal and Longitudinal Section

Figure 12 shows the imported geometry of the defective pipeline with a transverse and longitudinal section, Fig. 13 shows a photograph of the defective pipeline.

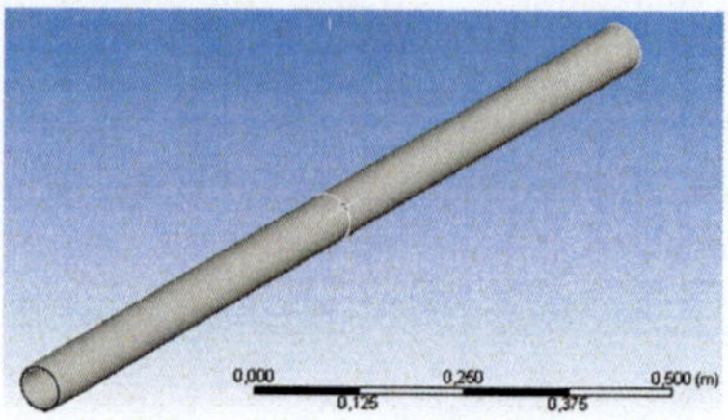

Fig. 12. Imported defective pipeline model.

Fig. 13. Photo of defective pipeline.

Table 6. Pipeline calculation with a longitudinal and transverse section.

Vibration mode number	Frequency, Hz	Vibration mode number	Frequency, Hz	Vibration mode number	Frequency, Hz	Vibration mode number	Frequency, Hz
1	176.82	13	4208.3	25	6521.1	37	7941.2
2	229.59	14	4319	26	6602.8	38	8230
3	630.94	15	4509.5	27	6746.5	39	8390.4
4	1046.8	16	4818.6	28	6826.4	40	8453.4
5	1190.1	17	4940.7	29	6989.3	41	8569.1
6	1865.4	18	5287	30	7120	42	8665.9
7	1925.5	19	5526.4	31	7203	43	8880.5
8	2597.2	20	5742.4	32	7422.7	44	8972.1
9	2640.3	21	5938.7	33	7513.2	45	9008
10	2975.3	22	6218.5	34	7611.1	46	9234.5
11	3620	23	6319.6	35	7775.8	47	9537.1
12	3772.7	24	6434.2	36	7929	48	9699.3

As a result of the study, 48 first modes of oscillation of the calculated pipe were determined. The results are summarized in Table 6.

As a result of the study, informative frequency ranges were determined to control the defective pipeline from the pipeline with a transverse section. The most characteristic frequencies are 6746.5; 7120; 7294.4; 7699.2; 8230 Hz.

The waveform of the defective cross-sectional pipeline at 7699.2 Hz is shown in Fig. 14.

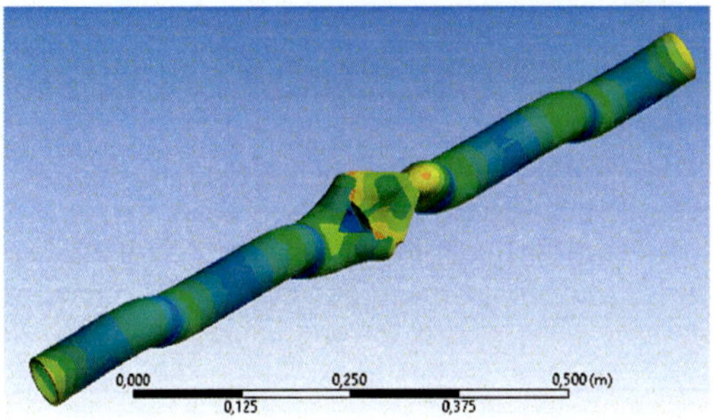

Fig. 14. The waveform of defective pipe with a longitudinal and transverse section at frequency 7699.2 Hz.

8 Conclusion

Using modal analysis, the natural frequencies of the pipeline were calculated with both defective and defects of various types and sizes. Theoretical studies have revealed the most characteristic frequencies for a defect of 7 mm: 5778.2; 5801.3; 6662.5; 9828.1; 10262 Hz; 8 mm: 5782.2; 6658.8; 6682.7; 7058.9; 7241.4 Hz; 10 mm: 5726.7; 6645.5; 7184; 7220.1; 9017.3 Hz; with a transverse section: 6712.8; 7114.4; 7195.6; 8969.2; 9496.1 Hz; with a transverse and longitudinal section: 6746.5; 7120; 7294.4; 7699.2; 8230 Hz. The remaining frequencies listed in Table 1 are common for both defective and defective pipelines and cannot serve as a basis for damage recognition.

References

1. Gaponenko, S.O., Kondratiev, A.E., Ibadov, A.A.: Mathematical modeling of vibrations of an elastic shell under the influence of ground. In: International Science and Technology Conference "EastConf".EastConf, p. 8725333 (2019)
2. Gaponenko, S.O., Kondratiev, A.E., Shakurova, R.Z.: Improving the efficiency of energy complexes and heat supply systems using mathematical modeling methods at the operational stage. E3S Web of Conferences **124**, 05029 (2019)
3. Zakharova, V.E., Gaponenko, S.O., Fedotova, A.O.: Mathematical modeling of low-frequency diagnostic vibration-acoustic vibrations of linear-extended energy objects of housing and communal services. E3S Web of Conferences 216, 01076 (2020)
4. Gaponenko, S.O., Shakurova, R.Z., Kondratiev, A.E.: Identification of diagnostic information for assessing the technical condition of power equipment. In: International Russian Automation Conference, pp. 115–119 (2021)

5. Nazarychev, S.A., Gaponenko, S.O., Zakharova, V.E.: Construction of a mathematical model of the vibrational process of pipelines under internal and external loads. Journal of Physics: Conference Series, p 012037 (2020)
6. Adegboye, M.A., Fung, W.K., Karnik, A.: Recent advances in pipeline monitoring and oil leakage detection technologies: principles and approaches. Sensors **19**(11), 2548 (2019)
7. Ullah, Z., Wang, X., Chen, Y., Zhang, T., Ju, H., Zhao, Y.: Time-domain output data identification model for pipeline flaw detection using blind source separation technique complexity pursuit. In Acoustics, Multidisciplinary Digital Publishing Institute **1**(1), 199–219 (2019)
8. Ju, H., Wang, X., Zhao, Y.: Variational specific mode extraction: a novel method for defect signal detection of ferromagnetic pipeline. Algorithms **13**(4), 105 (2020)
9. Xu, C., Du, S., Gong, P., Li, Z., Chen, G., Song, G.: An improved method for pipeline leakage localization with a single sensor based on modal acoustic emission and empirical mode decomposition with Hilbert transform. IEEE Sens. J. **20**(10), 5480–5491 (2020)
10. Ghavamian, A., Mustapha, F., Baharudin, B.T., Yidris, N.: Detection, localisation and assessment of defects in pipes using guided wave techniques: a review. Sensors **18**(12), 4470 (2018)
11. Yu, L., Li, S.Z.: Acoustic emission (AE) based small leak detection of galvanized steel pipe due to loosening of screw thread connection. Appl. Acoust. **120**, 85–89 (2017)
12. Hooi, M.P.: Fatigue crack detection. In: Aluminium Pipe By Using Non-Linear Vibro Acoustic Method (2020)
13. Shemshadi, M., Karimi, M., Veysi, F.: A simple method to design and analyze dynamic vibration absorber of pipeline structure using dimensional analysis. Fabio Rizzo **1**, 56–64 (2020)
14. Liu, G., Li, S., Li, Y., Chen, H.: Vibration analysis of pipelines with arbitrary branches by absorbing transfer matrix method. J. Sound Vib. **332**(24), 6519–6536 (2013)
15. Xu, Y., Jiao, Z.: Exact solution of axial liquid-pipe vibration with time-line interpolation. J. Fluids Struct. **70**, 500–518 (2017)
16. Zhang, D., Gao, W., Yan, X.: Determination of natural frequencies of pipes using white noise for magnetostrictive longitudinal guided-wave nondestructive testing. IEEE Trans. Instrum. Meas. **69**(6), 2678–2685 (2019)
17. Blachowski, B., Swiercz, A., Gutkiewicz, P., Szelazek, J., Gutkowski, W.: Structural damage detectability using modal and ultrasonic approaches. Measurement **85**, 210–221 (2016)
18. Leontiev, N.V.: Application of the ANSYS system to solving problems of modal and harmonic analysis. Educational and methodological material for the advanced training program "Information systems in mathematics and mechanics". Nizhny Novgorod, p 101 (2016)
19. Kudus, S.A., Suzuki, Y., Matsumura, M., Sugiura, K.: Damage assessment based on modal analysis of pipe structure. Jurnal Teknologi **80**(5) (2018)
20. Yi-min, H., Yong-shou, L., Bao-hui, L., Yan-jiang, L., Zhu-feng, Y.: Natural frequency analysis of fluid conveying pipeline with different boundary conditions. Nucl. Eng. Des. **3**, 461–467 (2018)
21. Yi-mina, H., GeSenga, W., Jieb, W.H.: A direct method of natural frequency analysis on pipeline conveying fluid with both ends supported. Nucl. Eng. Des. **1**, 12–22 (2017)

Analysis of Various Aspects in Metals Creation with Given Microheterogeneity Degree

A. Anikeev(✉), I. Chumanov, and V. Sedukhin

South Ural State University, 76. Lenin Avenue, Chelyabinsk 454080, Russia
anikeevan@susu.ru

Abstract. Automation of technological processes is one of the priority areas of industrial development. Areas in particular in need of automation and computerization are directions with not yet standardized process of obtaining, in particular, the direction of creating materials with a gradient of properties. The article discusses various aspects of the metals creation with given microheterogeneity degree, obtained by introducing dispersed particles into the melt. When creating such materials, it is important to take into account many aspects: the conditions for the dispersed particles separation from supersaturated solution, the particles interaction level with metal, the feed rate to the melt, distribution control, the particles number in the melt volume, the microheterogeneity influence on the properties' level, and other aspects. The article analyzes modern approaches for solving each aspect, proposes the methodology for calculating the crystallization rate and particle distribution when fed into the metal during centrifugal casting. Based on the developed methodology, computer model was built allowing to predict the particles distribution depending on many factors.

Keywords: Dispersed particles · Cast alloys · Wetting · Injection rate · Microheterogeneity · Distribution density

1 Introduction

Given the finiteness of natural resources and the new challenges posed by the technology development, the materials creation with given properties level, while reducing cost is the most important task of our time. One of the solutions to this problem can be economically alloyed met-als with given microheterogeneity degree. Microheterogeneity can be obtained by introducing carbides, oxides, or nitrides dispersed particles (or isolating them from supersaturated solution) and their predetermined distribution in the volumes that are most affected. The low cost of the resulting material is due to the fact that the main material (matrix) volume can consist of carbon steel, inexpensive steel, and a high level of mechanical characteristics in surface volumes is achieved by creating microheterogeneity.

When creating materials with given microheterogeneity degree, scientists face various problems:

A. A. Radionov and V. R. Gasiyarov (Eds.): RusAutoCon 2022, LNEE 986, pp. 147–156, 2023.
https://doi.org/10.1007/978-3-031-22311-2_15

1. the difference in the microparticles densities and the melt (uncontrolled uneven distribution);
2. microparticles interaction with metal (wetting level);
3. predicting the micro-particles distribution by the melt volume;
4. the microheterogeneity effect on properties level.

For controlled particles sedimentation in the melt by specific gravity, some researchers have proposed the method of introducing dispersed particles into the melt when casting on centrifugal casting machine.

The essence of the method: when pouring metal material into the centrifugal caster, solid refractory fine disperse particles of different densities are continuously fed into the metal jet, during the entire time of casting.

If the density of refractory disperse particles immersed in the melt differs from the density of the melt, then the force acting on the particle is not balanced by their own centrifugal force and gravity. Therefore, conditions arise for the particles to move to one side or the other, i.e., to the inner or outer surface of the blank to be formed. When the particle comes in contact with the crystallization front it is pressed by the melt to the crystallization front and no longer floats, but is captured by the growing dendrites. As a result surface hardening occurs.

To harden the outer surface of the tubular billet, refractory dispersed particles with a density higher than that of the melt are used. In this case the centrifugal force prevails over the Archimedean force, and the particle in the melt moves from the rotation axis to the crystallization front. It is pressed by the melt to the crystallization front, does not float, and is captured by the growing dendrites. As a result, the outer surface of the tube billet is hardened.

For the hardening of the inner surface of the tube billet refractory dispersed particles with a density less than the density of the melt are fed. In this case the value of the Archimedean force prevails over the centrifugal force and the particle in the melt moves to the rotation axis, floats to the free surface of the melt and is captured by the growing dendrites. This results in solidification of the inner surface of the tube billet.

For simultaneous hardening of the outer and inner surface of the tubular billet, particles of different densities (above and below the density of the melt) are simultaneously fed into the mixtures. When entering the mold, the particles are separated under the action of centrifugal forces and Archimedean floating force, and the inner and outer surfaces of the tubular billet are simultaneously hardened. By changing various casting parameters (workpiece size, volume and the supplied particles dispersion, crystallization rate, etc.), it is possible to obtain material with a given microheterogeneity degree [1, 2]. Khan's work [3] on the topic of "pushing" particles into the melt stream. Khan claims that such parameter as the melt jet's speed has special effect on refractory particles, since the low flow rate does not allow the particles to move to the calculated destination and can "get stuck" in the filling channel or intermediate tank. Here, the author operates with two main parameters of the interaction be-tween particles and the melt: the particles wettability and the free energy amount, which are directly dependent on each other. Next, based on these parameters, the influence of the flow velocity and the friction forces between the particles and the flow are calculated. Khan displays the jet's (flow) critical velocity parameter. The jet's speed is critical if, after passing through the certain calculated mileage of the casting channel with given cross section, the particles fraction ratio remaining in the channel due to crystallizing metal at the channel walls to the fraction of particles passing through the channel or to the total number of particles is certain number that satisfies the technological conditions and does not interfere further casting process due to the cross section reduction of the flow channel. In his research course, Khan derived formula for the critical flow velocity, which is directly proportional to the transported particle's radius, the acceleration of gravity, the difference in density between the particles substances and the carrier fluid, as well as the surface tension, expressed in terms of the sum of the fluid wetting angle and the correction distance, and vice versa proportional to the inverse wetting angle.

The short conclusion is the fact that the smaller the particle in size and density, the lower the speed necessary to maintain its buoyancy and prevent sedimentation, which is the critical flow rate. This formulation is quite plausible, because the author used the elementary laws of hydro-dynamics and the bodies's buoyancy physics. In addition, the experiment was carried out with properties scaling of the introduced particles and the liquid melt body, transferred into the ordinary water image and microscopic size polymer balls, but for clear comparison, additional aluminum and steel balls similar to polymer sizes were used [3].

To study the wetting dispersed metal particles parameters, leading research laboratories apply the still drop techniques and capillary technique. The distinctive feature in the second method is the high accuracy corresponding to real conditions. In this technique, the particles under investigation and the metal wetting them are heated in separate chambers, and the liquid metal is deposit-ed on particles substrate. This technique has already shown impressive results in number of studies, proving its accuracy [4]. The topic of the wettability was raised again by Catalina [5], supplementing Khan results with an additional unaccounted mathematical apparatus. Catalina's work was aimed to study the interaction dynamics of powder microparticles with moving fluid flow, as the result of which the model was developed for the dynamic dispersed particles inter-action with the melt body.

The author identified 3 states of interfacial interaction between the liquid and dispersed particles during crystallization: direct capture of dispersed particles during the passage through the crystallization front, particles ejection by the crystallization front, dispersed particles gradual complete absorption with their simultaneous ejection by the crystallization front, which subsequently forms the material's gradient structure.

The author experiments with three systems: aluminum (matrix) - zirconium oxide (inclusions), succinonitrile (succinic acid, matrix) - polystyrene (inclusions), biphenyl (preservative, matrix) - glass chips (inclusions). This work essence is to build such mathematical model of the dispersed particles behavior, which could give fairly accurate prediction of the dispersed particles distribution at given flow velocity, the crystallization front velocity, expressed in determining their ab-sorption / pushing potential / gradient distribution over the workpiece's body. The result of Cat-alina's work is more precise in relation to the Khan's work, and also creates more accurate description of particles interaction when exposed to the crystallization front. The particles interaction dynamic model with the melt proposed by Catalina is aimed to determine the value of the matrix buoyancy / absorption force based on information about the flow velocity.

The author claims that his model is valid only at speed range close to subcritical [5]. There are several mathematical models for prediction of the particles distribution by volume, in the world at the moment. Most of them are based on the steady state assumption: models describing the criteria for dropping particles by a growing crystallization front [6], models calculating the critical particle absorption rates by growing dendrites crystals [7].

2 Computer Modeling

Some models attempt to describe the system dynamic state [8] or to determine the criterion for capturing non-metallic inclusions by the solidification front during centrifugal metal casting [9]. Some researchers use existing models not to predict the introduced particles distribution, but to more efficiently remove existing oxide particles from the metal and increase its properties [10, 11].

Based on existing models, the article's authors developed mathematical and computer model of the particles distribution over the metal volume during centrifugal casting. The model takes into account various parameters: casting temperature, workpiece size, volume fraction and dispersion of particles, crystallization rate and others. The computer model was developed on the"Ansys" basis and allows predicting the dispersed particles input distribution in real conditions with a high degree of accuracy (Fig. 1, 2).

Fig. 1. Modeling the temperature fields dis-tribution during casting.

In designing the model, the following constraints were accepted:

1. The model describes some fixed volume of continuous medium without surrounding atmosphere and walls of working units and auxiliary devices;
2. Gravitational, centrifugal, centripetal forces, the force of surface tension in air, and thermodynamic interaction act on the melt for aggregate transformation;
3. Gravitational, centrifugal, centripetal, Archimedean forces act on particles except surface tension forces, thermodynamic interaction;

4. The behavior of particles in the volume of a metallic melt is described without taking into account the melt wetting phenomenon;
5. Values of density, viscosity, heat capacity of media, initial temperature values of boundary conditions, temperature of particles are assumed constant instead of polynomial dependencies
6. Disperse particles are combined into packages of 5–108 pieces each without reference to their geometric size.

The capabilities of the model are quite broad: it makes it possible to predict the movement of particles and trace the movement of each packet of particles at any moment of crystallization (Fig. 2, 3), and to present the results as a three-dimensional image (Fig. 4).

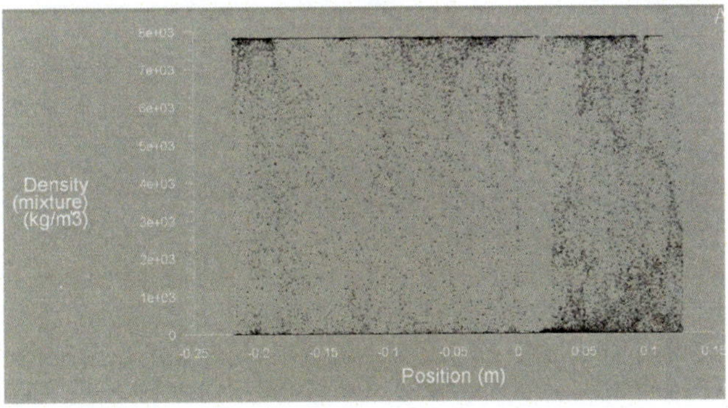

Fig. 2. Diagram of the distribution of all particles along the horizontal axis.

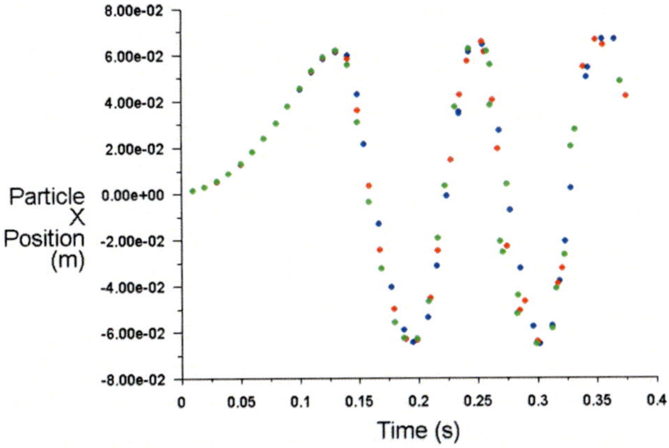

Fig. 3. Distribution diagrams of individual particle packets.

The introduced particles have significant effect on the metal properties. This is achieved by in-creasing the crystallization rate, particles incorporation into the crystal lattice and the "reinforcement" effect. The particles influence degree on the mechanical properties is determined by several factors: the metal chemical composition, the introduced particles type and dispersion, and the density of particles distribution in the crystallized metal volume [12].

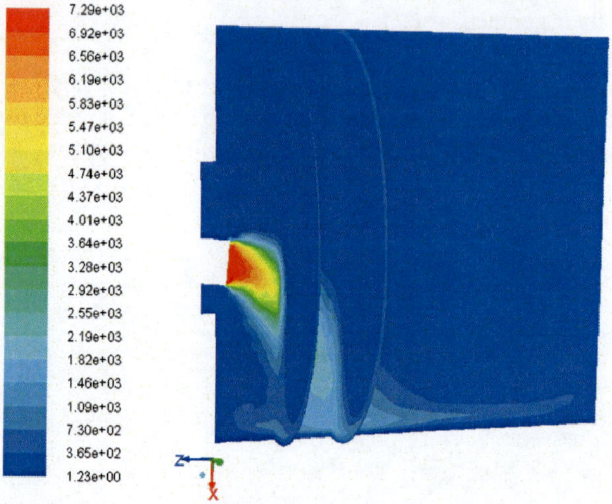

Fig. 4. Volume modeling of metal crystallization and particle distribution.

3 Experimental Results

The authors performed several simulation iterations with different parameters, as well as experimental melts of the metal. After the experiments, the obtained blanks were cut and prepared for metallographic studies. The metal samples were examined in the cast state without heat treatment and examined on an «Axio Observe»d microscope and a «JEOL» electron microscope in several sections along the length of the billet. Particular attention was paid to counting the number of particles per unit area (Fig. 5, 6).

The results of the experimental castings were compared with the results of mathematical and computer modeling (Fig. 7). Distribution curves at experimental castings.

Obtained, the results of computer simulation have differences from the results of experimental melting. The estimated deviation of the results is not more than 15%. In Fig. 6, the distribution lines on the curves show waviness with some increase or decrease in concentration. This phenomenon is observed due to rotation of the mold. When hitting the mold, the melt jet and the particles it contains succumb to a helical movement, which creates a twisting effect in the formation of the crystallizing layer. The absence of such a phenomenon in computer modeling can be explained by the insufficient number of introduced coefficients responsible for the thermal conductivity of the phases, the speed of the crystallization front, as well as the effect of the phases wetting each other.

Microhardness studies were carried out to confirm the uneven distribution of particles over the cross-section of the obtained blanks. Microhardness of samples was investigated by means of a stationary hardness meter according to the Micro-Vickers method. The load and holding time used were 4.9 N and 60 s, respectively. The ratio of this micro-hardness to the macro-hardness given by Rockwell C was 0.01. The measurement was performed in 10 iterations at 200 μm intervals from the inner edge to the outer edge of each specimen. The measurement results are presented in Table 1.

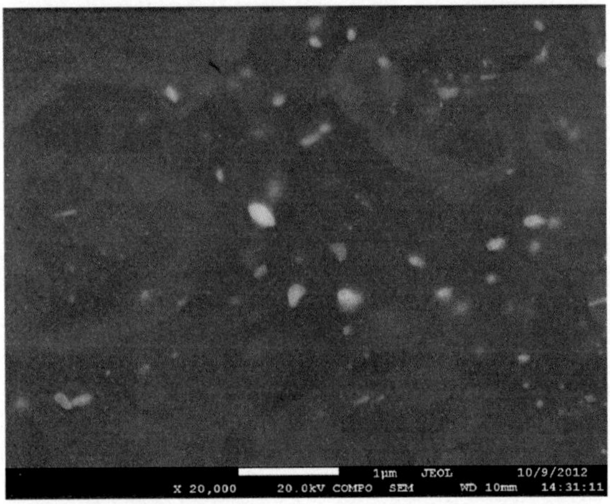

Fig. 5. Microstructure of experimental workpieces: tungsten carbides in the metal, x 20000.

Fig. 6. Microstructure of experimental workpieces: tungsten carbides in the metal, x 6000.

According to the data obtained, it can be noted that hardness and microhardness have maximum values at the outer edge, gradually decreasing to the inner edge. Introduction of dispersed WC particles increases values of these indicators. The maximum micro-hardness values are reached, respectively, at high concentrations of refractory particles.

Conducting a heat treatment reduces the hardness on a given corrosion-resistant steel, while reducing the hardness gradient over the volume of the castings. This is due to the alignment of the structure obtained by centrifugal casting.

For the experimental materials obtained, mechanical properties were tested in different sections from edge to center: impact toughness, ultimate strength, hardness, wear resistance. Thus, the introduction of 3.6% of tungsten carbide dispersed particles allows to increase the mechanical properties of low carbon steel, to increase the tensile strength by 36–38%, impact strength by 23–26% and wear resistance by 29–34%. According to the data obtained, diagrams were plotted, some of them are shown in the Fig. 8.

This approach feature is that when using different class steel (for example, stainless steels), the introduction effect will be different [13–20].

Thus, the article analyzes various aspects of creating metals with given microheterogeneity de-gree obtained by introducing dispersed particles into the melt, and modern approaches to solving each of the aspects. The computer model has been developed that has great application possibilities for materials with given microheterogeneity degree. This materials' class is one of the most promising, since it allows to significantly increase properties only in given volumes, reducing the scarce alloying elements consumption.

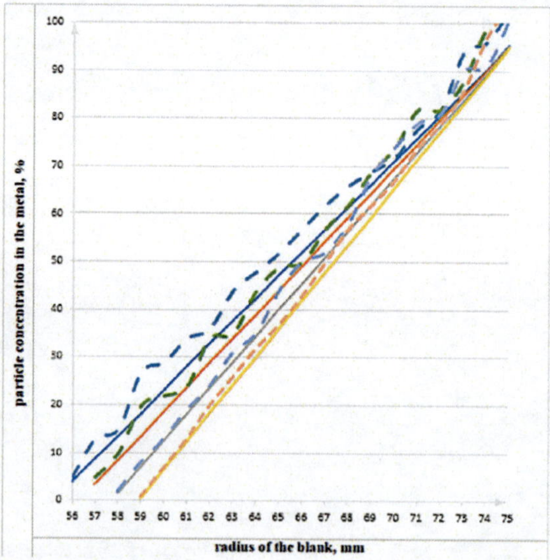

Fig. 7. Distribution diagram of WC carbides when simulating in the CFD-complex Ansys Fluent: 1, 2, 3, 4 - tungsten carbide particle distribution lines at different simulation iterations; 1', 2', 3', 4' - tungsten carbide particle.

Table 1. Results of microhardness measurements from inner edge (1) to outer edge (10), HV.

№	Sample Label					
	Reference	WC, 1 wt%	WC, 2 wt%	WC, 1 wt% after heat treatment	WC, 2 wt% after heat treatment	Reference after heat treatment
1	147.1	169.1	185.9	153.5	155.9	160
2	150.8	170.7	183.2	155.1	162.5	161.7
3	153.2	171.1	192.3	156	165.9	161.2
4	158.1	171.9	189.5	157.4	164.8	163.6
5	164	172.3	186.2	157	163.6	165.3
6	167.9	173.5	189.9	160.6	163.4	163.2
7	168.3	172.3	191.5	161	162.9	165.9
8	173.5	179.5	189.4	161	167.9	169.6
9	175.9	181.7	187.1	165.8	164.9	169.6
10	184.7	183.9	197.2	165.9	168.9	171.5

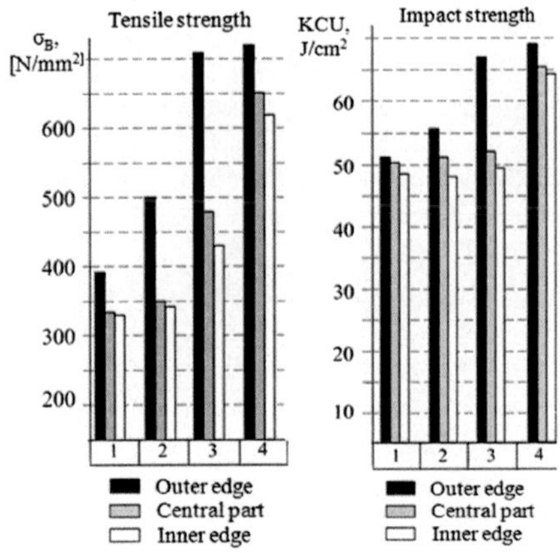

Fig. 8. Change diagrams of tensile strength and impact toughness of samples billets from steel AISI 1020.

References

1. Al-Mangour, B., Grzesiak, D.: Selective laser melting of TiC reinforced 316L stainless steel matrix nanocomposites: influence of starting TiC particle size and volume content. Mater. Des. **104**, 141–151 (2016)
2. Anikeev, A., Seduhin, V., Sergeev, D.: Increase in wear resistance by introduction of titanium carbide dispersed particles. Mater. Sci. Forum **843**, 269–273 (2016)
3. Han, Q., Hunt, D.: Particle pushing: critical flow rate required to put particles into motion. J. Cryst. Growth **152**(3), 221–227 (1995)

4. Chumanov, I., Anikeev, A.: Fabrication of functionally graded materials by introducing wolframium carbide dispersed particles during centrifugal casting and examination of FGM's structure. Procedia Engineering **2**, 816–820 (2015)
5. Catalina, A., Mukherjee, S., Stefanedcu, D.: A dynamic model for the interaction between a solid particle and an advancing solid/liquid interface. Metall. Mater. Trans. A **31**(10), 2559–2568 (2000)
6. Kiviö, M., Holappa, L., Louhenkilpi, S.: Studies on interfacial phenomena in titanium carbide/liquid steel systems for development of functionally graded material. Metall. Mater. Trans. B **47**, 2114–2122 (2016)
7. Kiviö, M., Holappa, M., Yoshikawa, T.: Interfacial phenomena in Fe-TiC systems and the effect of Cr and Ni. High Temp. Mater. Processes **31**(4–5), 645–656 (2012)
8. Li, H., Li, P., Chen, W.: Effect of WC and Co on the microstructure and properties of TiC steel-bonded carbide. Mater. Sci. Forum **898**, 1468–1477 (2017)
9. Martinez, E., Peaslee, D., Lekakh, S.: Calcium wire ladle treatment to improve cleanliness of centrifugally cast steel. Trans. American Foundry Society **119**, 513–520 (2011)
10. Sobczak, N., Nowak, R., Radziwill, W.: Experimental complex for investigations of high-temperature behaviour of molten metals in contact with refractory materials. Mater. Sci. Eng. A **495**(1–2), 43–49 (2008)
11. Türker, M., Çapan, L.: Effect of inclusions on mechanical properties of Nb stabilized austenitic stainless steels (316Nb) with centrifugal and sand casting techniques. Matériaux et Techniques **105**(3), 2017035 (2017)
12. Wang, Q., Zhang, L.: Detection of non-metallic inclusions in centrifugal continuous casting steel billets. Metall. Mater. Trans. B **47**(3), 1594–1612 (2016)
13. Wang, Q., Zhang, L.: Determination for the entrapment criterion of non-metallic inclusions by the solidification front during steel centrifugal continuous casting. Metall. and Mater. Trans. B. **47**(3), 1933–1949 (2016)
14. Watanabe, Y., Sato, H.: Review fabrication of functionally graded materials under a centrifugal force, in book "nanocomposites with unique properties and applications in medicine and industry". InTech., pp. 372–378 (2011)
15. Wilde, G., Perepezko, H.: Experimental study of particle incorporation during dendritic solidification. Mater. Sci. Eng. A **283**(1–2), 25–37 (2000)
16. Klimova, M., Shaysultanov, D., Semenyuk, A., Zherebtsov, S., Stepanov, N.: Effect of carbon on recrystallised microstructures and properties of cocrfemnni-type high-entropy alloys. J. Alloy. Compd. **851**, 156839 (2021)
17. Peng, J., et al.: The predicted rate-dependent deformation behaviour and multistage strain hardening in a model heterostructured body-centered cubic high entropy alloy. Int. J. Plast **145**, 103073 (2021)
18. Semenyuk, A., Klimova, M., Zherebtsov, S., Stepanov, N.: Effect of interstitial elements on the cryogenic mechanical behavior of fcc high entropy alloys. Mater. Sci. Forum **1016**, 1386–1391 (2021)
19. Yurchenko, N., Panina, E., Zherebtsov, S., Stepanov, N.: Design and characterization of eutectic refractory high entropy alloys. Materialia **16**, 101057 (2021)
20. Zherebtsov, S.N., Mikhailets, S.N., Zabegailo, I.V., Rogachev, E.A.: Use of a developed production process of centrifugal electroslag casting for manufacturing a reducer. Chem. Pet. Eng. **52**(11–12), 859–862 (2017)

Modeling of Integrated Internet of Things System on the Base of Semi-markov Processes

I. Lvovich[1], A. Preobrazhenskiy[1(✉)], and Ya. Lvovich[2]

[1] Voronezh Institute of High Technologies, 73A, Lenin Street, Voronezh 394043, Russia
komkovvivt@yandex.ru

[2] Voronezh State Technical University, 14, Moscow District, Voronezh 394026, Russia

Abstract. In this paper, to build a model for the functioning of the Internet of Things system, it is proposed to use the technology of semi-Markov processes. The main characteristics for the analyzed semi-Markov processes are indicated. It is shown how semi-Markov processes can be used in the course of decision making. Making a choice by a single action for time t is considered as a solution in semi-Markov processes. An illustration of a simulation-semantic model that can be used to describe the functioning of the Internet of things system is given. When considering the transmitted data in the Internet of Things system, the possibility of solving the problem of dynamic programming is considered. The results that were obtained on the basis of the proposed approach are given.

Keywords: Model · Internet of Things · Semi-Markov · Process · Decision making

1 Introduction

Internet of Things systems in many cases can be considered as structures in which processes are described on the basis of random approaches. It is proposed to use the apparatus of semi-Markov processes within the framework of the general statement of the problem associated with imitative-semantic modeling [1, 2]. This provides opportunities for handling incomplete information. The decision-making process under uncertainty will be supplemented by semi-Markov processes as well as algorithmic elements of dynamic programming. Then there are opportunities to implement the description of the search for the optimal strategy for the case with an infinite range of decision-making based on iteration methods over the criterion [3, 4].

The purpose of this paper is to develop an approach that allows simulating the operation of the Internet of Things system based on the apparatus of semi-Markov models.

A. A. Radionov and V. R. Gasiyarov (Eds.): RusAutoCon 2022, LNEE 986, pp. 157–165, 2023.
https://doi.org/10.1007/978-3-031-22311-2_16

2 Main Characteristics of Semi-markov Processes

To build a model of the functioning of the Internet of Things system, it is proposed to use the semi-Markov process (SMP). In this system, it is necessary to consider various states that correspond to some values and probabilistic transient characteristics. It is necessary to implement decision-making processes on the basis of data analysis [5, 6]. Transitions in the system are described on the basis of a probabilistic approach. The Markov process is considered in the form of SMP, in which deterministic time intervals are allocated between the phases of the process. In Fig. 1 we can see the illustration of two trajectory diagrams. They show how the process of serving users in the Internet of Things system can be carried out. The first trajectory diagram shows the semi-Markov process, that is, the probabilistic time intervals. The second trajectory diagram shows the Markov process, which corresponds to deterministic time intervals between transitions. There may be virtual transitions in the semi-Markov process, which is shown in the diagrams. In real transitions, a change of state will be implemented. Suppose there is a simple semi-Markov process. In Fig. 2 shows the information flow diagram that corresponds to it. On its basis, it is possible to describe a consistent decision-making process. For each of those states that are passed, the target values are summed up.

The person who makes the decision (PMD) for any point in time t analyzes the current state of s_t. After that, the $a_t = \mu_t(s_t)$ action is implemented, which will depend on the state. The v_t value is summed up by the average time.

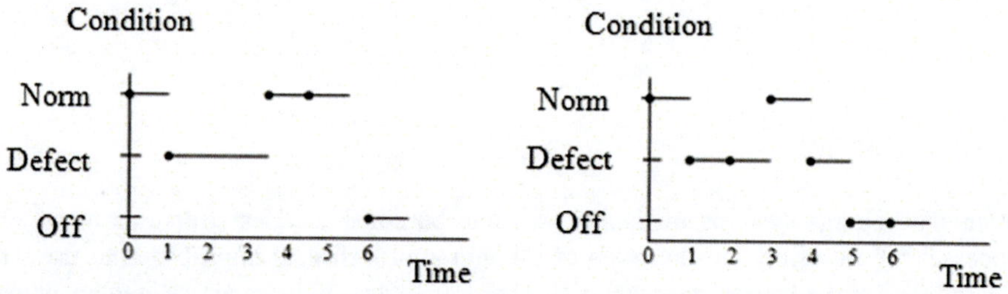

Fig. 1. Illustration of the diagram of process trajectories in the Internet of Things system: a) the semi-Markov process b) the Markov process.

We will indicate the main characteristics for the analyzed semi-Markov processes.

1. The set of moments of time $\boldsymbol{T}$;
2. There are random variables $\{D(t);\ t \in \boldsymbol{T}\}$, where $D(t) \in \boldsymbol{A} = \{1,2,\ldots,\alpha\}$. They are considered as a solution that corresponds to point in time t. They show the process of choosing a solution or a management process [7, 8].
3. The state of the process that corresponds to the time point t is considered as a set of random variables $\{S(t);\ t \in \boldsymbol{T}\}$, где $S(t) \in \boldsymbol{S} = \{1,2,\ldots\}$. They form a semi-Markov process that shows the final effect. This takes into account:

- Markov subchain. It is based on the set of random variables $\{S_k;\ k \geq 0\}$. It should be noted that $S_k = S(T_k)$, in the specified expression, the moments of time go in ascending order $T_1 < T_2 < T_3 < \ldots$ They are treated as random variables;
- set α that are related to the probabilities of transitions $\{P^{(a)}{}_{ij}(t);\ i \geq 0,\ j \geq 0,\ 1 \leq a \leq \alpha,\ t \in \boldsymbol{T}\}$. They will be between the states of the hook. What is the meaning of transition probabilities? They demonstrate the probability of transitions between state i and state and state j. This takes into account that during the implementation of activity a, the process for point in time t will enter the state i.

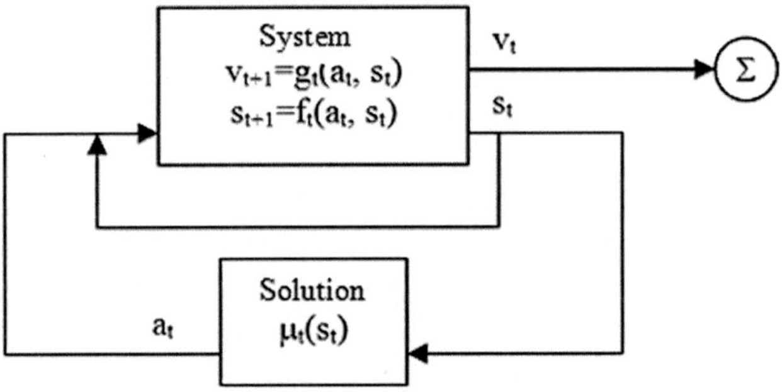

Fig. 2. Illustration of information flow a semi-Markov decision-making process.

We suppose that there is a certain action $a \in \boldsymbol{A}$, and states $i,j \in \boldsymbol{S}$:

$$\begin{aligned} P_{ij}^{(a)}(t) &= P\{S_{k+1} = j | S_k = i,\ d_k = a,\ T_k = t\} \\ &= P\{S(T_{k+1}) = j | S(T_k) = i,\ D(T_k) = a,\ T_k = t\}. \end{aligned} \tag{1}$$

It can be seen that the Markov property will be performed by:

$$\begin{aligned} P_{ij}^{(a)}(t) &= P\{S_{k+1} = j | S_k = i,\ d_k = a,\ T_k = t\} \\ &= P\{S_{k+1} = j | S_k = i,\ S_{k-1} = h, \ldots,\ D(T_k) = a,\ T_k = t\}; \end{aligned} \tag{2}$$

sets α that are correlated with implementation time intervals $\{\tau^{(a)}{}_{ij}(t);\ i \geq 0,\ j \geq 0,\ 1 \leq a \leq \alpha,\ t \in \boldsymbol{T}\}$. They will be sub-hook states correlated with some numbers that are random. They are characterized by probability distribution (PD):

$$\{h_{ij}^{(a)}(m, t);\ i \geq 0,\ j \geq 0,\ 1 \leq a \leq \alpha,\ m \geq 0,\ t \in \boldsymbol{T}\}. \tag{3}$$

They also have a corresponding probability distribution function (PDF):

$$\{H_{ij}^{(a)}(m, t);\ i \geq 0,\ j \geq 0,\ 1 \leq a \leq \alpha,\ m \geq 0,\ t \in \boldsymbol{T}\}. \tag{4}$$

What is the meaning of the time of stay? It demonstrates that the process will be in a certain state [9, 10] *i* during this time. The process enters the state *i* for time *t*.

If we analyze the possibility of its further transition to state *j*, the action will be *a*.

3 Use of Semi-markov Processes in Decision-Making

In the semi-Markov processes that allow decision-making in the course of analyzing Internet of Things systems [11, 12], the main components are highlighted. The set of time points *T*, the space of action ***A***, the space of states ***S***, the set of stepped transition functions from PD $q^{(a)}{}_{ij}(\cdot)$ and PDF $Q^{(a)}{}_{ij}(\cdot)$.

They are considered in terms of the probabilities of transition $P^{(a)}{}_{ij}(\cdot)$. There are also multiple intervals of residence with PD $h^{(a)}{}_{ij}(\cdot)$ and PDF $H^{(a)}{}_{ij}(\cdot)$.

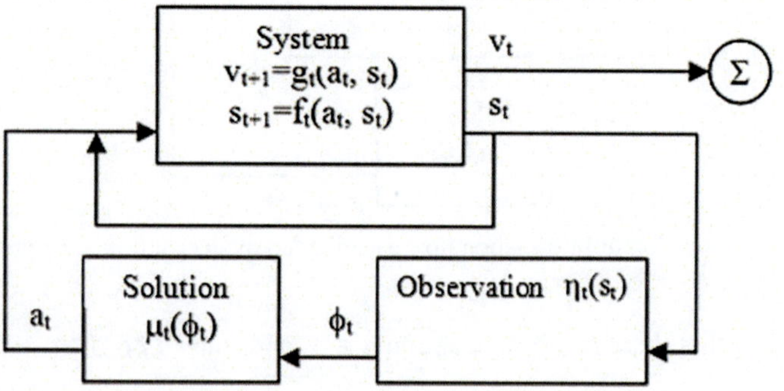

Fig. 3. The illustration of the information flow, correlated with a partial image of the observed semi-Markov process aimed at decision-making.

They can also be described on the basis of the transition probabilities $p^{(a)}{}_{ij}(\cdot)$. There are set of time intervals with PD $w^{(a)}{}_{i}(\cdot)$ and PDF $W^{(a)}{}_{i}(\cdot)$, as well as many estimated functions $\upsilon^{(a)}{}_{i}(\cdot)$. The Markov subchain $\{S(T_k); T_k \in \boldsymbol{T}\}$ is used to describe the corresponding stochastic processes [13, 14].

They have a decision-making process $\{D(t);\ t \in \boldsymbol{T}\}$, as well as a semi-Markov process that is associated with calculating the effect of $\{S(t); t \in \boldsymbol{T}\}$.The diagram of the information flow, which is related to a partially observed semi-Markov process related to decision-making, is illustrated in Fig. 3.

The decision maker (DM) will analyze the observation ω_t, which corresponds to any point in time *t*. The observation will be associated with the state of s_t based on the expression $\phi_t = \eta_t(s_t)$ and the action $a_t = \mu_t(\phi_t)$ is implemented [15, 16].

Only observation will have an impact on the action. Over the average time, summation of v_t values will be implemented.

The selection for a single action for point in time t is considered as a decision in the SMP. The random variable $D(t)$ for any m will affect uncertainty in decision-making. Solutions are implemented in a consistent manner [17, 18]. Solutions $D(t)$ may contain corresponding sets of alternative actions for different $\boldsymbol{A}_S$ states. Also, the states may correspond to different moments in time $\boldsymbol{A}_t$ or there may be different states for different points in time $\boldsymbol{A}_{St}$.

In the SMP, the state is considered to be an object of the Internet of Things at some point in time t. The random variable $S(t)$ shows the uncertainty of the status. There are state attribute variables $\{\theta_k(t);\ k = 1, 2, 3, \ldots, K,\ t \in \boldsymbol{T}\}$, where $\theta_k(t) \in \Theta_{k,t} = \{0,1,2,3,\ldots\}$. Then, on the basis of their product, the space of states is determined:

$$S_t = \prod_k \Theta_{k,t}. \tag{5}$$

Using the Cartesian product, from a set of variable attributes of states, determine the state s for any t:

$$\sigma_\tau \in \Theta_{1,\tau} \times \Theta_{2,\tau} \times \Theta_{3,\tau} \times \ldots \times \Theta_{K,\tau}. \tag{6}$$

As a result, in the problem we get:

- Many actions $\boldsymbol{A}$.
- the semi-Markov process on the basis of which it is possible to calculate the effect at multiple time points $\boldsymbol{T} \subseteq \{0, 1, 2, \ldots\}$, when a dynamic influence diagram and a cyclic Markov tree are analyzed, and $\boldsymbol{T} \subseteq [0, +\infty]$ if a stochastic tree is analyzed. To define a semi-Markov process, we must specify:
- The Markov subchain in which there is a state space $\boldsymbol{S}$;
- Three sets of probabilities of transitions $P^{(a)}{}_{ij}$ between states $\boldsymbol{S}$, depending on which actions from $\boldsymbol{A}$;
- Constant intervals of residence time with PDF $H^{(a)}{}_{ij}(m,t) = \mathbf{1}(m-1)$, where $\mathbf{1}(m-1)$ is considered as a step function with a duration of $m = 1$, if there is a dynamic influence diagram and a Markov cyclic tree, as well as exponential intervals of residence time with PDF $H^{(a)}{}_{ij}(\mathrm{m,t}) = 1-^{-\mu \mathrm{im}}$, in the specified expression μ_i are considered as transition frequencies. They depend on what states are for the stochastic tree;

4. Three sets of evaluation functions $\nu^{(a)}{}_i(\mathrm{m})$. They are related to the magnitude of the coefficient of the effect of information transfer that is expected for any state from $\boldsymbol{S}$, depending on which actions from $\boldsymbol{A}$.

   In Fig. 4 we can see the simulation-semantic model.

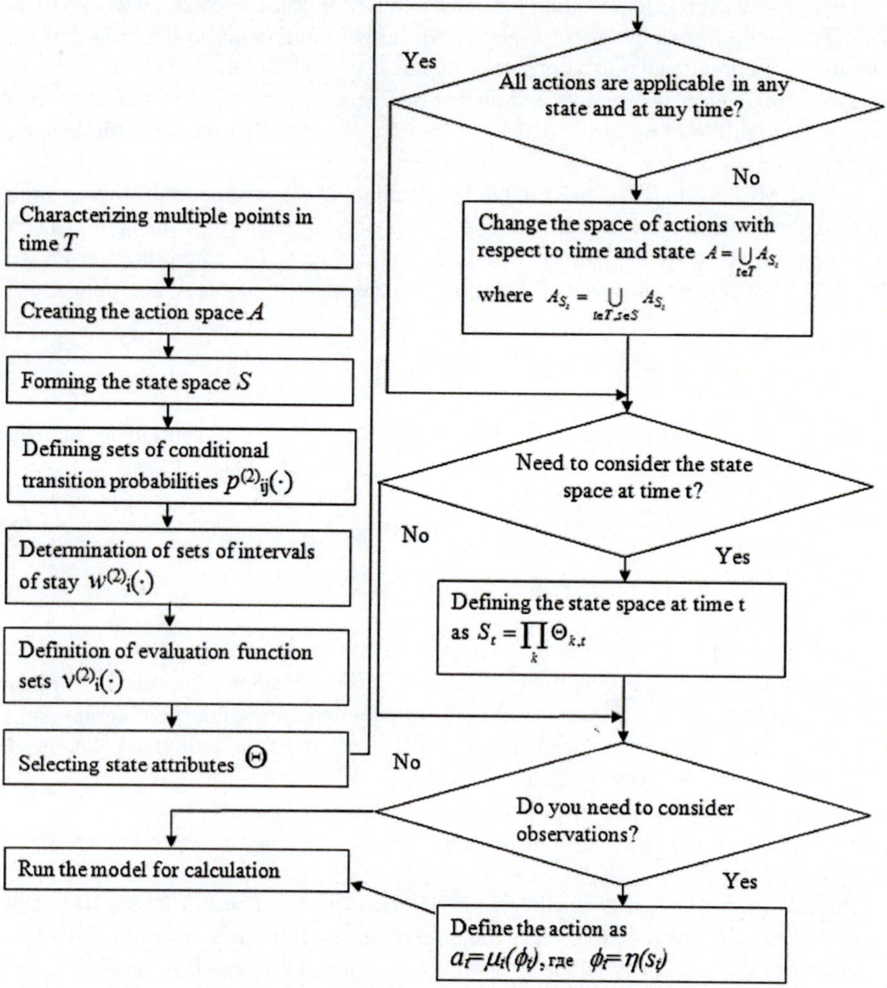

Fig. 4. Illustration of the simulation-semantic model.

4 Using Simulation-Semantic Modeling in Decision-Making in the System

When considering the amount of data transmitted, we can come to the task of dynamic programming [19, 20]. Then, if there is a range of solutions of n steps, with some coefficient of $0 \leq \beta \leq 1$, you can get the optimal value of V_i^*, which will be for each state $i \in \boldsymbol{S}$, when there is an initial value of $V_i(0)$ and a time point $t \in \boldsymbol{T}$:

$$V_i^*(n,\beta,t) = \max_a \left\{ \sum_j \sum_{m=n+1}^{\infty} q_{ij}^{(a)}(m,t) \left[\sum_{l=0}^{n-1} \beta^l y_i^{(a)}(l) + \beta^n V_i(0) \right] + \sum_j \sum_{m=1}^{n} q_{ij}^{(a)}(m,t) \left[\sum_{l=0}^{m-1} \beta^l y_i^{(a)}(l) + \beta^m b_{ij}^{(a)} + \beta^m V_j^*(n-m,\beta,t+m) \right] \right\}; \quad (7)$$

In this expression, $n \geq 0$, $i,j \in \boldsymbol{S}$, $a \in \boldsymbol{A}$, $t \in \boldsymbol{T}$, $q_{ij}{}^{(a)}(m,t) = P_{ij}{}^{(a)}(t) \cdot h_{ij}{}^{(a)}(m,t)$ is treated as a step function of transition from state i to j. This is determined by action a at interval m after entering state i for time t.

$$\nu_{\iota\varphi}{}^{(\alpha)}(\mu) = [\textstyle\sum_{l=0}^{m-1} \psi_\iota{}^{(\alpha)}(\lambda)] + \beta_{\iota\varphi}{}^{(\alpha)}. \quad (8)$$

is considered as an estimated transition function with an effect coefficient of $y_i{}^{(a)}(\cdot)$ and an incremental correction $b_{ij}{}^{(a)}$ for the transition $\xi(a,i,j,t)$. In this case, the expression

$$\nu_\iota{}^{(\alpha)}(\mu) = \sum_j \theta_{\iota\varphi}{}^{(\alpha)}(\cdot)\, \nu_{\iota\varphi}{}^{(\alpha)}(\mu). \quad (9)$$

expresses the value of the objective function, which is obtained for each state $i \in \boldsymbol{S}$ at interval m, depending on the action of $a \in \boldsymbol{A}$. The first term in (7) demonstrates the value that is expected when the next transition from the state i is realized after the time interval n. The second term in (7) demonstrates that the transition [21, 22] will occur before the expiration of the time interval n. In a particular case, the MP, (7) will move to (5), thus:

$$V_i^*(n,\beta,t) = \max_a \{ v_i^{(a)}(1) + \beta \sum_j P_{ij}^{(a)}(t) \cdot V_j^*(n-1,\beta,t+1) \};\ n \geq 0,\ i,j \in \boldsymbol{S},\ a \in \boldsymbol{A},\ t \in \boldsymbol{T}. \quad (10)$$

Based on the method of iteration by criterion, it is possible to analyze the optimality Eq. (7) or (10). In the equation, the optimal strategy $\pi^* = \{\mu_0{}^*, \mu_1{}^*, \mu_2{}^*, \mu_3{}^*, \ldots\}$, where $\mu_t{}^*: \boldsymbol{S} \to \boldsymbol{A}$ and $t \in \boldsymbol{T}$ shows the solution.

In this case, there will be a maximization of $V_{init}{}^*(N,\beta,0)$, which is considered as the optimal value of the effect from the initial state after a period of time N at time $t = 0$. The time in the decision-making range that remains is $n = N - t$.

If we analyze the time interval N, then the complexity of the iteration algorithm for the SMP is calculated – $O(N^2 \cdot |A| \cdot |S|^2)$. This is determined by the fact that for each stage of decision-making it is required to analyze all moments of time t' such that $t \leq t' \leq t + n$.

Calculation of complexity by algorithm, if considered MP, is implemented in the form of $O(N \cdot |A| \cdot |S|^2)$. Then we can see the unambiguous correspondence [23, 24] between the duration n and the current time t: $n = N - t$.

5 Results

The analysis of the Internet of things system, which included 109 objects, was carried out. The system was monitored for 20 h. The description of processes based on semi-Markov models was considered to correspond to 14 h. The set of actions included 17 elements. The probabilities of transitions between states were estimated.

During the analysis of the system, two possible strategies were identified. The action space was changed every 3 h. Sequences of control operations were determined by imitating semantic modeling. When using optimization approaches, it was possible to reduce the control time by 27%.

6 Conclusion

It is shown that it is possible to integrate the mathematical apparatus of semi-Markov processes and observation spaces in order to carry out structural modeling of dynamic decision-making. It is demonstrated that the decision-making process for uncertainty conditions can be supplemented by semi-Markov processes with algorithmic elements of dynamic programming. This makes it possible to describe the search for the optimal strategy when using simulation-semantic models.

References

1. Neittaanmäki, P., Repin, S., Tuovinen, T.: Mathematical Modeling and Optimization of Complex Structures; Series: Computational Methods in Applied Sciences. Springer International Publishing AG, Switzerland (2016). https://doi.org/10.1007/978-3-319-23564-6
2. Liu, C., Fan, J., Branch, J., Leung, K.: Toward QoI and energy-efficiency in Internet-of-Things sensory environments. IEEE Trans. Emerg. Top Comput. **2**(4), 473–487 (2014)
3. Koedinger, K.R., Matsuda, N., MacLellan, C.J., McLaughlin, E.A.: Methods for evaluating simulated learners: Examples from simstudent. In: 17th International Conference on Artificial Intelligence in Education AIED **5**, 45–54 (2015)
4. Ramesh, A., Goldwasser, D., Huang, B., Daume, H., Getoor, L.: Modeling learner engagement in moocs using probabilistic soft logic. NIPS Workshop on Data Driven Education (2013)
5. Gubbi, J., Buyya, R., Marusic, S.: Palaniswami M. Internet of Things (IoT): a vision, architectural elements, and future directions. Fut. Gen. Comput. Syst. **29**(7), 1645–1660 (2013)
6. Jianyong, L., Xiaobo, Z.: On average reward semi-Markov decision processes with a general multichain structure. Math Oper Res **29**(2), 339–352 (2004)
7. Orlova, D.: Stability of solutions in ensuring the functioning of organizational and technical systems. Modeling, Optimization and Information Technologies **1**(20), 325–336 (2018)
8. Almeida, V., Doneda, D., Monteiro, M.: Governance challenges for the Internet of Things. IEEE Internet Comput, 4б(б), 56–59 (2015)
9. Hu, Q., Yue, W.: Optimal replacement of a system according to a semi-Markov decision process in a semi-Markov environment. Optim. Methods Softw. **18**(2), 181–196 (2003)
10. Rodriguez-Molina, J., Martinez, J.-F., Castillejo, P., Lopez, L.: Combining wireless sensor networks and semantic middleware for an internet of things-based sportsman/woman monitoring application. Sensors **13**, 1787–1835 (2013)
11. Sorokin, S.O.: Optimization modeling of the functioning of the system of homogeneous objects in a multidimensional digital environment. Modeling, Optimization and Information Technol. **3**(22), 153–164 (2018)
12. Lvovich, I., Lvovich, Y., Preobrazhenskiy, A., Preobrazhenskiy, Y., Choporov, O.: Managing developing internet of things systems based on classification and predictive models and algorithms. International Seminar on Electron Devices Design and Production SED 2021 - Proceedings 9444493 (2021)

13. Lvovich, I., Preobrazhenskiy, A., Lvovich, Y., Choporov, O.: Modeling and optimization of processing large data arrays in information systems. In: Proceedings of ITNT 2021 - 7th IEEE International Conference on Information Technology and Nanotechnology **7**, p. 9649229 (2021)
14. Odu, G.: Review of multi-criteria optimization methods - theory and applications. IOSR Journal of Eng. **3**(10), 01–14 (2013)
15. Barnaghi, P., Wang, W., Henson, C., Taylor, K.: Semantics for the internet of things: early progress and back to the future. International Journal on Semantic Web and Information Systems, **8**(1), 121 (2012)
16. Wang, Y.T., Zhang, X.: Cooperative caching placement in cache-enabled D2D underlaid cellular network. IEEE Commun Lett **21**(5), 1151–1154 (2017)
17. Wortmann, F., Flüchter, K.: Internet of Things. Bus. Inf. Syst. Eng. **57**(3), 221–224 (2015)
18. Chen, Z., Pappas, N., Kountouris, M.: Probabilistic caching in wireless D2D networks: cache hit optimal versus throughput optimal. IEEE Commun Lett **21**(3), 584–587 (2017)
19. Caire, J.M., Molisch, G.: Wireless device-to-device caching networks: basic principles and system performance. IEEE J Sel Area Comm **34**(1), 176–189 (2016)
20. Miao, A.A.: Network lifetime maximization for cellular-based M2M networks. IEEE Access **5**, 18927–18940 (2017)
21. Lutakamale, A.S., Kaijage, S.: Wildfire monitoring and detection system using wireless sensor network: a case study of Tanzania. Wirel. Sens. Netw. **9**, 74–289 (2017)
22. Stankovic, J.A.: Research directions for the internet of things. IEEE Internet Things J. **1**(1), 3–9 (2014)
23. Minerva, R., Biru, A., Rotondi, D.: Towards a definition of the Internet of Things (IoT). IEEE Internet Initiative Torino Italy (2015). https://iot.ieee.org/images/files/pdf/IEEE_IoT_Towards_Definition_Internet_of_Things_Revision1_27MAY15.pdf
24. Broadband Internet Technical Advisory Group. Internet of Things (IoT) Security and Privacy Recommendations, BITAG website (2016). https://www.bitag.org/documents/BITAG_Report_-_Internet_of_Things_(IoT)_Security_and_Privacy_Recommendations.pdf

Modeling of the Heat and Mass Transfer Process in Assessing the Thermal Effect When Processing an Oil Well with a Hot Coolant Through Hollow Rods

E. Goltsov(✉), N. Kostarev, and N. Trufanova

National Research Polytechnic University, 29, Komsomolsky Avenue, Perm 614990, Russia
evgeny.goltsov@yandex.ru

Abstract. The article considers numerical simulation of the processes of heat and mass transfer in an oil well complicated by asphalt-resin-paraffin deposits, taking into account the heating of the oil fluid flow supplied by the coolant. The results of numerical experiments were obtained by implementing a two-dimensional axisymmetric mathematical model based on the laws of conservation of mass, energy and momentum vector. The numerical investigation was performed in the ANSYS Fluent finite element analysis environment using the finite volume method. The solutions of the equations of stationary turbulent heat and mass transfer in an oil well have been obtained when estimating the thermal effect of hot oil treatment through the hollow rods of the downhole pump. The adequacy of the proposed model has been evaluated. Fields of temperatures and velocities, temperature distribution in the borehole by depth at different modes of coolant feeding are obtained. We calculated the thermal state of the well structure when flushing with liquid through hollow rods with thermal insulation and evaluated the efficiency of the technology and its effectiveness at a given temperature of paraffin melting (60 °C).

Keywords: Heat transfer · Mass transfer · Mathematical model · Ansys Fluent · Coolant · Boreholes · Hollow rods

1 Introduction

When operating oil wells, there are a number of problems associated with the processes of asphalt-resin-paraffin (ARPD) deposition on the walls of the equipment. Efficiency of combating this phenomenon depends on the region, technological parameters of production and the chosen method of thermal effect assessment. It is of practical interest to assess the efficiency of the methods for controlling the ARPD for a particular well. The method of mathematical modeling allows evaluating the efficiency of this or that approach with minimum costs. Thus, the works [1] and [2] consider the wells, where the heating cable is used as a method of combating paraffin deposits, presented a numerical solution of non-stationary problem of heat and mass transfer in an oil well with a heating

A. A. Radionov and V. R. Gasiyarov (Eds.): RusAutoCon 2022, LNEE 986, pp. 166–175, 2023.
https://doi.org/10.1007/978-3-031-22311-2_17

cable. The articles [3] and [4] investigate the influence of electromagnetic radiation on asphalt-resin-paraffin deposits, analyze the dynamics of paraffin deposition and the rate of its removal. The use of hot oil pumped into the well to warm it up to a temperature above the paraffin crystallization temperature is considered in [5]. The paper presents a method for determining the temperature regime in direct and reverse evaluation of the thermal effect when treating a well with a hot thermal fluid based on the thermal balance of the fluid being in the elevator pipes and annulus. Of scientific and practical interest is the method presented in this work, in which the assessment of the thermal effect in the treatment with hot thermal oil well is carried out through the hollow rods of the deep well pump.

2 Main Part

The following study examines a method for estimating the thermal effect of treating an oil well with hot coolant through hollow rods located in a tubing string. Moving along the rods, the coolant heats the wellbore fluid flow, exits through the bypass sleeve, mixes with the main oil flow and rises to the wellhead. Figure 1 shows a general schematic of the equipment for evaluating the thermal effect of treating the well with hot coolant.

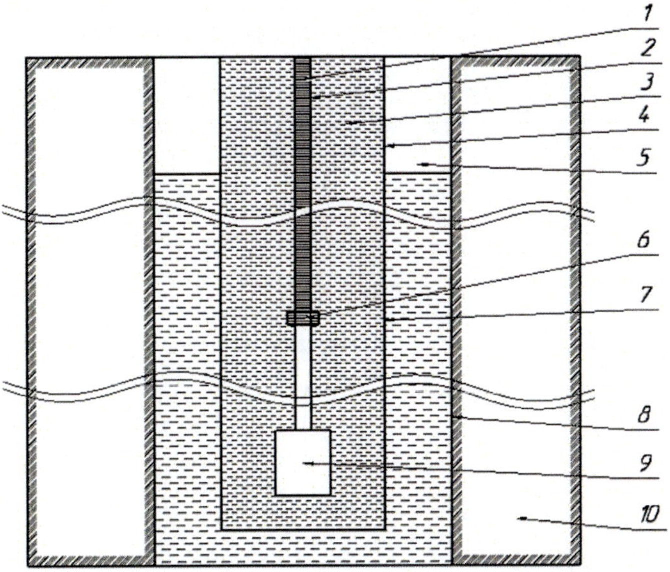

Fig. 1. General scheme of the well and equipment to assess the thermal effect of hot coolant treatment. Structure of equipment: 1 - coolant; 2 - hollow rods; 3 - oil in tubing; 4 - tubing string; 5 - associated oil gas; 6 - transfer flushing sleeve; 7 - tubing; 8 - production string; 9 - pump; 10 - ground mass.

The work considers the section from the wellhead to the flushing sleeve, which is 800 m long. The hollow rod string consists of two parts of different diameters and is located inside the tubing. The following assumptions are made in describing the heat and mass transfer processes: the thermal properties of solid materials do not depend on temperature, the infinite earth mass is replaced by a limited area, the oil fluid is considered as a single-phase medium. We consider a stationary, axisymmetric problem of turbulent heat and mass transfer of oil in the tubing, in the hollow rods and in the annular space. A realizable k-epsilon model from the RANS (Reynolds averaged Navier Stokes equations) family of models is used to describe the turbulent transport. To simulate the turbulent flow in the boundary layer, we use near-wall functions.

The mathematical model of motion and heat transfer in an oil well is based on the laws of conservation of mass, amount of motion and energy, taking into account the assumptions made, has the following form:

Transport equations (Navier–Stokes equations averaged by Reynolds):

$$\rho_i(\bar{\upsilon}_{ir}\frac{\partial\bar{\upsilon}_{ir}}{\partial r}+\bar{\upsilon}_{iz}\frac{\partial\bar{\upsilon}_{ir}}{\partial z})=-\frac{\partial\bar{P}_i}{\partial r}+\mu_i\left(\frac{\partial^2\bar{\upsilon}_{ir}}{\partial r^2}+\frac{1}{r}\frac{\partial\bar{\upsilon}_{ir}}{\partial r}+\frac{\partial^2\bar{\upsilon}_{ir}}{\partial z^2}-\frac{\bar{\upsilon}_{ir}}{r^2}\right)+ \\ +\frac{1}{r}\frac{\partial}{\partial r}\left(-r\rho_i\overline{\upsilon_{ir}'^2}\right)+\frac{\partial}{\partial z}\left(-\rho_i\overline{\upsilon_{ir}'\upsilon_{iz}'}\right), \tag{1}$$

$$\rho_i(\bar{\upsilon}_{ir}\frac{\partial\bar{\upsilon}_{iz}}{\partial r}+\bar{\upsilon}_{iz}\frac{\partial\bar{\upsilon}_{iz}}{\partial z})=-\frac{\partial\bar{P}_i}{\partial z}+\mu_i\left(\frac{\partial^2\bar{\upsilon}_{iz}}{\partial r^2}+\frac{1}{r}\frac{\partial\bar{\upsilon}_{iz}}{\partial r}+\frac{\partial^2\bar{\upsilon}_{iz}}{\partial z^2}\right)+ \\ +\frac{1}{r}\frac{\partial}{\partial r}\left(-r\rho_i\overline{\upsilon_{ir}'\upsilon_{iz}'}\right)+\frac{\partial}{\partial z}\left(-\rho_i\overline{\upsilon_{iz}'^2}\right) \tag{2}$$

The transport equation for the kinetic energy of turbulence:

$$(\frac{\partial\rho_i k\bar{\upsilon}_{ir}}{\partial r}+\frac{\partial\rho_i k\bar{\upsilon}_{iz}}{\partial z})=\frac{\partial}{\partial r}\left[\left(\mu_i+\frac{\mu_t}{\sigma_k}\right)\frac{\partial k}{\partial r}\right] \\ +\frac{\partial}{\partial z}\left[\left(\mu_i+\frac{\mu_t}{\sigma_k}\right)\frac{\partial k}{\partial z}\right]+G_k+G_b-\rho_i\varepsilon-Y_M \tag{3}$$

The transfer equation for the rate of dissipation of kinetic energy of turbulence:

$$(\frac{\partial\rho_i\varepsilon\bar{\upsilon}_{ir}}{\partial r}+\frac{\partial\rho_i\varepsilon\bar{\upsilon}_{iz}}{\partial z})=\frac{\partial}{\partial r}\left[\left(\mu_i+\frac{\mu_t}{\sigma_\varepsilon}\right)\frac{\partial\varepsilon}{\partial r}\right]+\frac{\partial}{\partial z}\left[\left(\mu_i+\frac{\mu_t}{\sigma_\varepsilon}\right)\frac{\partial\varepsilon}{\partial z}\right]+\rho_i C_1 S\varepsilon \\ -\rho_i C_2\frac{\varepsilon^2}{k+\sqrt{\nu_i\varepsilon}}+C_{1\varepsilon}\frac{\varepsilon}{k}C_{3\varepsilon}G_b \tag{4}$$

Turbulent viscosity:

$$\mu_t=\rho_i C_\mu\frac{k^2}{\varepsilon} \tag{5}$$

The incompressibility equation:

$$-\rho_i\left(\frac{\partial(r\overline{\upsilon}_{iz})}{\partial z}+\frac{\partial(r\overline{\upsilon}_{ir})}{\partial r}\right)=0 \tag{6}$$

Energy equation for liquid elements:

$$\begin{aligned}&\frac{\partial}{\partial r}\left[\overline{\upsilon}_{ir}(\rho_i E+p)\right]+\frac{\partial}{\partial z}\left[\overline{\upsilon}_{iz}(\rho_i E+p)\right]=\frac{\partial}{\partial r}\left(k_{eff}\frac{\partial T}{\partial r}+\overline{\upsilon}_{ir}(\tau)_{eff}\right)+\\&+\frac{\partial}{\partial z}\left(k_{eff}\frac{\partial T}{\partial z}+\overline{\upsilon}_{iz}(\tau)_{eff}\right)\end{aligned} \tag{7}$$

Energy equation for solid elements:

$$\lambda_j\left(\frac{\partial^2 T}{\partial r^2}+\frac{\partial^2 T}{\partial z^2}\right)=0 \tag{8}$$

The system of differential Eqs. (1–8) was closed by the following boundary conditions: at the inlet to the pump pipe, a velocity plot for oil was set corresponding to the well's flow rate equal to 20 m3/day and the oil temperature at the coupling level equal to 15 °C; a coolant flow velocity plot corresponding to a given flow rate of 150 m3/day and the coolant temperature of 120 °C; at the wellhead, the convective heat exchange condition:

$$q=\alpha\left(T_0-T_{up}\right) \tag{9}$$

A temperature distribution corresponding to a geotherm with a geothermal gradient equal to 0.2 °C/10 m is set on the surface bounding the earth's region:

$$T_{zm}=T(z). \tag{10}$$

The condition of ideal thermal contact was set at the interface of heterogeneous media:

$$T^{(n)}|_{r_i}=T^{(n+1)}|_{r_i},\ \lambda_n\frac{\partial T^{(n)}}{\partial r}\bigg|_{r_i}=\lambda_{n+1}\frac{\partial T^{(n+1)}}{\partial r}\bigg|_{r_i}. \tag{11}$$

For speeds in the center of hollow rods and tubing - the maximum speed condition, and on the walls - the conditions of adhesion and non-penetration.

Differential equations of heat and mass transfer in an oil well can be solved analytically or numerically [6–9]. Problem (1)– (11) was solved numerically by the finite volume method in the ANSYS Fluent software product.

Table 1 describes the properties of materials used in modeling the heat and mass transfer process.

Table 1. Properties of the materials used.

Material	Density ρ, kg/m^3	Heat capacity C, J/(kg·K)	Thermal conductivity λ,W/(m·K)	Viscosity μ, mPa·s
Ground	1900	1680	1.82	-
Steel	8030	502.48	16.27	-
Oil	800	2000	0.15	10
Cement	1400	900	0.85	-
Thermal insulation (compound)	840	1200	0.3	-

3 Results

As a result of a numerical study of the processes of heat and mass transfer of oil and coolant in an oil well with hollow rods, temperature and velocity fields were obtained throughout the entire volume of the studied area. The flow rate of the well in question is 20 m3/day. The melting point of the ASPO varies linearly from the temperature at the mouth equal to 52 °C, to the temperature at a depth of 800 m equal to 59.3 °C. The results are shown in Figs. 2–7.

Figure 2 shows the velocity field in the tubing at 799 m depth of the flow-through socket. The figure shows that the coolant in the center of the rod channel has maximum velocity values, then, leaving the socket, it mixes with the oil flow and continues moving along the tubing. The speed of oil flow with coolant is 2 times lower than the speed of coolant in hollow rods, despite the fact that the flow of coolant is added to the flow of produced oil. The drop in velocity is due to the fact that the cross-sectional area in tubing is significantly larger than in hollow rods.

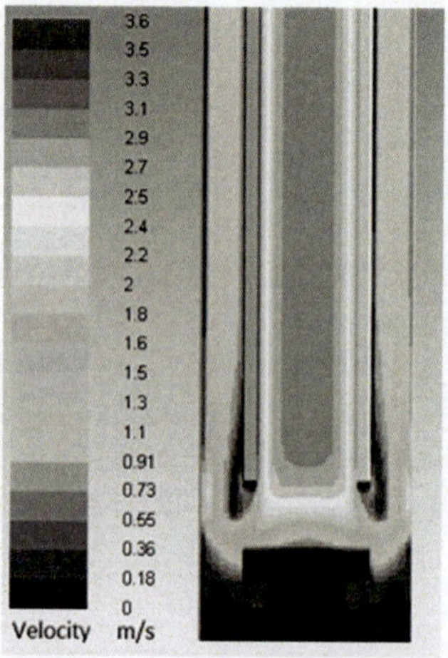

Fig. 2. Velocity field in the tubing for the well section where the coolant exits the flow coupling. The coolant is supplied with a flow rate of 150 m3 / day with a temperature of 120 °C.

Figure 3 shows the temperature fields of a well Sect. 800 m long with different temperatures of the injected coolant. The figure shows the intensity of the change in the temperature of the coolant as it moves along the hollow rods. The higher the initial temperature of the coolant, the higher the temperature of the oil fluid moving through the tubing. The temperature of the coolant itself and the temperature of the metal rods decreases with increasing depth of the well. The maximum oil temperature of 158 °C is reached at the mouth of the tubing channel at a coolant temperature of 210 °C, while the average oil temperature at the height of the tubing is 89 °C.

The effect of coolant flow on the temperature of oil in the tubing is shown in Fig. 4. It is obvious that with an increase in coolant flow, the intensity of well heating increases and the average temperature of oil in the tubing increases from 55 °C to 89 °C with an increase in flow from 150 m^3/day to 250 m^3/day.

The influence of the coolant temperature on the efficiency of assessing the thermal effect when treating a well with a hot coolant is shown in Fig. 5., Fig. 6 shows that even at the temperature of the injected oil equal to 210 °C (curve 1), the temperature at the coupling level (800 m) falls below the melting point of paraffin and is 55 °C. While at a coolant temperature equal to 120 °C, the oil temperature at the same depth is 40 °C (curve 4). Hence, it can be concluded that an increase in the temperature of the coolant at a given flow rate of 150 m^3/day and a well flow rate of 20 m^3/day does not prevent the process of deposition of paraffin on the walls of the tubing or to carry out a full assessment of the thermal effect when treating an oil well with a hot coolant.

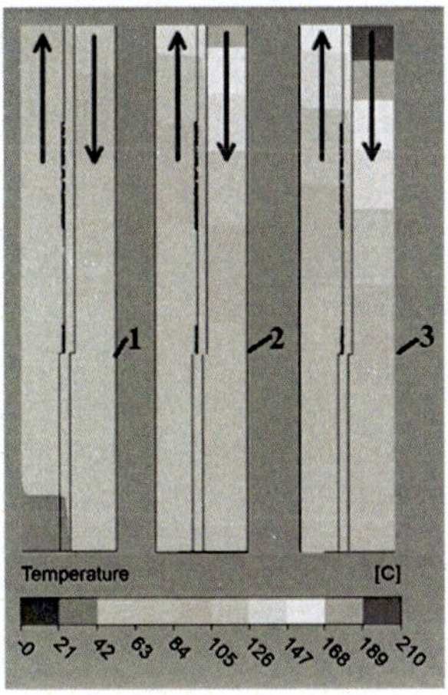

Fig. 3. Temperature fields in tubing for wells with different coolant temperatures: 1 – 120 °C; 2 – 150 °C; 3 – 210 °C.

Analysis of the processes of heat and mass transfer during the flow of oil and coolant at different values of coolant flow allowed us to determine the value of the latter, when using which (250 m^3/day) the oil temperature along the entire length of the well is greater than the melting point of paraffin (curve 5), despite the fact that the temperature of the coolant was set equal to 120 °C Fig. 6., curve 2. In this in this case, the oil temperature at

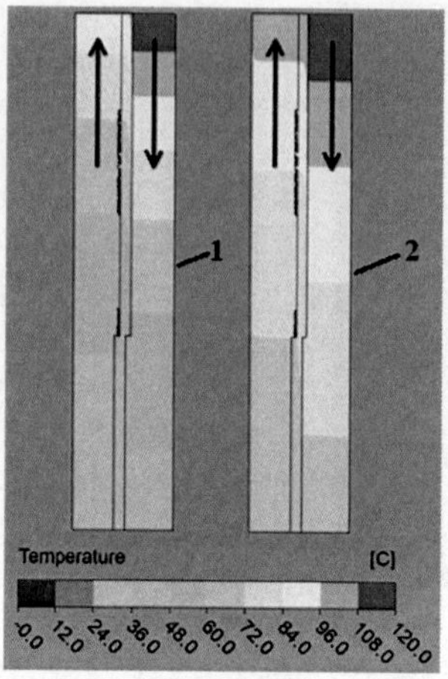

Fig. 4. Temperature fields in the tubing for wells with different coolant flow rates: 1 – 150 m^3/day; 2 – 250 m^3/day.

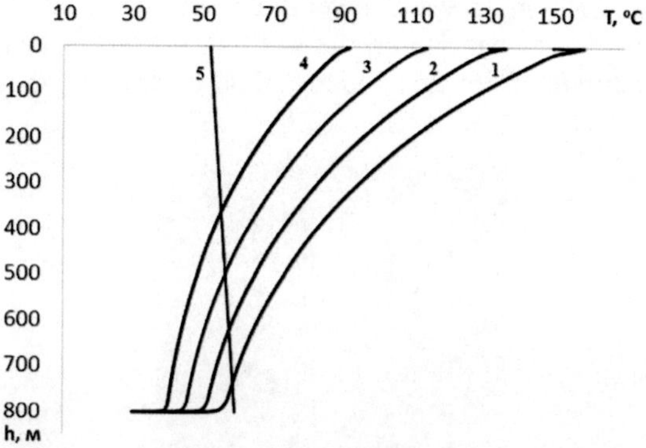

Fig. 5. Temperature distribution on the inner wall of the tubing along the depth at different temperatures of the coolant. The coolant is oil supplied with a flow rate of 150 m^3/day. The flow rate of the well is 20 m^3/day. 1 – the temperature is 210 °C; 2 – the temperature is 180 °C; 3 – the temperature is 150 °C; 4 – the temperature is 120 °C; 5 – the melting point of paraffin.

the coupling level is 62 °C. Thus, under such conditions, the assessment of the thermal effect when treated with a hot coolant, there is a complete removal of paraffin deposits from the walls of the well.

The most effective assessment of the thermal effect of a well with hot oil can be achieved by changing both the temperature of the coolant and its flow rate. The paper considers the case when the temperature of the coolant was set equal to 150 °C, and the flow rate was 200 m^3/day. The temperature distribution on the tubing wall in this mode for assessing the thermal effect during treatment with a hot coolant is shown in Fig. 6. (curve 1). It can be seen from the figure that the temperature along the considered

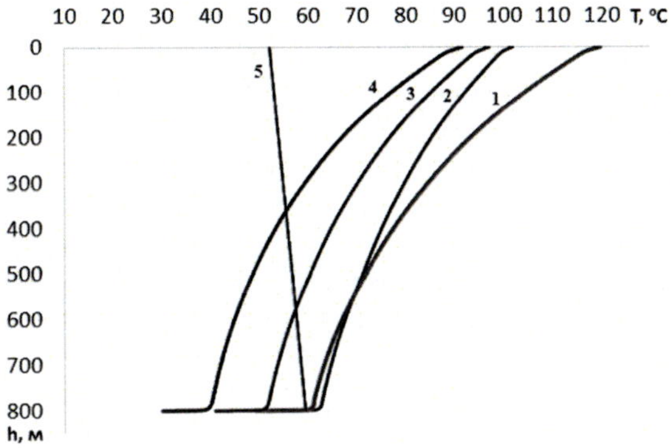

Fig. 6. Temperature distribution on the inner wall of the tubing from the depth at different coolant flow rates. The flow rate of the well is 20 m^3/day. 1 – coolant oil heated to 150 °C and supplied with a flow rate of 200 m^3/day; 2 – coolant oil heated to 120 °C and supplied with a flow rate of 250 m^3/day; 3 – coolant oil heated to 120 °C and supplied with a flow rate of 200 m^3/day; 4 – coolant oil heated to 120 °C and supplied with a flow rate of 150 m^3/day; 5 – the melting point of paraffin.

section does not fall below the melting point of paraffin, therefore, the removal of ASF will occur throughout the entire section.

In order to heat the well more effectively, a thermal insulation layer is applied to the inner surface of the rods Fig. 7. To evaluate the effectiveness of the thermal insulation layer when flushing the well through the hollow rods of a downhole pump, two well models with and without the thermal insulation layer on the inner surface of the hollow rods were implemented. As a result, temperature distribution curves on the inner tubing wall were obtained for both cases.

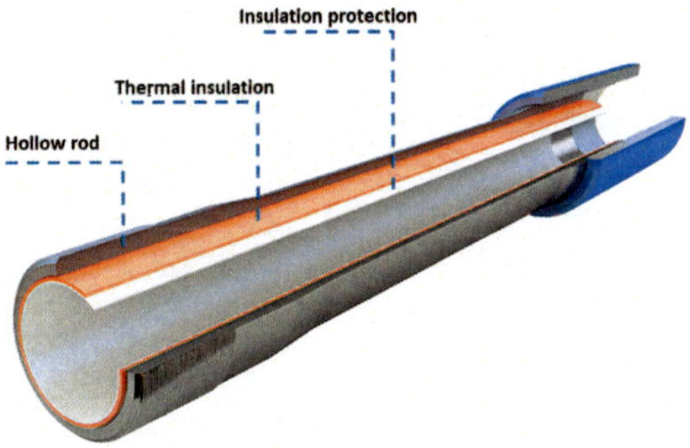

Fig. 7. General view of the boom design.

The calculation results are shown in Fig. 8 and Table 2.

Figure 8 shows that the presence of thermal insulation layer on the inner surface of hollow rods significantly affects the temperature distribution on the tubing wall in the well. With a thermal insulation layer in place, the temperature at the coupling level (1,200 m) reaches 56 °C, an increase of 36 °C compared to the case without thermal insulation. This is due to the fact that the coolant has a higher temperature at the outlet of the coupling equal to 63 °C. Whereas, moving along the rods without thermal insulation, the coolant has time to cool down to 20.7 °C at the coupling level.

According to the results of the conducted studies, the following conclusion can be drawn: when assessing the thermal effect when treating a well with a coolant supplied through hollow rods, the coolant flow has a more significant effect than its temperature, however, for the most effective assessment of the thermal effect, it is necessary to vary the values of both of these parameters.

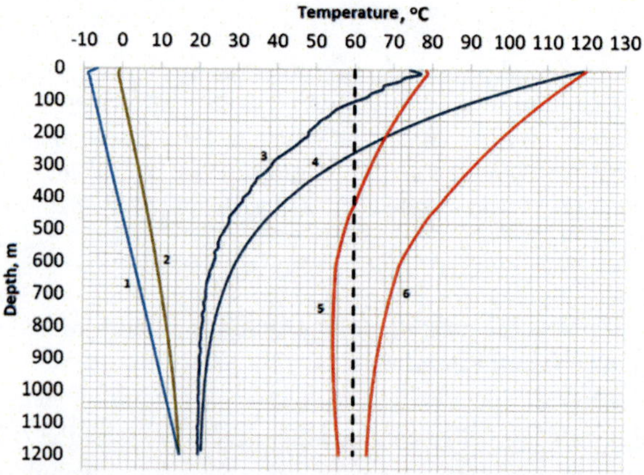

Fig. 8. Temperature distribution on the inner wall of the tubing and inside the rods by depth for the cases with and without thermal insulation. The heat-transfer agent is oil, supplied at a flow rate of 150 m^3/day and a temperature of 120 °C for 5 h. Paraffin crystallization temperature is assumed to be 60 °C.1 – geotherma; 2 - without flushing; 3 - without tubing insulation; 4 – without insulation rod; 5 - with tubing insulation; 6 - with rod insulation.

Table 2. Comparison of flushing performance with and without thermal insulation.

Boom design	Indicators		
	Flow temperature at the coupling level, °C	Depth of complete removal of ARPD, m	Depth of the most probable ARPD removal area, m
With insulation	56	424	1200
Without insulation	20	109	424

4 Conclusion

This paper shows how the efficiency of heat treatment of an oil well with a coolant through hollow rods of a deep pump depends on the parameters of the coolant supply. The influence of the flow rate and temperature of the coolant on the temperature distribution in the well column is estimated. The presence of a thermal insulation layer significantly increases the efficiency of flushing the well with hot coolant through the hollow rods.

References

1. Kostarev, N., Trufanova, N.: Thermal process control in oil well with the help of heating cable. Electr. Eng. **11**, 60 (2017)
2. Kovrigin, L., Kukharchuk, I.: Automatic control system for removal of paraffin deposits in oil well in permafrost region by thermal method. Chem. Eng. Res. And Des 115(Part A):**116** (2016)
3. Balakirev, V., Sotnikov, G., Tkach, Y., Yatsenko, T.: Heating and melting of asphalt–paraffin plugs in oil-well equipment using an electromagnetic radiation source operating in a periodic mode. J. of Applied Mech. and Tech. Phys. **42**(4), 680 (2001)
4. Fatykhov, M.: Heating and melting of paraffin in a coaxial system under the effect of high-frequency electromagnetic radiation. High Temp. **40**(5), 746 (2002)
5. Kamenshchikov, F.: Thermal dewaking of wells. J. Clerk Maxwell, A Treatise on Electricity and Magnetism, Izh.: Research Center "Regular and chaotic dynamics", Moscow (2005)
6. Musakaev, N., Borodin, S., Romanyuk, S.: Methods for solving the one-dimensional radial problem of heat transfer to the frozen rocks surrounding the borehole. Vestnik of the Tyumen State University **7**, 19 (2014)
7. Gubaidullin, A., Musakaev, N., Borodin, S.: Computer simulation of processes in an oil and gas well equipped with electric centrifugal pumps. Izvestia of Higher Educational Institutions. Oil and Gas **5**, 69 (2010)
8. Lekomtsev, A., Ilyushin, P., Yu, C., Korobov, G.: Determination of technological parameters of the thermal treatment of the well through the hollow rods of the deep pumping equipment. Exposition Oil Gas **4**, 50–54 (2020)
9. Musakaev, N.: A mathematical study of temperature situation in a well in the presence of a source of electrical heating. Izv. Vysh. Uch. Zav. Neft and Gas **6**, 43 (2010)

A Method for Planning the Supply of Petroleum Products to Filling Stations Based on Multi-agent Resource Conversion Processes

K. Aksyonov(✉), A. Antonova, and O. Aksyonova

Ural Federal University, 32, Mira Street, Ekaterinburg 620078, Russia
k.a.aksenov@urfu.ru

Abstract. The paper proposes a method for planning the supply of fuel to gas stations by fuel trucks. The considered routing problem has several practical aspects obstructing using methods for solving the transport problem. These aspects include the presence of several sections in the tanks of fuel trucks with the requirement not to mix different types of fuel in the tank, the absence of fixed start and end times for loading and unloading fuel, the presence of a sequence for draining fuel from sections of a fuel truck. The article proposes a hybrid method based on multi-agent modeling, which allows considering all the features of the considered routing problem. Agents in a multi-agent system correspond to vehicles, as well as logistics points of loading and unloading. The proposed method is implemented in a decision support system based on the use of software products of the BPsim family. The method has been tested in solving the problem of fuel supply to a gas station network.

Keywords: Supply planning · Simulation modeling · Multi-agent modeling · Transport problem

1 Introduction

One of the ways to save resources when transporting goods is to use decision support systems in the field of logistics. Development of software packages that solve the logistic problems is in demand. The main function of such systems is the ability to calculate and build efficient routes for various purposes in a transport network. The paper considers one of such problems, which is the optimal routing of a fleet of vehicles to perform a given set of orders. The problem of optimal routes finding, and the problem of optimal vehicle loading are studied in a theory of combinatorial optimization [1]. In practice, such problems are complicated using a fleet of vehicles with limited payload capacity, irregular average vehicle speeds during the day, and the need for timely delivery. Dedicated routing software must meet requirements for flexibility and scalability also.

Relevance of the topic studied is related to development of practice-oriented algorithms for solving problems of cargo transportation routing considering the optimal distribution of the transport resources along the delivery routes, as well as the possibility

A. A. Radionov and V. R. Gasiyarov (Eds.): RusAutoCon 2022, LNEE 986, pp. 176–185, 2023.
https://doi.org/10.1007/978-3-031-22311-2_18

of choosing several distribution options. Objects of the study are gas stations located on the map of the study area, which require the supply of petroleum products for operation and functioning.

The purpose of the paper is to develop a scheduling method with certain petroleum transportation rules based on multi-agent resource conversion processes.

2 Literature Review

The problem of finding routes of products transportation with minimal cost from one warehouse to another is a generalization of a traveling salesman problem and belongs to the class of NP-complete problems [2]. In the second half of the twentieth century, existing methods for solving the traveling salesman problem were actively used and modified to solve the routing problem. Among them are branch and bound methods [3, 4], dynamic programming [5], and integer linear programming [6, 7]. However, starting from the 1960s, the main efforts were directed to development of approximate heuristic algorithms. Approximate heuristic algorithms can be divided into three large groups: constructive [8, 9], two-phase [10, 11], and improvement [12] ones.

Metaheuristic algorithms began to develop actively after the 1990s. Metaheuristic algorithms arose from observing the processes of animate and inanimate nature and include, for example, evolutionary modeling and ant colonies [13, 14].

The factors limiting the application of the routing problem methods to the fuel delivery problem are the following [15].

- Depending on the type of fuel, each cargo can be transported in one section of a tank truck.
- There is a certain sequence of draining fuel from a tanker truck depending on design of the drain devices.
- There is no time component in the form of start and end times of truck runs, loading and unloading time.
- No division into types of cargo.
- Presence of several sections of gasoline in tankers is not considered.
- Physical limitations of tank trucks are not considered.
- Preferences of tanks at gas stations are not considered.
- Possibility of serving nearby gas stations with one gasoline tanker per trip is not considered.

These factors are proposed to be considered using a model of a multi-agent resource conversion process [16]. The model allows you to simulate various fleet management schemes considering the current situation (urgency and volume of cargo transportation, condition and location of vehicles, fuel residues in vehicles, location of gas stations), analyze various scenarios, and choose the most effective solution for a given point in time.

Let us consider a hybrid method for planning the oil product supply of a filling station network based on integration of the transport method, the multi-agent approach and simulation modeling.

3 Method for Planning the Supply of Petroleum Products to a Filling Station Network

Let us consider integration of the methods for solving the transport problem and multi-agent modeling.

3.1 Application of Methods for Solving the Transport Problem

The transport problem is a mathematical linear programming problem about finding the optimal distribution of homogeneous objects from a battery to receivers while minimizing the cost of movement.

Let us consider the main methods for solving the transport problem, such as a north-west corner method, a minimum cost method and a distribution method. The following situation has been taken as an example: suppliers (sections of tanks with gasoline fuel) A1, A2, A3 and A4 contain respectively 5960, 4440, 6010 and 4420 L of fuel, which must be delivered to consumers B1, B2, B3 in the required amount of 5000, 10,500 and 4330 L. A classical 'cost' parameter in the transport problem is defined through the priority of transportation and is given by a matrix where the priority of transportation is the number of days, during which a filling station will be able to work without new supplies for each type of fuel. The lower the priority, the greater the request for urgent fuel delivery. This concept of cost differs from the concept in the transport problem. However, from the point of view of the fuel delivering problem, the priority of transportation is the main factor affecting the cost since the lack of fuel at a gas station leads to its stop.

The northwest corner method involves delivering in the upper left corner of the table until the consumer's demand is satisfied, or the supplier's capacity is not exhausted, otherwise this consumer is ignored. The plan received may not be optimal as the cost of transportation is not considered. The result is shown in Table 1.

The minimum cost method proposes that at each step the delivery with the lowest priority (cost) of transporting a unit of cargo among all empty cells is selected. This method allows you to find a plan close to optimal in the case of a small dimension of the problem. The result of the algorithm is shown in Table 2.

The distribution method involves checking the optimality of the plan obtained by another method using a matrix of estimates. If this plan is not optimal, an optimization loop is performed, and the next optimality check is fulfilled. The method is also called the potential method and is the most accurate however at the same time the most time-consuming. The result of the algorithm is shown in Table 3. This result is the most accurate of the three methods considered and is obtained using four optimization cycles.

Analysis of the above methods allows us to conclude that the most suitable algorithm for further development is the method with minimum costs.

Table 1. Result of the northwest corner method.

Section volume, liter	The required volume, liter		
	6000	10,500	4330
	Transportation priority/Transportation volume, liter		
5960	2.1/5960	2/0	2/0
4440	1/40	2.5/4400	1/0
6010	1/0	2.8/6010	2/0
4420	2.8/0	2.1/90	1/4330

Table 2. Result of the minimum cost method.

Section volume, liter	The required volume, liter		
	6000	10,500	4330
	Transportation priority/Transportation volume, liter		
5960	2.1/0	2/5960	2/0
4440	1/0	2.5/110	1/4330
6010	1/6000	2.8/10	2/0
4420	2.8/0	2.1/4420	1/0

Table 3. Result of the distribution method.

Section volume, liter	The required volume, liter		
	6000	10,500	4330
	Transportation priority/Transportation volume, liter		
5960	2.1/0	2/5960	2/0
4440	1/110	2.5/0	1/4330
6010	1/5890	2.8/120	2/0
4420	2.8/0	2.1/4420	1/0

3.2 Application of Multi-agent Modeling Methods

Analysis of decision support systems (DSS) could not allow us to identify specialized software for fuel transportation planning in open sources. Some systems are used to simulate in logistics, such as AnyLogic [17], ARIS [18], Business Studio [19], G2 [20], Magenta [21], BPsim [22].

Over the past decades, the scope of simulation systems has expanded significantly.

Firstly, simulation systems or simulation models are embedded the management scheme of an enterprise or organization, they are integrated through data exchange interfaces with sensors, controllers, and corporate information systems, and they receive data on the current situation at the management level.

Secondly, most simulation maps during a simulation experiment allow you to change or update the original data and adapt the solution based on a situation.

Thirdly, in most applied decision support systems that use a modeling subsystem, formalization or heuristic-based decision subsystems are used to formalize the knowledge of decision makers and control algorithms.

Thus, we can talk about open modules or simulation systems. It should also be noticed that elements of expert systems (knowledge bases), inference machines and ontologies are actively used in decision support systems.

Let us consider the advantages and disadvantages of simulation systems in relation to the problem being solved. Modeling software evaluation is presented in Table 4.

The software systems listed in Table 4 have several disadvantages that limit or exclude the possibility of their use in solving the problem of developing a system for planning and managing fuel traffic. All the systems are not problem-oriented in the field of fuel transportation planning and a problem of redevelopment of the vehicle route in the case of a new order has not been resolved.

In the Magenta system, there is no possibility to adjust and coordinate the plan with the user, In ARIS, AnyLogic, Business Studio systems, the knowledge of specialists is not considered when planning a route. The greatest opportunities for software implementation of the fuel supply planning method are provided by the BPsim family and the Magenta modeling system.

Let us consider development of the method for planning the oil product supply of a gas station network using multi-agent modeling via products of the BPsim family and the minimal costs method for transport problem solving. Agents in a multi-agent system will correspond to vehicles and logistics nodes (loading and unloading nodes - tank farms, gas stations, garages).

3.3 Hybrid Fuel Delivery Scheduling Method

The hybrid fuel delivery scheduling method includes the following steps.

1. Calculation of fuel demand at filling stations. Solving the transport problem in relation to planning supplies from oil depots to gas stations without solving the problem of dense packing of fuel trucks.
2. Ranking of all demands, determining the most urgent demands according to their priority. Definition a supplier (warehouse or logistics center) for each order and calculating a delivery route.
3. Creation of a transportation plan. Assigning a vehicle to each order and evaluating the delivery time. At this stage, the intelligent scheduling agent uses a knowledge base with statistics of fuel sales from a gas station, physical limitations of gas stations and fuel trucks (their compatibility and serviceability), as well as preferences for using.
4. Verification of the logistic plan by a logistic specialist or a dispatcher.
5. Checking and adjusting the delivering plan via the multi-agent simulation model of the resource conversion process.
6. Plan implementation.

Table 4. Software evaluation.

Parameter	G2	AnyLogic	Magenta	BPsim	ARIS	Business Studio
Development of a structure of a conceptual model (ontology)/loading of geodata and control rules	NO/+	NO/+	+ / +	+ / +	+ /NO	NO/NO
Resources, objects, converters	+	+	+	+	+	+
Hierarchical process model	+	+	NO	+	+	+
Agents	NO	+	+	+	NO	NO
Using heuristics to describe a model node	+	NO	+	+	NO	NO
Reacting to an external event	+	NO	+	+	NO	NO
Dynamic scheduling	+	NO	+	+	NO	NO
Rescheduling	+	NO	+	+	NO	NO
Coordination of plans in real time	NO	NO	+	+	NO	NO
Plan execution control	+	NO	+	NO	NO	NO
Coordinating plans with the user	NO	NO	NO	+	NO	NO
Manual tuning of plans	NO	NO	NO	+	NO	NO
Interface of a specialist (dispatcher)	NO	NO	+	+	NO	NO

7. Correction of the delivering plan by an expert under conditions of external influences. The main criterion for the success of the plan is to ensure the uninterrupted operation of the filling station network.

4 Implementation of the Program of the Hybrid Method of Fuel Delivery Scheduling

Based on a hybrid decision-making method described, a decision support system was developed using integration of the BPsim.MAS simulation system and the BPsim.DSS decision support system. The BPsim.DSS system is a problem-oriented system that supports manual and automated updating of the distribution plan.

According to the InterRRaP architecture [23], the dispatcher agent model consists of the four levels.

1. The environment model corresponds to the following elements of the multi-agent resource conversion process: converters, resources, reactive intelligent agents, mechanisms, parameters, and goals. The external environment performs the following functions – generates tasks, transfers messages between agents, processes agent commands, performs resource conversion, changes the current state of the external environment. The external environment is a control object model built from the elements of the resource conversion process. The control object model describes the processes of fuel consumption at filling stations, the workflows of filling stations, oil depots and fuel supply processes.
2. The components of the interface with the outside world and reactive behavior are implemented in the form of a tactical base of the agent's production rules and an inference mechanism (simulator algorithm).
3. The local planning level is intended to find effective solutions. The local planning component is implemented based on a strategic database that uses frames.
4. The subsystem of cooperation with other agents is intended to integrate the hybrid agents with each other, as well as with external information systems.

Design of a conceptual model of the subject area and the agent's knowledge bases are implemented based on the extension of a UML class diagram [24]. The inference mechanism is implemented using a solution search scheme built based on a sequence diagram. Each decision of the strategic knowledge base is an agent's action plan. Each plan has a set of rules from the base of the reactive component. Based on the found solution, the agent's current plan is updated. Search for all options contained in the agent's strategic knowledge base forms the agent's plan library.

5 Application of the Developed Method and Software

The model implements the process of consumption and supply of fuel to the network of filling stations of the Bashneft company in Sverdlovsk region. The model parameters have been taken as follows.

- Number of nodes in total is 179, among them operations are 59 and intelligent agents are 120.
- Total number of agent rules is 941.
- Number of resources is 703.
- Number of fuel trucks is 12, number of requests for delivery during a day is 22.

The visual interface of the Scheduler agent manager is shown in Fig. 1.

The following indicators of filling station capacities have been taken as a basis: type of fuel, capacity, minimum balance, average daily consumption. The average daily consumption is calculated based on the consumption statistics of the product from each container. For each station, the average night consumption values were calculated - as average values for the interval from 10 pm to 6 am and average daily consumption values for the interval from 6 am to 10 pm. Initial data also includes data on the cost of transporting fuel both by own fuel trucks and by freelancers, fuel supply prices from oil depots, and a distance matrix.

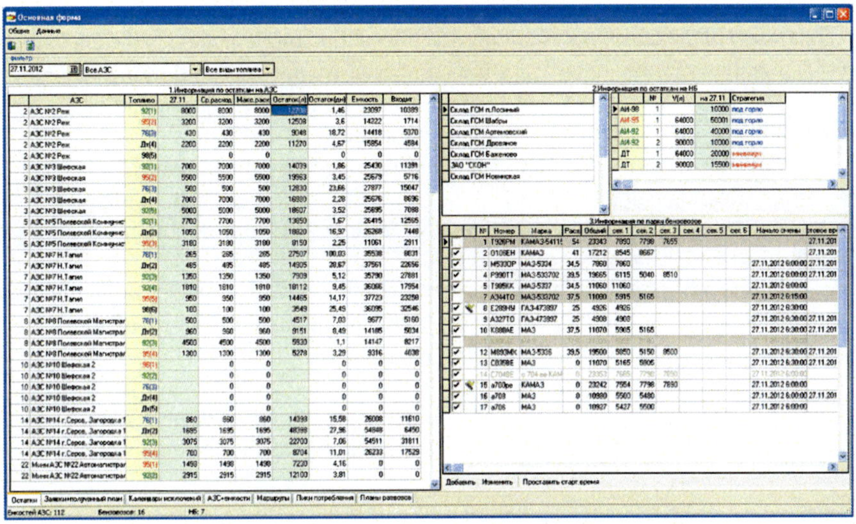

Fig. 1. The visual interface of the Scheduler agent manager.

The developed hybrid method and a set of models have been used to analyze the operation of the Bashneft gas station network.

The results of computational experiments are compared with the actual delivery data. For the experiments, real data of the plan and fact of the distribution of petroleum products have been taken.

Table 5 shows a comparison of the calculated and actual plans for daily delivery with the same number of truck runs.

Table 5. Comparison of daily transportation volumes.

Integrated analysis of runs	Own fuel trucks	Other trucks
Difference in the number of runs (calculated and actual)	NO/ +	NO / +
The ratio of the number of runs (own and others), calculated	+	+
The ratio of the number of runs (own and others), actual	+	+
The volume of transportation (own and others) in liters, calculated	NO	+
The volume of transportation (own and others) in liters, actual	+	NO

The experiments results show that the proposed method gave more requests for execution to own fleet of fuel trucks (the ratio is 10 to 5) compared to the dispatcher solution (the ratio is 8 to 7) so the volume of transportation of fuel by own trucks increased.

The results of the table analysis show that the method implemented in the DSS system shows a greedier strategy in terms of the number of truck's runs and traffic volume (on average, 13% higher). However, the method is inferior to the dispatcher actions in terms of the volume of fuel specified by the delivery strategy by 1.4%.

The results of experiments on the data have shown the following results.

- The experimental plan gives consistency with the results of the actual plan drawn up by the specialist dispatcher.
- The use of the automatic scheduling algorithm allows us to consider the preferences of the decision maker in terms of distribution strategies, the priority of servicing of gas stations, restrictions on vehicles and gas stations.

Thus, the new hybrid method for planning the supply chain of petroleum products for filling stations is flexible and meets all the requirements of the subject area.

6 Conclusion and Future Work

The paper presents a hybrid method for planning the transportation of fuel to a network of gas stations. The method has been developed using integration of transport problem solving methods, a multi-agent simulation model of the resource conversion process, and expert systems. The method has been implemented in the software that integrates the developed Scheduler system and products of the BPsim family.

The method and the software developed have been tested at the current network of filling stations in Sverdlovsk region. When solving the problem of planning the fuel delivery, the problem of integrating the DSS with the corporate information system of the enterprise has been solved, which makes it possible to further apply the DSS to solving problems of real-time control.

Further work is related to the adjustment of the proposed method in solving applied problems of planning the oil product supply of the filling station network.

References

1. Korte, B., Vygen, J.: Combinatorial Optimization: Theory and Algorithms. Springer, Berlin, Heidelberg (2006)
2. Lawler, E.L., Lenstra, J.K., Rinnooy Kan, A.H.G., Shmoys, D.B.: The Traveling Salesman Problem: A Guided Tour of Combinatorial Optimization. Wiley (1985)
3. Augerat, P., Naddef, D., Belenguer, J.M., Benavent, E., Corberan, A., Rinaldi, G.: Computational results with a branch and cut code for the capacitated vehicle routing problem. France: N. p (1995)
4. Melamed, I.I., Sigal, I.K.H.: The linear convolution of criteria in the bicriteria traveling salesman problem. Comput. Math. Math. Phys. **37**, 8 (1997)
5. Christofides, N., Mingozzi, A., Toth, P.: Exact algorithms for the vehicle routing problem, based on spanning tree and shortest path relaxations. Math Prog **20**, 255–282 (1981). https://doi.org/10.1007/BF01589353
6. Desrochers, M., Desrosiers, J., Solomon, M.: A new optimization algorithm for the vehicle routing problem with time windows. Oper. Res. **40**(2), 342–354 (1992). https://doi.org/10.1287/opre.40.2.342

7. Laporte, G., Nobert, Y., Desrochers, M.: Optimal routing under capacity and distance restrictions. Oper. Res. **33**, 5 (1985)
8. Clarke, G., Wright, J.R.: Scheduling of vehicle routing problem from a central depot to a number of delivery points. Oper. Res. **12**, 568–581 (1964). https://doi.org/10.1287/opre.12.4.568
9. Solomon, M.: Algorithms for the vehicle routing and scheduling problem with time window constraints. Oper. Res. **35**, 254–265 (1987). https://doi.org/10.1287/opre.35.2.254
10. Gillett, B.E., Miller, L.R.: A heuristic algorithm for the vehicle-dispatch problem. Oper. Res. **22**(2), 340–349 (1974). https://doi.org/10.1287/opre.22.2.340
11. Fisher, M., Jaikumar, R.: A generalized assignment heuristic for vehicle routing. Networks **11**(2), 109–124 (1981). https://doi.org/10.1002/net.3230110205
12. Funke, B., Grunert, T., Irnich, S.: Local search for vehicle routing and scheduling problems: review and conceptual integration. J Heur **11**(4), 267–306 (2005). https://doi.org/10.1007/s10732-005-1997-2
13. Dorigo, M., Gambardella, L.M.: Ant colony system: a cooperative learning approach to the traveling salesman problem. IEEE Trans Evol Comp. **1**(1), 53–66 (1997). https://doi.org/10.1109/4235.585892
14. Glover, F., Laguna, M.: Tabu Search. Springer, Boston (1997)
15. Aksyonov, K., Aksyonova, O., Ayvazyan, H.: Identification of problems and limitations of the planning of fuel distribution through the gas station network and their solution. In: Proceedings of the International Multi-conference on Engineering, Computer and Information Sciences, Ural Hi-Tech Park, Yekaterinburg, 25–27 October 2019
16. Aksyonov, K., Bykov, E., Aksyonova, O., Goncharova, N., Nevolina, A.: The architecture of the multi-agent resource conversion processes. In: Proceedings of the UKSim 11th European Modelling Symposium on Mathematical Modelling and Computer Simulation, Manchester, 19–21 November 2017
17. AnyLogic Cloud modeling system. Available from http://www.anylogic.com. Accessed 15 Jul. 2022
18. ARIS modeling system. Available from https://ariscloud.com/aris-basic/. Accessed 15 Jul. 2022
19. Business Studio modeling system. Available from https://www.businessstudio.ru/demo/business_studio/. Accessed 15 Jul. 2022
20. G2 real-time expert system. Available from https://ignitetech.com/softwarelibrary/gensym. Accessed 15 Jul. 2022
21. Himoff, J., Rzevski, G., Himoff, J., Skobelev, P.: Magenta technology multi-agent logistics i-Scheduler for road transportation. In: Nakashima, H., Wellman, M. (eds.) AAMAS '06: 5th International Joint Conference on Autonomous Agents and Multi-agent Systems, Hakodate Japan, May 2006. Association for Computing Machinery, New York, United States (2006)
22. Aksyonov, K. et al.: Decision support systems application to business processes at enterprises in Russia. In: Jao, C. (ed.) Efficient Decision Support Systems: Practice and Challenges in Multidisciplinary Domains. Intech Open (2011)
23. Muller, J.P., Pischel, M.: The Agent Architecture InteRRap: Concept and Application. German Research Center for Artificial Intelligence, Germany (1993)
24. UML official website. Available from http://www.uml.org. Accessed 15 Jul. 2022

Multi-factorial Analysis of Environmental Safety and Optimization of Oily Waste Recycling System

M. Derevyanov(✉), Yu. Pleshivtseva, and A. Afinogentov

Samara State Technical University, 244, Molodogvardeyskay Str., Samara 443100, Russia
mder2007@mail.ru

Abstract. The article is devoted to the urgent problem of reducing the influence of anthropogenic factors in the form of the formation of various oil-containing wastes (OCW) during the extraction, recycling, and transportation of hydrocarbon resources on the example of one of the regions of Russia. The algorithm of multifactorial analysis of environmental safety and optimization of the OCW recycling system based on the Data Envelopment Analysis (DEA) method is proposed. The algorithm uses well-known CCR and Super-efficiency models of the DEA method. The interconnected recycling system under consideration includes storage facilities for OCW, technologies for their recycling, and the corresponding infrastructure. Solution of optimization problem allows determining the best combinations of storage facilities and technologies as part of the OCW recycling system, ensuring maximum environmental safety of the recycling process. The novelty of the proposed approach is the formulation and solution of the corresponding optimization problem of mathematical programming for evaluating combinations of storage facilities and technologies for their environmental safety and comprehensive analysis of the results obtained by average estimates in the analyzed system. The approbation of the algorithm shows that the obtained comparative estimates make it possible to develop management decisions to obtain the most environmentally safe functioning of the OCW recycling system.

Keywords: Oily waste · Recycling · Multi-factorial analysis · Environmental safety · Optimization

1 Introduction

The activities of enterprises in almost all industries are associated with oil and gas usage, because of which various oil-containing waste (OCW) is formed [1]. Russia is the absolute world leader in the pollution of the natural environment of various hydrocarbon structures and their constituent high-molecular resinous asphaltene substances, drilling fluids, and chemical reagents widely used in the extraction and preparation of oil, gas, and gas condensate [2]. This fact is associated with the problems of soil pollution, reservoir, and wastewater, the formation of salts and sludge, and the use of poorly equipped sludge barns, oil collectors, oil pollution storages of various compositions

A. A. Radionov and V. R. Gasiyarov (Eds.): RusAutoCon 2022, LNEE 986, pp. 186–193, 2023.
https://doi.org/10.1007/978-3-031-22311-2_19

for storing hydrocarbon waste. According to the Ministry of Natural Resources of the Russian Federation [3], up to 510 million tons of soil are polluted annually, up to 25 million tons of associated and secondary hydrocarbon raw materials are burned, spilled, and buried in the ground. With the doubling of the number of OCWs every 5–7 years, the problems of treatment, recycling, and their disposal reach a critical level of impact on the environment and economy, because, unlike our country, where the share of processing and recycling of industrial waste is no more than 30%, in developed countries of Europe, North America, and Asia, this share reaches 50% [4]. The main reasons for the current situation are the low efficiency of technological processes of recycling and the lack of an integrated approach to the organization of the OCW recycling system.

The OCW recycling system refers to a complex of waste storage facilities (barns, oil sludge storage facilities, settling ponds, etc.) and technologies for OCW recycling located in the same region, including stationary and mobile installations. There are various approaches to solving the environmental problems of the processing of OCWs and their further use. A comparative assessment of the life cycle of traditional (incineration and burial) and two new approaches to energy recovery, including solvent extraction and pyrolysis, to the purification of oily sludge is carried out in [5]. Other innovative technologies for processing OCWs include ultrasonic irradiation, electrokinetic,

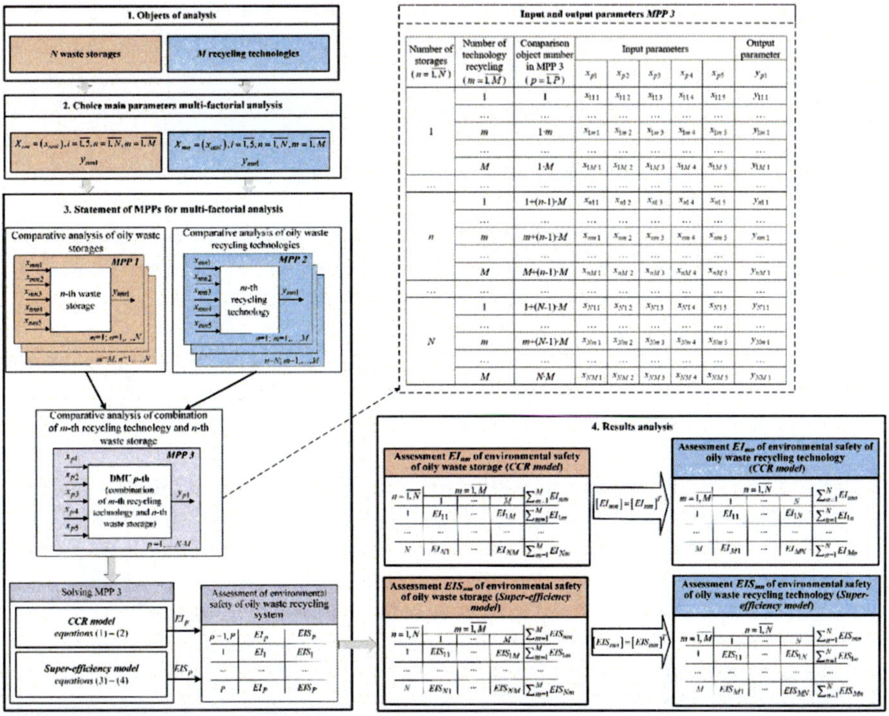

Fig. 1. Algorithm for multi-factorial analysis of environmental safety and optimization of oily waste recycling system

freezing/thawing, and extraction of hydrocarbons using surfactants and their combinations [6–9]. These developed technologies are related to various cleaning mechanisms, resource recovery efficiency, energy consumption, and environmental impact [6]. However, the works [5–9] do not sufficiently assess environmental safety, which, for example, for Russia with its vast territories and the scale of production in the oil and gas complex is crucial for decision-making in the OCW recycling system.

One of the approaches to an adequate assessment of environmental safety can be a multifactorial analysis based on the Data Envelope Analysis (DEA) method [10] and proposed by the authors in various versions [11–13].

The article proposes an algorithm for multi-factorial analysis of environmental safety and optimization of the OCW recycling system based on the DEA method, extending the approach presented in [11] to a broader class of analyzed objects in the waste recycling system of oil and gas complex enterprises (Fig. 1). Solution of optimization problem provides the best combinations of storage facilities and technologies as part of the OCW recycling system, ensuring maximum environmental safety of the recycling process.

2 Algorithm of Multi-factorial Analysis of Environmental Safety and Optimization

For a multifactorial analysis of environmental safety and optimization of the OCW recycling system, it is proposed to apply the DEA method, widely used to evaluate the effectiveness of complex technical systems. The DEA method makes it possible to determine the efficiency of objects in the analyzed OCW recycling system based on a comparative analysis of their input and output parameters by solving the corresponding mathematical programming problem (MPP) [14], which can be solved by standard methods [15]. The method allows obtaining comparative integral relative multifactorial estimates and visualization of the comparison results in a visual and convenient form for analysis [10]. The developed algorithm includes four stages:

(1) **Selection of analysis objects.** The OCW recycling system, including N waste storage facilities and M recycling technologies, is analyzed considering that the issue of quality control of the information available about them has been resolved at the previous stages of the study.
(2) **Choice of the main characteristics affecting the environmental safety of the recycling process.** The physical and chemical composition characterizes the OCW placed in the particular storage. For assessments of the OCW recycling technologies, the special coefficients are used that characterize the formation of various substances in the recycling process, the secondary use of the components of the OCW recycling, and the rationing of the useful recycling product obtained from the components of the OCW recycling [11].
(3) **Statement of the problem for the multifactorial analysis of environmental safety and optimization.** In the well-known approach to the multivariate analysis described in [11], it is proposed to solve the relevant problems MPP 1 and

MPP 2 separately and assess the environmental safety of storage facilities and technologies for OCW recycling. The objects of analysis of MPP 1 and MPP 2 are the n-th storage of OCW, $n = \overline{1, N}$, and the m-th recycling technology of OCW, $m = \overline{1, M}$, , respectively (Fig. 1).

Under step 2 of the algorithm, the following parameters that negatively affect environmental safety are proposed to be considered as components of the input vector, $X_{nm} = (x_{nmi}),\ i = \overline{1,\ 5},\ n = \overline{1,\ N},\ m = \overline{1, M}$, in MPP 1: x_{nm1}- gross emission of greenhouse gases (carbon dioxide and methane); x_{nm2} - gross emission of sulfur dioxide; x_{nm3}- gross emission of soot; x_{nm4} - the mass of the unprocessed part of mineral and mechanical impurities; x_{nm5} - the mass of the resulting contaminated water. As an output parameter $y_{nm1},\ n = \overline{1,\ N},\ m = \overline{1, M}$, positively influencing the assessment of environmental safety, it is proposed to consider a lot of useful recycling products [11].

The input $X_{mn} = (x_{mni}),\ i = \overline{1,\ 5}$ and output y_{mn1} parameters in MPP 2 are determined in the same way, using the same formulas and tables [11] as for MPP 1 according to the environmental safety assessment of the n-th OCW storage.

Thus, in the analyzed waste recycling system, $P = N \times M$ combinations of N storage facilities and M technologies for their processing can be theoretically considered (Fig. 1). The novelty of the approach proposed in this article is MPP 3, formulated to assess environmental safety and optimize the system as a whole, where a combination of m-th recycling technology, $m = \overline{1, M}$ and n-th OCW storage, $n = \overline{1,\ N}$, is considered as the p-th object of comparison, $p = \overline{1,\ P}$. The input parameters $X_p = \left(x_{pi}\right),\ i = \overline{1,\ 5}$, $p = \overline{1,\ P}$, of each p-th object of comparison represent the corresponding components of the vectors $X_{nm} = (x_{nmi}),\ i = \overline{1,\ 5}$ in MPP 1 and $X_{mn} = (x_{mni}),\ i = \overline{1,\ 5}, n = \overline{1,\ N}$, $m = \overline{1, M}$, „ in MPP 2 (Fig. 1).

The output parameters $y_{p1}, p = \overline{1,\ P}$, of each p-th object of comparison represent the corresponding parameters $y_{nm1}, m = \overline{1, M}, n = \overline{1,\ N}$ in MPP 1 and $y_{mn1}, m = \overline{1, M}$, $n = \overline{1,\ N}$, „ in MPP 2 (Fig. 1). The specified input $x_{pi},\ i = \overline{1,\ 5}$ and output $y_{p1}, p = \overline{1,\ P}$ parameters represent the initial data for the formulation and solution of MPP 3.

The MPP 3, formulated on the basic CCR model of the DEA method of [15], allows to obtain estimates $EI_p, p = \overline{1, P}$ of the environmental safety of the p-th combination of the m-th recycling technology and the n-th OCW storage. This problem can be written in the following form:

$$EI_p(X_p, Y_p) = \frac{u_{p1}y_{p1}}{\sum_{i=1}^{5} v_{pi}x_{pi}} \to \max_{(U,V)\in G}; \quad (1)$$

$$\frac{u_{p1}y_{p1}}{\sum_{i=1}^{5} v_{pi}x_{pi}} \le 1,\ \forall \left(p = \overline{1, P};\ u_{p1} \ge 0;\ v_{pi} \ge 0,\ i = \overline{1,\ 5}\right), \quad (2)$$

where $U = u_{p1};\ \ V = (v_{pi}),\ i = \overline{1,\ 5}$ are the vectors of the weighting coefficients for each comparison object in the analyzed system.

The estimates $EI_p, p = \overline{1, P}$ are distributed over a single interval [0,1], while several comparison objects can have the maximum score $EI_p = 1$ in the analyzed system at once. Nevertheless, in many cases it is necessary to determine which of the P objects of comparison is the best combination of storage facilities and technologies as part of the OCW recycling system that ensure maximum environmental safety of the recycling process. Then it is advisable to formulate the MPP using the Super-efficiency model of the DEA method in the following form [16]:

$$EIS_p(X_p, Y_p) = \frac{\overline{u}_{p1} y_{p1}}{\sum_{i=1}^{5} \overline{v}_{pi} x_{pi}} \rightarrow \max_{(\overline{U},\overline{V}) \in \overline{G}}; \tag{3}$$

$$\frac{\overline{u}_{p1} y_{p1}}{\sum_{i=1}^{5} \overline{v}_{pi} x_{pi}} \leq 1, \ \forall \begin{pmatrix} p = \overline{1, P};\ p \neq k;\ \overline{u}_{p1} \geq 0; \\ \overline{v}_{pi} \geq 0,\ i = \overline{1, 5} \end{pmatrix}. \tag{4}$$

Here $\overline{U} = \overline{u}_{p1}$; $\overline{V} = (\overline{v}_{pi})$, $i = \overline{1, 5}$ are the vectors of the unknown weighting coefficients for each p-th object being compared, $p = \overline{1, P}$, excluding the weighting coefficients $\overline{u}_{k1}$ and $\overline{v}_{ki}$, $i = \overline{1, 5}$; $\overline{u}_{k1}$ and $\overline{v}_{ki}$, $i = \overline{1, 5}$, are the weighting coefficients obtained for the output and input parameters of the k-th object of comparison with the value of the efficiency estimates $EI_k = 1$, , respectively, calculated based on the solution of the problem (1)–(2); k is the ordinal number of the comparison object with the value of the assessment in the analyzed OCW recycling system, which are excluded from the solution of the MPP 3 in the form (3)–(4).

The MPP 3 solution in the form (3)–(4) makes it possible to determine the environmental safety assessments EIS_p, $p = \overline{1, P}$ of the p-th combination of the m-th recycling technology and the n-th OCW storage, obtained on the Super-efficiency model of the DEA method. The estimates EIS_p are distributed over the interval $[0,\infty)$. As a result, the object of comparison, representing the optimal combination of processing technology and storage of OCW, having the highest assessments, can be considered the most environmentally friendly in the analyzed OCW recycling system.

(4) **Analysis of the results**

In contrast to the approach proposed in [11], which is associated with the complexity of determining environmental safety assessments for a large number of objects, the approach under consideration (paragraph 3) is practically free from restrictions on the number of analyzed objects. However, in turn, the analysis and interpretation of the resulting estimates for P objects presents considerable complexity, to overcome which the following approach is used.

It is proposed to transform matrix-columns of size $1 \times P$ containing estimates EI_p and EIS_p, $p = \overline{1, P}$, into matrices of size $N \times M$, denoting the estimates EI_{nm} and EIS_{nm}, $n = \overline{1, N}$, $m = \overline{1, M}$, respectively. This transformation is proposed to be carried out by recalculating the indices $p = m + (n - 1) \cdot M$, $n = \overline{1, N}$, $m = \overline{1, M}$, of the initial estimates, in the same way as shown in Fig. 1. Assessments EI_{nm} and EIS_{nm}, $n = \overline{1, N}$, $m = \overline{1, M}$, allow to rank the analyzed storages according to the degree of environmental safety of recycling the waste stored in them.

To obtain estimates EI_{mn}, EIS_{mn}, $m = \overline{1, M}$, $n = \overline{1, N}$, allowing to rank the analyzed technologies according to their environmental safety during the OCW recycling in each n-th storage, it is necessary to convert the corresponding estimates EI_{nm}, EIS_{nm}, $n = \overline{1, N}$, $m = \overline{1, M}$, of the environmental safety of OCW storage according to the following formulas:

$$[EI_{mn}] = [EI_{nm}]^T, m = \overline{1, M}, n = \overline{1, N}; \tag{5}$$

$$[EIS_{mn}] = [EIS_{nm}]^T, m = \overline{1, M}, n = \overline{1, N}. \quad (6)$$

3 Results of the Algorithm Testing

The developed algorithm for multi-factorial analysis of environmental safety and optimization was tested on the example of the OCW recycling system, consisting of $M = 8$ recycling technologies and $N = 15$ OCW storage facilities located in the Samara region. Calculations are made using the specialized software developed in MATLAB [17].

Figures 2 and 3 present the results of the analysis (step 4 of the algorithm) of the assessments EIS_{nm} and EIS_{mn}, $m = \overline{1, 8}$, $n = \overline{1, 15}$, of the environmental safety of the OCW recycling system based on the Super-efficiency model of the DEA method.

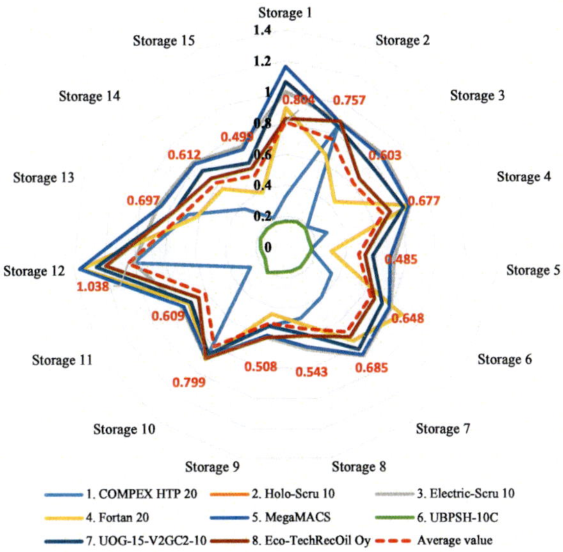

Fig. 2. Comparative assessment of storage facilities according to the degree of environmental safety of waste based on estimates EIS_{nm}.

The results presented in Figs. 2 and 3 show that the best combination is "storage No. 12 - MegaMACS technology" with an assessment $EIS_{12\ 5} = EIS_{5\ 12} = 1.356$, that ensures maximum environmental safety of OCW recycling. It is most unsafe for the environment to recycle waste in storage No. 6 using Liquid Drilling Sludge Processing Plant (UBPSH -10C) technology, since this combination has a minimal assessment of environmental safety in the analyzed OCW recycling system.

Comprehensively evaluating the analysis results in Fig. 2, it is environmentally safe to process OCWs in storage No. 12, using almost all the analyzed technologies for OCW recycling in the system, except for technology No. 6 (UBPSH -10C), with an assessment $EIS_{12\ 6} = 0.166$. Similarly, the analysis of the assessments in Fig. 3 shows

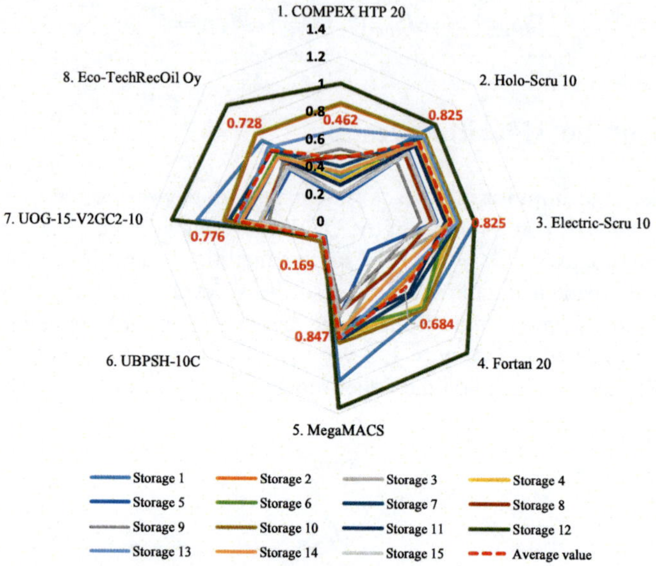

Fig. 3. Comparative assessment of the environmental safety of the use of waste recycling technologies in the storage facilities of the analyzed system based on the assessments EIS_{mn}.

that technology No. 6 (UBPSH-10C) is the most environmentally unsafe in the analyzed system and has minimal estimates for all analyzed OCW storage facilities.

The average values of the environmental safety assessments of the OCW recycling system make it possible to assess environmental risks during the complex processing of OCW by all technologies in every storage (Fig. 2) and by each technology of all waste in storage (Fig. 3). For example, it is environmentally safe to recycle OCW in storage No. 12 using any of the technologies in the system (the average assessment is equal to 1.038).

The analysis of the average values of environmental safety assessments in Fig. 3 is practically significant since the prospects for waste processing from the technology side are traditionally assessed. So, it is environmentally safe to recycle all OCW in storage using MegaMACS technology in the analyzed system (the average score is 0.847).

Acknowledgements. The research and reported study were funded by the Russian Foundation for Basic Research (projects No. 20-08-00240 and 20-08-00353).

References

1. Ossai, I., Ahmed, A., Hassan, A., et al.: Remediation of soil and water contaminated with petroleum hydrocarbon: a review. Environ. Technol. Innov. **17**, 100526. https://doi.org/10.1016/j.eti.2019.100526 (2020)
2. Bulatov, V., Igenbaeva, N., Nanishvili, O.: Oil and gas complex waste as a technological indicator of the geoecological state of Russian regions. Bull. Sci. Pract. **7**(8), 46–55 (2021). https://doi.org/10.33619/2414-2948/69/05

3. Krapivsky, E.: Oil sludge: destruction, recycling, decontamination. Vologda: Infra Eng., p. 430 (2021)
4. Waste management - problems and solutions: an analytical review.: Federal State Budgetary Scientific Institution Research Institute – Republican research scientific and consulting center of expertise. https://www.extech.ru/files/anr_2015/anr_5.pdf. Accessed 21 Sept. 2021 (2015)
5. Hu, G., Feng, H., He, P., et al.: Comparative life-cycle assessment of traditional and emerging oily sludge treatment approaches. J. Clean Prod., **251**, 119594. https://doi.org/10.1016/j.jclepro.2019.119594 (2020)
6. Hu, G., Li, J., Zeng, G.: Recent development in the treatment of oily sludge from petroleum industry: A review. J. Hazard. Mater. **261**, 470–490 (2013). https://doi.org/10.1016/j.jhazmat.2013.07.069
7. Da Silva, L., Alves, F., de França, F.: A review of the technological solutions for the treatment of oily sludges from petroleum refineries. Waste Manag. Res. **30**(10), 1016–1030 (2012). https://doi.org/10.1177/0734242X12448517
8. Hu, G., Li, J., Huang, S., Li, Y.: Oil recovery from petroleum sludge through ultrasonic assisted solvent extraction. J. Environ. Sci. Health **51**(11), 921–929 (2016). https://doi.org/10.1080/10934529.2016.1191308
9. Hochberg, S., Tansel, B., Laha, S.: Materials and energy recovery from oily sludges removed from crude oil storage tanks (tank bottoms): A review of technologies. J. Environ. Manag. **305**, 114428. doi:https://doi.org/10.1016/j.jenvman.2022.114428 (2022)
10. Charnes, A., Cooper, W., Rhodes, E.: Measuring the efficiency of decision-making units. Eur. J. Oper. Res. **6**(2), 429–444 (1978)
11. Derevyanov, M., Pleshivtseva, Y., Afinogentov, A., et al.: Optimization of the disposal system of oily waste according to the criterion of environmental safety. Atlantis Highlights Mater. Sci. Technol. **1**, 33–38. https://doi.org/10.2991/isees-19.2019.8 (2019)
12. Derevyanov, M., Pleshivtseva, Y., Afinogentov, A.: Multi-criteria analysis of resource saving and energy efficiency in the technogenic waste processing system of the oil and gas industry. Vestn. Samara State Tech. Univ. Tech. Sci. Ser. **29**(1), 19–35. https://doi.org/10.14498/tech.2021.1.2(2021)
13. Derevyanov, M., Pleshivtseva, Y., Afinogentov, A., et al.: Application of data envelopment analysis for multi-criteria evaluation of system for technogenic waste recycling in oil refining industry. In: IOP Conference Series: Materials Science and Engineering, vol. 1089, p. 012023. https://doi.org/10.1088/1757-899X/1089/1/012023 (2020)
14. Cooper, W., Seiford, L., Tone, K.: Data Envelopment Analysis: A Comprehensive Text with Models, Applications, References and DEA-Solver Software, 2nd edn., p. 490. Springer, Boston, MA (2008)
15. Todd, M.: The many facets of linear programming. Math. Program. **91**(3), 417–436 (2002). https://doi.org/10.1007/s101070100261
16. Chen, Y., Du, J.: Super-efficiency in data envelopment analysis. In: Data Envelopment Analysis. International Series in Operations Research & Management Science, vol. 221, Springer, Boston, MA, p. 34. https://doi.org/10.1007/978-1-4899-7553-9_14(2015)
17. Pleshivtseva, Y., Derevyanov, M., Mandra, A., et al.: Optimization of a complex-structured system for the complex processing of technogenic waste from oil and gas enterprises. Certificate for registration of a computer program No. 2021664227 Russian Federation, applicant and copyright holder Samara State Tech. University, claim. 24 Aug. 2021, registered 1 Sept. 2021 (2021)

Research and Analysis of Computing Cluster Configuration Management Systems

A. I. Martyshkin[1](✉) and R. A. Biktashev[2]

[1] Penza State Technological University (PSTU), Baydukov Proyezd/Gagarin Street, 1a/11, Penza 440039, Russia
alexey314@yandex.ru
[2] Penza State University (PSU), Krasnaya Street, 40, Penza 440026, Russia

Abstract. The article discusses the issues of research and analysis of computer cluster configuration management systems. The question of the evolution of cluster systems, the features of its architecture and possible configurations are touched upon. The main purpose of the article is to study and analyze the configuration management systems of a computing cluster. The main tasks considered in the article are: research, classification of parallel computing systems and determination of the place of cluster systems among them; study of the features of cluster architecture and its configurations.

The theory of configuration management systems is considered, the four most currently used configuration management systems are investigated: Puppet, Chef, SaltStack and Ansible. Based on the configuration management systems comparison methodology, the study and comparison of these systems were carried out. Such indicators as properties related to the input specification, with the implementation of configurations and with the management of specifications are evaluated. Recommendations on the choice of a configuration management system for a computing cluster are proposed. The Ansible system was chosen as an example, and based on it, an overview of basic information about the operation of configuration management systems was conducted.

The result of the study is an analysis of the configuration management systems of the computing cluster. Recommendations on the choice of a configuration management system for a computing cluster are proposed. In conclusion, the article presents the main results and conclusions obtained during the study.

Keywords: Architecture · Cluster · Cluster configuration · Cluster system · Memory · Parallel system · Processor · Reconfigurable system

1 Introduction

The active use and implementation of cluster technologies is an integral part of the modern world. The development of computing clusters (CC) allows you to expand the scope of high-performance (including reconfigurable systems) computing. In almost all spheres of life (energy, mechanical engineering, etc.), many technological and engineering processes are carried out based on CC [1].

A. A. Radionov and V. R. Gasiyarov (Eds.): RusAutoCon 2022, LNEE 986, pp. 194–205, 2023.
https://doi.org/10.1007/978-3-031-22311-2_20

Cluster systems (ClS) differ in architecture, type and number of computing nodes (CN), the composition of the operating systems (OS) used, the type of RAM, the class of tasks performed, but they all require centralized management. With the increase in the number of cluster nodes, the problem of configuration management (CM) in such systems becomes more and more urgent. In such cases, CM-systems come to the administrator's aid to ensure that the functional and physical characteristics of the cluster nodes meet the necessary requirements.

The purpose of the work is to study and analyze CC CM-systems. To achieve the above goal, the study considers a number of tasks:

- Research, classification of parallel computing systems (CSs) and determination of the place of ClSs among them.
- Study of the features of cluster architecture and its configurations.

2 Theoretical Analysis

There are two major approaches to reducing power consumption. On the hardware side: energy-efficient power supplies, the use of low-power element base, etc. The second approach is from the software side [2, 3].

Clusters belong to parallel computing systems (CS). In order to show the place of ClS among many other types of parallel architectures, it is necessary to give its classification. Throughout the history of computers, many CS have been created with various ways of organizing serial and parallel computing [2]. If we take into account such characteristics as the purpose of the system, the type and number of processors, the topology of their relationship, the way memory is organized, methods of managing system elements, methods of performing operations, etc., then the number of different types of structures will be voluminous. As a result, the task of systematization of CS has become the most urgent.

Michael Flynn's classification of CSs [4, 5] is not completely adequate, because it generates ambiguity in grouping existing architectures. For a detailed understanding of the features of a particular architecture, it is necessary to introduce a more detailed classification. Today there are several such classifications [6].

From a hardware point of view, all the variety of existing architectures can be divided into symmetric multiprocessor (SMP), massively parallel (MPP) and hybrid architecture (NUMA) [7].

2.1 Evolution of ClS

In 1962, J. K.R. Licklider from MIT published a series of papers presenting his concept of "Galactic Networks", which considered the possibility of a global interconnected set of computers with access to data from any node. In the same year, independently of MIT, the RAND Corporation developed the concept of a packet switching network. In 1965, Lawrence J. Roberts and Thomas Merrill paired the TX-2 computer located in Massachusetts with the Q-32 computer located in California, using a telephone line for this, demonstrating for the first time a large-scale computer network [8]. Today, clusters of the Beowulf type are widespread, widely used, therefore, they are developing and improving.

2.2 Features of Cluster Architecture

All computing ClSs are formed according to the principle: hierarchy of calculators, memory, switching environments. However, each level of the computing hierarchy can have its own method of accessing memory, a method of processing and transmitting data. In addition, different schemes for constructing calculators, memory, and switching environments can be applied in each aircraft. That is why ClSs are called multiarchitectural. There are three types of hierarchy: the hierarchy of calculators (core, processor, node, CS); hierarchy of memory (RAM, multi-level Cache memory, external media); hierarchy of communication environments (buses, controllers, switches, networks). Each level of the hierarchy is characterized by its performance, bandwidth of communication channels and memory.

Most of the CNs of modern ClSs are built on the basis of AMD Opteron and Xeon platforms. For Xeon-based ClSs all calculators are connected to a single memory controller, as a result of which conflicts may arise when simultaneously requesting access. In the cluster nodes based on AMD Opteron, each ClS processor has its own memory controller.

When choosing a communication environment, the speed of information transmission over communication channels (bandwidth) and the initial delay time when sending messages (latency) are usually taken into account. Performance, cost, and scalability are also important.

All CNs of a homogeneous cluster run under the same OS, have the same configuration and, accordingly, the same performance. The CNs of a heterogeneous cluster run under different operating systems, have different configurations and, accordingly, have different performance. Due to the heterogeneity of a heterogeneous cluster, problems arise with the distribution of tasks to be performed on different nodes, there is a need to create different executable files, as well as balancing the load of nodes.

There are clusters with shared memory (SM) and distributed memory (DM) [9, 10]. By OS type, clusters are classified into Windows and UNIX clusters.

Depending on the tasks solved on the cluster, the following types of ClS can be distinguished:

- High-speed clusters.
- High availability/high reliability clusters.
- GRID distributed computing systems.
- Load balancing clusters.
- Clusters of servers organized programmatically.

The performance of clusters is determined by the performance of nodes and the speed of communication between them. Classic areas where clusters are used: science, industry, image and video processing.

When any of the CN fails, tasks are automatically redistributed between the serviceable nodes of the high-availability ClS. Figure 1 shows a diagram of a high availability cluster with no failure points [11].

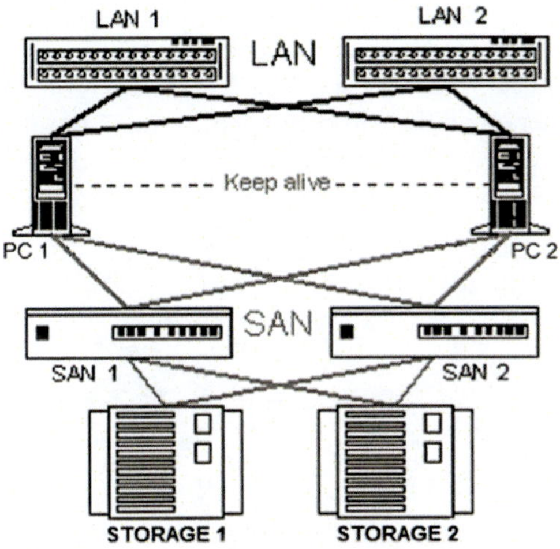

Fig. 1. High availability cluster.

There are also mixed systems that combine the features of HP-clusters and HA-clusters [12] (Fig. 2). But these systems lose out in speed and possible downtime to conventional systems.

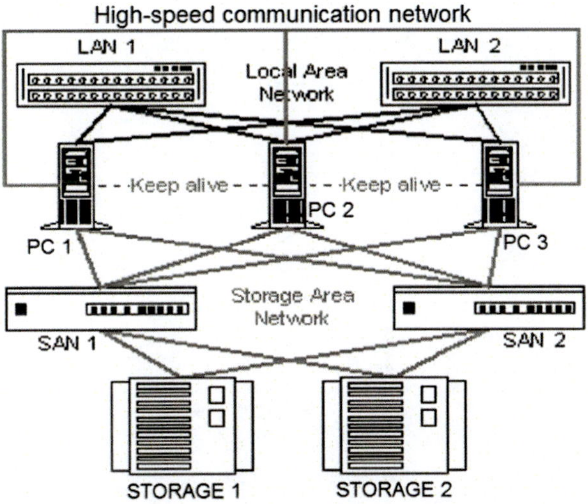

Fig. 2. High-speed fault-tolerant cluster.

GRID computing networks are designed to perform distributed computing and represent a complex of various types of calculators [13].

Network load balancing clusters (NLB-cluster) [14] are designed to regulate performance and improve reliability (Fig. 3). Requests in NLB-clusters are distributed through one or more input nodes and redirected to other nodes for processing.

Server clusters are designed to process identical requests on a group of servers that are logically combined at the level of software protocols (Fig. 4). A group of servers is characterized by greater reliability and performance compared to a single server.

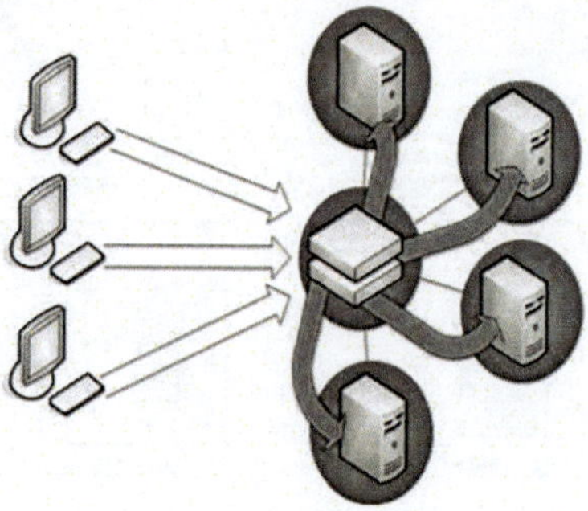

Fig. 3. Load balancing cluster.

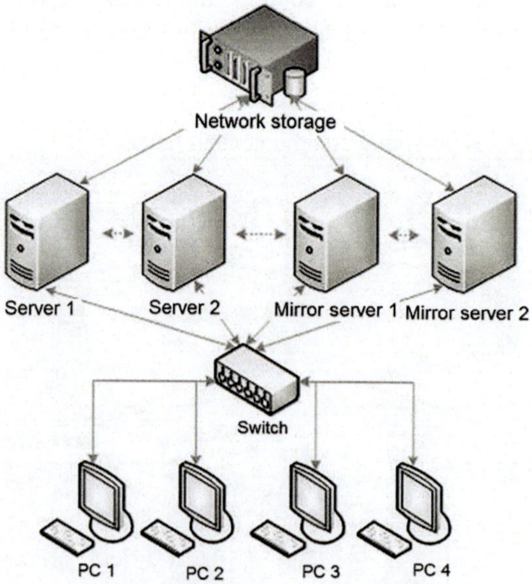

Fig. 4. Server cluster.

3 CC CM-Systems

The "active-active" configuration [15] is applicable in cases when the priority of tasks is equally high, the launch of low-priority tasks is unacceptable. The user has at his disposal a highly accessible system with server duplication. If one of the servers fails, applications are redirected to the rest, which adversely affects the overall performance of the system [16].

CM is a system development process that ensures the consistency of the functional and physical characteristics of a product with the requirements imposed on it. CM is carried out throughout the entire life cycle of the CIS, it contributes to the orderly management of system information and system changes in order to improve performance, reliability, prolong life, reduce cost, reduce risk and liability, as well as to correct defects.

CM-systems allow you to automate the configuration of servers programmatically. The use of CM-systems allows you to reduce the server setup time and perform it correctly (if control scripts are written without errors).

Currently, CM-systems are striving to implement the Infrastructure as a Code (IaC) paradigm [17]. Manual operations on nodes are minimized or reduced to zero.

3.1 Research of CC CM-Systems

During the period of development of CM-systems, a large number of proprietary CM-tools have been created from leading computer companies such as IBM, Hewlett-Packard, Microsoft and others. Puppet, Chef, SaltStack and Ansible CM-systems are popular today [18].

3.2 Puppet CM-System

Puppet is a platform written in Ruby and represents a set of tools for managing the configuration of computer systems. This CM-tool can be used for configuration management on Unix, Linux and Microsoft Windows platforms. In addition, it can be used to control both homogeneous and heterogeneous CC.

3.3 Chef CM-System

Chef is a CM-systems, the client part of which is written in Ruby, and the server part is written in Erlang. Chef is used to simplify the configuration and maintenance of multiple servers, can be integrated into cloud platforms. It is used to manage both Linux clusters and Windows clusters. It can be used to control both homogeneous and heterogeneous CC [19]. The Chef diagram is shown in Fig. 5.

3.4 SaltStack CM-System

The SaltStack diagram is shown in Fig. 6.

This CM-system is built on two basic principles: remote execution and configuration management. In a remote execution system, Salt uses Python to perform complex tasks with single-function calls. The CM-system in SaltStack is based on the foundation of remote execution to create repeatable configurations for minions [20].

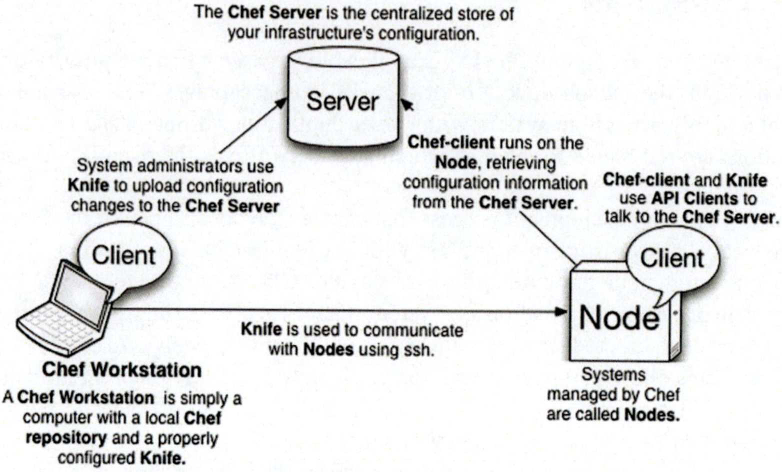

Fig. 5. Chef diagram.

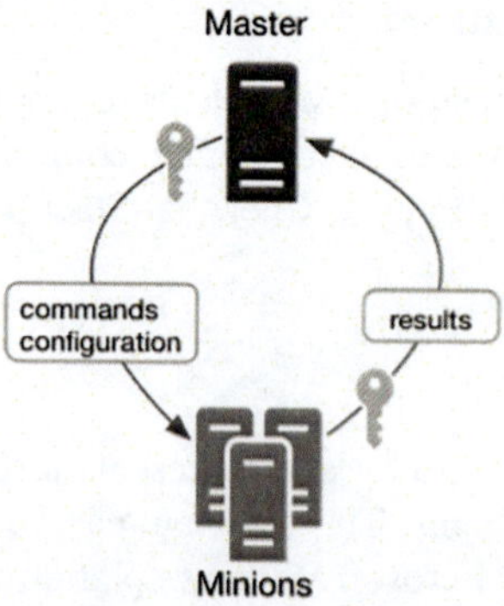

Fig. 6. SaltStack diagram.

4 Methodology for Comparing Cluster CM-Systems

The study is based on a methodology for comparing CM-systems [21], proposed and published in 2010. The structure of the comparison of configuration and configuration tools consists of the following categories:

- Properties related to the input specification.
- Properties related to the implementation of configurations.
- Properties related to specification management.
- Properties related to product support.

One of the main reasons for the need to automate cluster configuration is repetitive tasks that are economically inefficient and increase the likelihood of errors. The solution to these problems depends on the level of modularity in the system. Three modularity mechanisms described in [22] are defined.

To minimize the likelihood of a conflict, the access rights to the specification are delimited. Many version control systems can provide access control at the authorization level. In most programming languages and system configurations, the relationship between a file name and its contents is very limited or does not even exist. There is an option to create your own access control lists (ACLs) [23].

5 Experimental Part

Ansible system was chosen as an example to study the principles of deployment and operation of CM-systems. This decision is justified by the fact that this CM-tool is the easiest to deploy and use, it does not create a database after installation, accordingly, it does not launch daemons running constantly – the utility is launched only when necessary. Ansible is built on an agentless principle, cluster nodes are managed via SSH protocol. When installing Ansible, the minimum system requirements for the Control Machine are given in [24]. To install the necessary Python libraries on the Control Machine, the command is executed (Fig. 7).

```
$ sudo apt-get install python-yaml python-jinja2 python-paramiko python-crypto
```

Fig. 7. Installing Python libraries.

We will install Ansible from the source: clone the Git repository [25], with further loading of the environments with the commands shown in Fig. 8.

```
$ git clone git://github.com/ansible/ansible.git
$ cd ./ansible
$ source ./hacking/env-setup
```

Fig. 8. Installing Ansible by cloning a Git repository.

Executing single commands on remote nodes is just a basic Ansible tool. Playbooks are a set of Ansible commands designed for configuring, deploying and managing CC. It is possible to describe it in the form of policies intended for remote PCs, or in the form of a set of steps [26]. As an example, consider installing Apache on selected cluster nodes. Figure 9 shows an inventory file called hosts, where all hosts are Debian OS-based systems.

```
[web]
host1.primer.org
```

Fig. 9. Inventory file.

```
- hosts: web
tasks:
- name: Installs apache web server
apt: pkg=apache2 state=installed update_cache=true
```

Fig. 10. Playbook for installing Apache on web class machines.

Figure 10 shows a Playbook for installing Apache on web class machines. The launch of the Playbook is shown in Fig. 11.

```
ansible-playbook -i step-04/hosts -l host1.example.org step-04/apache.yml
```

Fig. 11. Launching the Playbook.

In Fig. 11, the notation is adopted: step-04/hosts – inventory file; -l restricts the launch by host1.example.org; apache.yml – Playbook.

6 Results

An approximate response to the launch of the Playbook is shown in Fig. 12.

```
PLAY [web] *********************
GATHERING FACTS *********************
ok: [host1.primer.org]
TASK: [Installs apache web server] *********************
changed: [host1.primer.org]
PLAY RECAP *********************
host1.example.org          : ok=2   changed=1   unreachable=0   failed=0
```

Fig. 12. The result of launching the Playbook.

Ansible has a method to account for the possibility of an error – stoping all processing in case something went wrong. Therefore, before restarting the server, it is necessary to check the validity of the configuration. Adding a virtual host must be started before deleting the default virtual host [26]. The launch of the Playbook and the result of its execution are shown in Fig. 13.

Apache2ctl returns error code 1. Ansible understands this and stops working, but the virtual host is still added, therefore, with each next restart attempt, Apache will "swear" at the configuration and shut down. There is a problem of returning to a working state after catching an error. The result of executing the modified Playbook configuration file is shown in Fig. 14.

```
$ ansible-playbook -i step-06/hosts -l host1.primer.org:step-06/apache.yml
PLAY [web] *********************
GATHERING FACTS *********************
ok: [host1.primer.org]                    primer
TASK: [Installs apache web server] *********************
ok: [host1.primer.org]
TASK: [Push future default virtual host configuration]
*********************
changed: [host1.primer.org]
TASK: [Activates our virtualhost] *********************
changed: [host1.primer.org]
TASK: [Check that our config is valid] *********************
failed: [host1.primer.org]=> {"changed": true, "cmd": ["apache2ctl",
"configtest"], "delta": "0:00:00.045046", "end": "2022-02-16
19:09:32.002063", "rc": 1, "start": "2022-02-16 19:09:31.957017"}
stderr: Syntax error on line 2 of /etc/apache2/sites-enabled/awesome-app:
Invalid command 'RocumentDoot', perhaps misspelled or defined by a
module not included in the server configuration
stdout: Action 'configtest' failed.
The Apache error log may have more information.
FATAL: all hosts have already failed -- aborting
PLAY RECAP *********************
host1.primer.org: ok=4   changed=2   unreachable=0   failed=1
```

Fig. 13. The result of launching the Playbook, taking into account the occurrence of an error.

```
$ ansible -i step-07/hosts -m service -a
'name-apache2 state-restarted' host1.primer.org
host1.primer.org | success >> {
    "changed": true,
    "name": "apache2",
    "state":"started"
}
```

Fig. 14. The result of running the Playbook with a return in case of errors.

7 Recommendations for Choosing a CC CM-System

The choice of a CC CN-system is quite complicated and depends on many factors. When selecting a CM-system, it is necessary to take into account the main characteristics of the cluster: the type of CPU, the type of RAM used, the type and composition of the OS used, the number of cluster nodes, the cluster architecture, the tasks performed on the cluster and the configuration to be achieved. Depending on the goals of implementing the CM-tool in an existing or new cluster architecture, it is necessary to take into account the properties associated with the input specification, implementation and management of specifications, as well as the degree of product support. Based on the conducted research, it is possible to choose a CM-system suitable for certain requirements.

Puppet and Chef are more developer-oriented, while SaltStack and Ansible are more suitable for the needs of system administrators. Chef is well suited for managing clusters on Debian/Ubuntu, but requires more debugging under CentOS/RHEL. The most suitable platform for working with Puppet is CentOS/RHEL, and for Salt – all versions of Linux. Ansible works well over SSH with almost all OS except SmartOS. Puppet is the safest for heterogeneous environments, while Ansible and Salt are more suitable for large homogeneous infrastructures. In addition, SaltStack is suitable for those who care about the speed of messaging and implementation of changes on cluster nodes, and Puppet and Chef – a declarative approach and description of states [22].

8 Conclusions

The result of the study is an analysis of CC configuration management systems. The question of the evolution of the ClS, the features of its architecture and possible configurations are studied.

The theory of CM-systems is studied, the four most currently used CM-systems are considered: Puppet, Chef, SaltStack and Ansible. Based on the CM-systems comparison methodology, the study and comparison of these systems were carried out. Such indicators as properties related to the input specification, with the implementation of configurations and with the management of specifications are evaluated. Recommendations on the choice of a CC CM-system are proposed. The Ansible system was chosen as an example, and based on it, an overview of basic information about the operation of CM-systems was conducted.

References

1. Voevodin, V.V., Zhumatiy, S.A.: Computing and Cluster Systems. Publishing House of Moscow State University, Moscow, p. 150 (2007)
2. Kalyaev, I.A., Levin, I.I., Semernikov, E.A., Shmoilov, V.I.: Reconfigurable Multiconveyor Computing Structures. Southern Scientific Center of the Russian Academy of Sciences Publishing House, Rostov-on-Don, p. 393 (2008)
3. Pashchenko, V.D., Martyshkin, I.A., Trokoz, D.A., Pashchenko, T.Y., Babich, M.Y., Butaev, M.M.: Parallel decomposition of control algorithms for computational processes based on the use of nondeterministic automaton logic. Nexo revista cientifica **34**(1), 1–12 (2021)
4. Flynn, M.J.: Very high speed computers. Proc. IEEE **54**, 1901–1909 (1966)
5. Flynn, M.J.: Some computer organizations and their effectiveness. IEEE Trans. Comput. **21**(9), 948–960 (1972)
6. Padua, D. (ed.).: Encyclopedia of Parallel Computing. Springer. https://doi.org/10.1007/978-0-387-09766-4 (2011)
7. Introduction to SMP, NUMA, and MPP architecture—Programmer Sought.: https://programmersought.com/article/3458509649/. Accessed 24 Jan. 2022
8. Wikipedia. History of computer clusters.: https://en.wikipedia.org/wiki/History_of_computer_clusters. Accessed 14 Feb. 2022

9. Sukhikh, A.V., Vasyaeva, N.S.: Research classification of cluster systems. Cybern. Program. **2**, 20–27 (2016)
10. Bormotov, A., Gorokhova, A.: Modeling the clustering of dispersed systems using dynamic models. Lect. Notes Civ. Eng. **180**, 61–70 (2022)
11. High-availability cluster – Wikipedia.: https://en.wikipedia.org/wiki/High-availability_cluster. Accessed 28 Jan. 2022
12. The division into High Availability and High Performance systems is Linux Cluster systems. https://vuzlit.ru/1032391/razdelenie_high_avalibility_high_performance_sistemy. Accessed 10 Feb. 2022
13. Foster, I., Kesselman, C., Tuecke, S.: The anatomy of the grid: enabling scalable virtual organizations. Int. J. Supercomput. Appl. **15**(3) (2001). http://www.globus.org/research/papers/anatomy.pdf. Accessed 4 Feb. 2022
14. Network Load Balancing | Microsoft Docs. https://docs.microsoft.com/en-us/windows-server/networking/technologies/network-load-balancing. Accessed 14 Feb. 2022
15. Hannifin, D., Alpern, N.J., Joey Alpern Microsoft. : Windows Server 2008 R2. The Administrator's Essential Reference, pp. 399–459 (2010). https://doi.org/10.1016/B978-1-59749-578-3.00009-8
16. Virtual Computer Museum. Clusters. Borzenko A. http://www.computer-museum.ru/technlgy/klaster.htm. Accessed 14 Feb. 2022
17. What is Infrastructure as Code? - Azure DevOps | Microsoft Docs. https://docs.microsoft.com/en-us/devops/deliver/what-is-infrastructure-as-code. Accessed 12 Feb. 2022
18. Chef vs Puppet vs Ansible vs Saltstack: What's Better for 2022 – Merehead. https://merehead.com/ru/blog/cchef-vs-puppet-vs-ansible-vs-saltstack-better-in-2022/. Accessed 6 Feb. 2022
19. Joshi, M.: Mastering Chef. Packt Publishing, p. 315 (2015)
20. Myers, C.: Learning SaltStack. Packt Publishing Ltd, p. 174 (2015)
21. Delaet, T., Joosen, W., Vanbrabant, B.: A survey of system configuration tools. In: Proceedings of the 23rd Large Installations Systems Administration (LISA) Conference, San Jose, CA, USA, pp. 1–14 (2010)
22. Zhitinsky, S., Chistyakov, A.: Configuration management systems. Syst. Adm. **6**(139), 35–38 (2014)
23. Access Control Lists – Win32 apps | Microsoft Docs. https://docs.microsoft.com/en-us/windows/win32/secauthz/access-control-lists?redirectedfrom=MSDN. Accessed 5 Feb. 2022
24. Prakhin, D.: Ansible project. Syst. Adm. **3**(148), 4–9 (2015)
25. WikipediA. Computer cluster. https://en.wikipedia.org/wiki/Computer_cluster. Accessed 25 Jan. 2022
26. WikipediA. Configuration management. https://en.wikipedia.org/wiki/Configuration_management. Accessed 16 Jan. 2022

The Possibility of Using the Any Logic Simulation System in the Study of Models of Embedded Reconfigurable Computing Systems Based on Queuing Theory

A. I. Martyshkin(✉)

Penza State Technological University (PSTU), Baydukov Proyezd/Gagarin Street, 1a/11, Penza 440039, Russia
alexey314@yandex.ru

Abstract. The article discusses the possibility of using the AnyLogic simulation system in the study of models of embedded reconfigurable computing systems based on queuing theory. The most effective method of analyzing queuing systems (QS) is simulation modeling. The article substantiates the choice of the development environment, provides an overview of AnyLogic tools for the development of QS models. AnyLogic implements a special library that allows you to create QS models from ready-made blocks. Within the framework of the article, simulation modeling is used to create models in order to study the characteristics and assess the quality of service of reconfigurable computing systems, which allow analyzing their basic properties and identifying patterns inherent in the processes of priority processing of incoming tasks in various service disciplines, load and structural parameters. Studies of the influence of priorities on task processing, analysis of a system with different laws of distribution of intervals between tasks, analysis of systems with heterogeneous load have been carried out. With the help of AnyLogic, models have been developed that allow to study QS with various combinations of input parameters. As a result of the simulation, the characteristics are calculated: the load of the reconfigurable computing system (RCS) processor, the probability of losing tasks, the average and maximum waiting time and stay of tasks in the RCS and the average, current and maximum queue length, confidence intervals for loading, waiting time, stay time. The results obtained during the study will be useful to developers of embedded systems (for example, RCS).

Keywords: Algorithm · Analytical modeling · Priority · Queuing theory · Reconfigurable computing system · Service discipline · Simulation modeling

1 Introduction

1.1 Theoretical Analysis

The purpose of this work is to study queuing systems with a different number of servicing devices and different queue management algorithms using the AnyLogic program. To achieve this goal, it is necessary to solve a number of tasks:

A. A. Radionov and V. R. Gasiyarov (Eds.): RusAutoCon 2022, LNEE 986, pp. 206–215, 2023.
https://doi.org/10.1007/978-3-031-22311-2_21

- Consider the known means of simulation modeling.
- Study the AnyLogic simulation environment and gain basic skills in it.
- Develop RCS models based on QS in AnyLogic environment.

The need for this study is due to the opportunity to study the real objects represented by the QS, and to obtain information about the properties, modeling results and to analyze the obtained characteristics, to identify the dependence on the input parameters. Also, the developed models will allow to investigate the mechanisms of priority maintenance for various input parameters of reconfigurable computing systems.

Modeling is the process of replacing one source object with another, called a model, for conducting experiments, the purpose of which is information about the system, based on the study of its properties. The objects of modeling are systems and processes occurring in them. Two conditions are required to obtain practical benefits from modeling: 1. the model corresponds to the original object and has the necessary set of properties for the study of this object; 2. the model reduces the complexity of the process of studying the original object in comparison with the study of the object without the use of modeling.

When analyzing the processes occurring in technical systems, the following mathematical modeling methods are most common [1–14]:

- analytical;
- numerical;
- simulation (statistical);
- combined.

QS is a dynamic system containing one or more devices serving incoming tasks. Tasks enter the system and wait for their service in the queue (storage). The simplest QS can be represented as follows (Fig. 1).

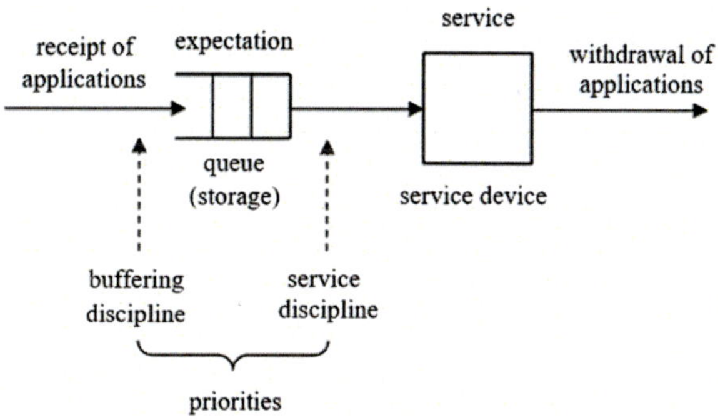

Fig. 1. The simplest QS.

The QS study involves determining the system parameters necessary and sufficient to obtain results that determine the purpose of the study. To describe the QS, the following parameters must be set: structural; load; functional [15]. The buffering discipline (BD) is a rule that determines the order in which incoming tasks of different classes are queued for service. Service discipline (SD) is a rule that determines the choice of a task from the service queue.

So, a priority QS is a system that provides preferred processing of some tasks over others in accordance with some queue management strategy.

1.2 Classification of QS

Known and used for research QS can be classified by the number of servicing devices, as well as by the number of classes of tasks [13, 15]. Since the parametrization of the QS involves the use of probabilistic distributions of random variables, we present specific distributions used in queuing models: uniform distribution, exponential distribution, Erlang distribution, triangular distribution [16–18].

1.3 Problem Statement

Models for conducting experiments are considered, the purpose of which is to study the characteristics of a reconfigurable computing system. It is necessary to provide a visualization of the process of functioning of the systems and the results of modeling. The values of structural functional and load parameters can be passed to the models as input data.

1.4 Description of AnyLogic

Today, software products are known and widely used to study various characteristics of computing systems, including MatLab, Mathcad, AnyLogic, GPSS World, etc. They are all good in their own way. In this paper, we will pay special attention to the use of the AnyLogic [19] software product in the study of queuing systems and possible models that can be used to represent reconfigurable computing systems. AnyLogic is a Russian software that has been known on the market for a long time and is used in various fields. The difference from the well-known modeling systems is as follows: AnyLogic is a comprehensive tool with which you can solve a wide range of modeling tasks. AnyLogic allows you to model a wide class of systems using implemented blocks [20–22]. Each implemented block has a specific purpose and functionality. Each block in the process diagram has a set of parameters that can be specified depending on the logic of the simulated system. With the help of the built-in animation, which allows you to visualize the process of processing tasks, you can explore the current state of the model: the number of generated tasks, the length of the queue, the number of lost tasks, the number of processed tasks (Fig. 2).

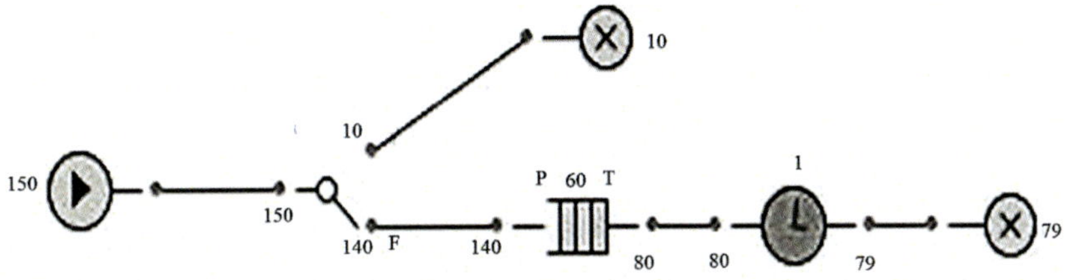

Fig. 2. Standard animation of blocks.

Figure 2 shows that 150 tasks have been generated by the source, 79 of them have been processed by the processor (currently one task is being serviced), 60 tasks are in the queue waiting for the processor to be released in the queue and 10 tasks are lost when the queue overflows.

AnyLogic allows you to create stochastic models, for which it provides 29 different random variable distribution laws [21].

2 Experimental Part

2.1 Research of QS

The relevance and at the same time the complexity of solving queuing theory problems from different fields of technology [6] made them the object of attention of programmers relatively early. The purpose of most programs is to study basic QS models and only a few allow calculations for priority maintenance of tasks.

As an example, consider the application. the Stset program, developed and implemented in the educational process of the Programming Department of the Penza State Technological University [23]. The Stset program is designed for analytical calculation of the necessary characteristics, including single- and multi-channel QSs with heterogeneous load and priorities. Thus, the presented program implements analytical methods for calculating QS. The use of GPSS World and AnyLogic systems is supposed to be used for the study of QS by the method of simulation modeling in [20]. This requires some time spent on mastering specific software products that have a lot of tools [24].

The models synthesized with the help of AnyLogic [25] are designed to conduct simulation experiments in order to study the characteristics and assess the quality of service indicators of a reconfigurable system with queues. The models visualize the process of RCS functioning and simulation results.

2.2 The Structure of the Model

When building a model, the following principles are laid down in the basis of its structure and functioning: hierarchy, exchange of tasks. The structure of the model is a hierarchical structure of objects.

During the simulation, it is possible to perform a number of actions to change the course of the experiment process through the control panel by selecting the appropriate option [21]. The visualization and logic of the priority data processing process is shown in Fig. 3.

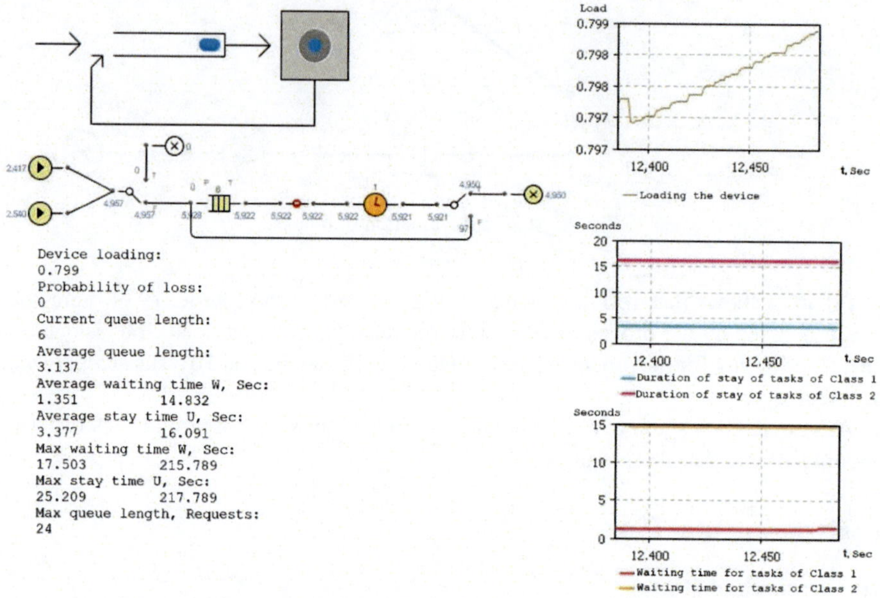

Fig. 3. Visualization and logic of the priority data processing process.

Setting the values of the input parameters begins with selecting the configuration of the model. Next, you need to enter the values of the average intensity of the receipt of tasks and the average duration of maintenance, as well as the storage capacity. The next step is to choose the distribution law of the intensity of receipt of tasks of each class. After the end of setting the values of all the necessary parameters, you need to click the "Run" button to proceed to the stage of the experiment.

The second stage of working with the model is the execution of an experiment, during which the QS operation is simulated in time based on the values of the parameters set at the first stage of working with the model.

Statistics on device loading, average waiting time for maintenance tasks and average time spent by tasks of different classes in the system are displayed on time graphs (Fig. 4).

According to the graphs shown in Fig. 4, it is possible to determine the values of the stay time and waiting time for different classes and the load values of the devices. According to the schedule, it is easier to visually understand for which class of tasks the waiting time or stay is longer/shorter. The experiment can continue until the user stops it, or indefinitely.

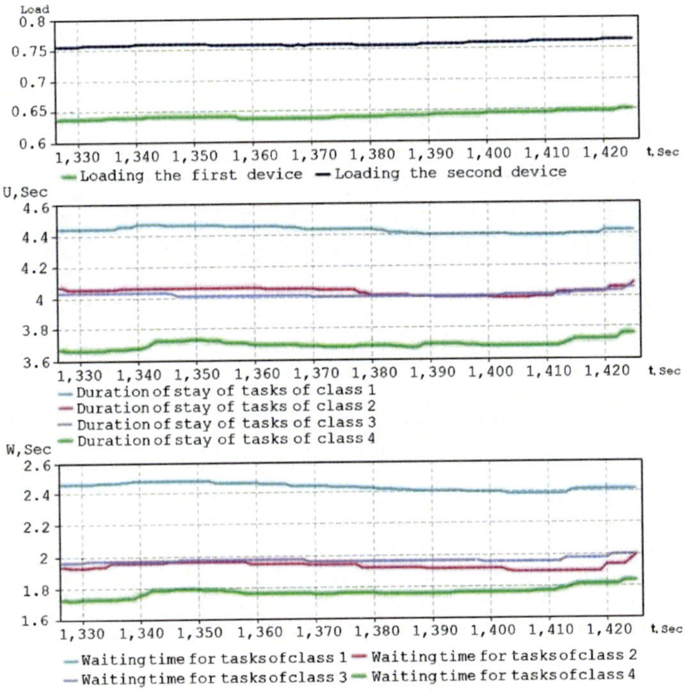

Fig. 4. Graphs showing the collected statistics.

3 Result

3.1 Investigation of the Impact of Priorities on Task Processing

When studying processes in systems with priority maintenance, two–channel QS with a heterogeneous flow of tasks having a non-priority service discipline or a service discipline with relative priorities, where the priority of tasks decreases with increasing task class (task class 1 is the maximum priority), are compared.

To study priority algorithms, QS is used to calculate such characteristics of the system operation as processor load p, delay in the system U, waiting time for service W, queue length l and the probability of losing tasks π, confidence intervals for loading, waiting time, stay time.

The initial data for the study are:

λi (i = 1, 2,…, n) – the intensity of the receipt of tasks of the i-th class;
bi (i = 1, 2,…, n) – the average duration of servicing tasks of the i-th class.

The graphs show the waiting time and the time spent in different service disciplines (Figs. 5 and 6).

From the graph (Fig. 5) it can be seen that with relative priorities, the waiting time increases with a decrease in the priority of the task (for tasks of grades 1–3, the

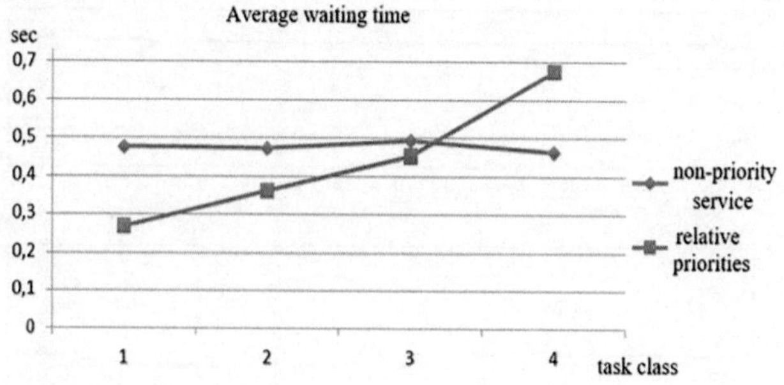

Fig. 5. Schedule of waiting time for maintenance of tasks.

average waiting time decreased with the introduction of priorities). With non-priority maintenance, the waiting time, taking into account errors, is considered to be equal.

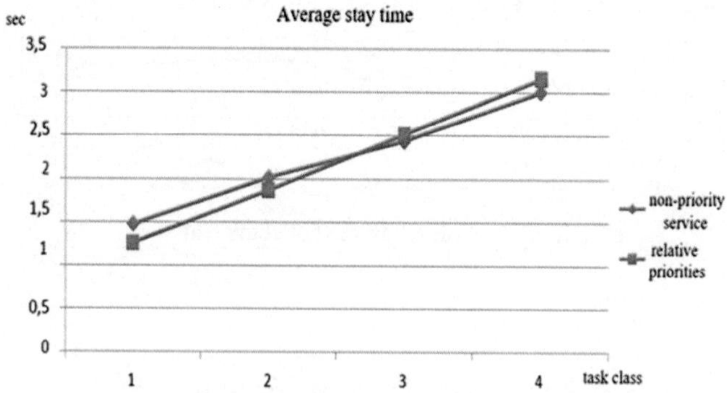

Fig. 6. Time schedule of tasks in the system.

It can be seen from the graph (Fig. 6) that with relative priorities, the stay time increases with decreasing priority of the task. With non-priority service, the stay time increases, because with an increase in the class, the duration of service increases.

Experiments were conducted with priority and non-priority service, analyzing the results of which it can be concluded that when the priority is lowered (service discipline with relative priority), the waiting time increases compared to the non-priority discipline. The highest priority has the minimum waiting time, which is significantly different from the waiting time of this class of task at non-priority service. This can be used when transferring data that requires high service speeds.

3.2 Analysis of Systems with Heterogeneous Load

In the study of a heterogeneous flow of tasks, a two-channel QS with a homogeneous flow of tasks will be analyzed (a QS with 4 classes of tasks and a non-priority service discipline was simplified to a homogeneous flow for the experiment) and a two-channel QS with a heterogeneous flow of tasks.

The initial data of the study are:

- n is the number of classes of various types (here: 3 for non-stationary task flow and 1 for stationary);
- λi (i = 1, 2,..., n) – the intensity of the receipt of tasks of the i-th class;
- bi (i = 1, 2,..., n) – the average duration of servicing tasks of the i-th class.

The graphs show the waiting time and the residence time of homogeneous and heterogeneous task flows (Figs. 7 and 8).

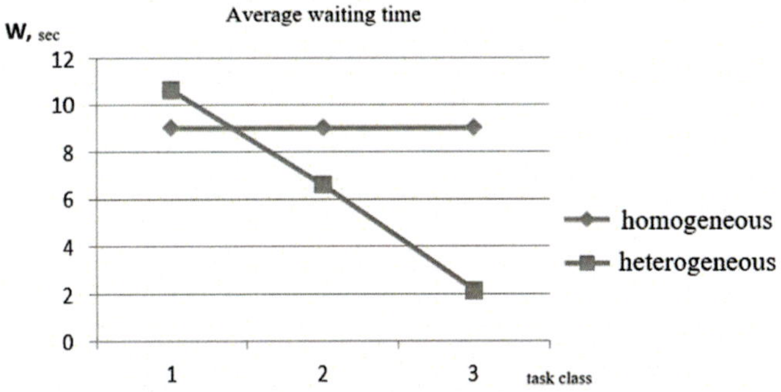

Fig. 7. Schedule of waiting time for tasks.

Figure 7 shows that with a homogeneous flow of tasks, the waiting time is equal, and with a heterogeneous one, due to a decrease in the intensity of the receipt of tasks, the waiting time by class 2 and 3 decreases.

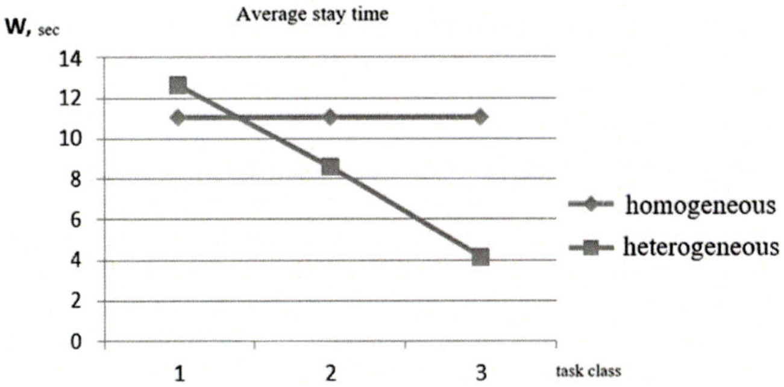

Fig. 8. Time schedule of tasks.

Figure 8 shows that with a homogeneous flow of tasks, the stay time is equal, and with a heterogeneous one, due to a decrease in the intensity of the receipt of tasks, the stay time of class 2 and 3 decreases. Analysis of the results of the experiments showed that when the flow of tasks is homogeneous, the waiting time and the stay time have equal values, with a heterogeneous flow of tasks, the waiting time and the stay time decrease with a decrease in the intensity of the receipt of tasks.

4 Conclusions

The study of modeling methods has shown that for the study of real computing systems (for example, RCS), presented in the form of queuing systems, it is most effective to use simulation modeling. In the AnyLogic environment, simulation models of RCS based on QS from ready-made blocks have been created. The study of the influence of priorities on task processing, the analysis of systems with a heterogeneous load and the analysis of a system with different laws of distribution of intervals between tasks. The results obtained will be useful for developers of embedded systems (for example, RCS).

Acknowledgements. The research was conducted at the expense of a grant from the Russian Science Foundation № 21-71-00110, https://rscf.ru/project/21-71-00110/.

References

1. Kleinrock, L.: Queueing Systems: Volume I—Theory. Wiley Interscience, New York, p. 417 (1975)
2. Kleinrock, L.: Queueing Systems: Volume II—Computer Applications. Wiley Interscience, New York, p. 576 (1976)
3. Taha, H.A.: Chapter 17. Queuing Systems. Introduction to operations Research = Operations Research: An Introduction. 7th ed. “Williams”, Moscow, pp. 629–697 (2007)
4. Martyshkin, A.I.: Software package for determining characteristics of task managers of reconfigurable computer systems using priority queueing networks. Revista Inclusiones **6**(SI), 463–474 (2019)
5. Ushenina, I.V.: Multi-channel FXLMS filter as an array of processing blocks and a method to maintain its performance when increasing the number of channels. Cybern. Autom. Control. Theory Methods Intell. Algorithms **986**, 315–324 (2019)
6. Ushenina, I.V.: Multi-input multiplexers and selectors as basic shifter blocks in arithmetic devices operating in floating point: comparative analysis. In: Journal of Physics: Conference Series, vol. 1399, no. 3 (2019)
7. Gnedenko, B.V., Kovalenko, I.N.: Introduction to the theory of queuing. Nauka Publishing House, Main Editorial Office of Physical and mathematical literature, Moscow, p. 432 (1966)
8. Matveev, V.F., Ushakov, V.G.: Queuing Systems, p. 240. MSU, Moscow (1984)
9. Melikov, A.Z., Ponomarenko, L.A., Ismailov, B.G.: Analysis of queuing system with dynamic priorities. J. Autom. Inf. Sci. **47**(9), 23–33 (2015)
10. Marigodov, V.K.: Queuing system with absolute priorities as a model of diversity radio-reception. Radioelectron. Commun. Syst. **52**(12), 638–642 (2009)
11. Jayswal, N.: Queues with Priorities. Mir, Moscow (1973)
12. Tabelskaya, A.S., Guseva, E.N.: Modeling of queuing systems. Alley Sci. **2**(16), 508–512 (2017)
13. Aliyev, T.I.: Fundamentals of Modeling Discrete Systems. St. Petersburg State University ITMO, St. Petersburg (2009)
14. Bocharov, P.P., Pechinkin, A.V.: Theory of Queuing, p. 530. RUDN, Moscow (1995)

15. Mayorov, S.A., Novikov, G.I., Aliyev, T.I., et al.: Edited Fundamentals of the Theory of Computing Systems, p. 408. Higher School, Moscow (1978)
16. Wentzel, E.S.: Operations research. "Soviet radio", Moscow, p. 552 (1972)
17. Wentzel, E.S.: Probability Theory: Textbook for Students. Universities. 10th ed. Publishing Center "Academy", Moscow, p. 576 (2005)
18. Prokhorov, Yu.V. (ed.): Mathematical Encyclopedic Dictionary, p. 847. Soviet Encyclopedia, Moscow (1988)
19. Boev, V.D.: Conceptual Design System with AnyLogic, and GPSS World, 2nd edn., p. 542. INTUIT, Moscow (2016)
20. Kupriyashkin, A.G.: Fundamentals of Systems Modeling. Norilsk industrir. in-T. Research Institute, Norilsk, p. 135 (2015)
21. Boev, V.D.: Conceptual Design of Systems in Anylogic 7 and GPSS World: Textbook, 2nd edn., p. 555. INTUIT, Moscow (2016)
22. Paley, A.G., Pollak, G.A.: Simulation Modeling. Development of Simulation Models by Means of Powersim and AnyLogic: Textbook. Lan, St. Petersburg, p. 208 (2019)
23. Biktashev, R.A., Martyshkin, A.I., Vostokov, N.G.: The program complex for determining the characteristics of the task managers multiprocessor systems using priority stochastic queueing networks. Fundam. Res. **10**(1), 13–20 (2013)
24. Karpenko, O.S.: A software package for the study of queuing networks with group service requirements. Inf. Technol. Educ. **4**, 108–111 (2021)
25. Official website of AnyLogic. www.anylogic.ru. Accessed 16 Jan. 2022

Agent-Based Model Assessing the Quality of the Cyber-Physical System

A. N. Varnavsky(✉)

HSE University, 20, Myasnitskaya St, Moscow 101000, Russia
avarnavsky@hse.ru

Abstract. Currently, cyber-physical systems are actively developing. Therefore, research aimed at creating efficient systems is relevant. One of the tasks of such systems is the redistribution of functions between the human and the system. In the case of intensive activity of the employee and the accumulation of fatigue, it is necessary to support his activity with a robot. Despite the existence of some models, this issue has not been fully studied especially concerning cyber-physical systems. The work aims to develop an agent-based model for assessing the quality of the cyber-physical system, consisting of a worker performing intensive work and a robot to support the worker. The model involves the parameters of the worker and the parameters of the labor activity of the worker and robot. As a result of the work, a model was built, and modeling was performed. The simulation result is a series of curves and their analysis.

Keywords: Cyber-physical system · Fatigue · Robot · Worker · Performance

1 Introduction

Currently, the concepts of creating cyber-physical production systems are being actively developed. Such systems integrate industrial, technical, biological objects, networks, smart devices, sensor systems, and analytical platforms with cloud computing. Many processes in cyber-physical systems are performed autonomously and decentralized. It can be noted that cyber-physical production systems are developing within the framework of the Industry 4.0 strategy [1, 2]. An example is the German platform SmartFactory KL (https://smartfactory.de/). Therefore, the development of new approaches and models for the design, construction, and operation of cyber-physical systems is relevant.

As biological objects of cyber-physical production systems are operators and production workers, without which some types of technological processes are impossible. The task of technical and smart devices is to help a human in fulfilling the production tasks facing him, i.e. distribution of functions and joint execution of tasks, operations, and processes [2].

When considering the task of distributing functions between humans and machines, it is necessary to consider the capabilities of man and machine. With the development of technology, many human functions have been taken over by machines, but it is currently not possible to completely replace a person for many tasks.

A. A. Radionov and V. R. Gasiyarov (Eds.): RusAutoCon 2022, LNEE 986, pp. 216–225, 2023.
https://doi.org/10.1007/978-3-031-22311-2_22

It is possible to compare the capabilities of a person and a machine in terms of such characteristics as receiving, storing, reproducing, processing information, making decisions, and exercising influences on external objects. A machine is much superior to a human in the speed of perception and processing. Also, a machine, unlike a man, can process information very accurately and quickly. The advantage of the machine is the ability to quickly perform control actions on external objects. However, in terms of the ability to perceive the most diverse information, a person surpasses a machine. A man, in comparison with a machine, has more diverse possibilities for processing information, and the ability to analyze and synthesize it in many ways. Also, human actions are more diverse and plastic; he can easily and quickly reorganize depending on the conditions, and he can find new ways to perform the assigned tasks and operations. Although artificial intelligence systems are capable of learning and gaining experience, a person is more effective in assessing the performance of the system based on a wide variety of factors, taking into account the experience gained in managing the system.

There are also subjective factors to consider. A machine, unlike a man, is not able to forget the program and be distracted. A person, under the influence of various subjective factors, can reduce the efficiency and quality of the activity. This can especially happen under the influence of fatigue and reduced performance in tasks and operations with a high load and intensity.

Thus, the task of improving the quality of interaction between machines and humans is important, in particular, in the tasks of changing the working capacity and increasing human fatigue. Despite the existence of some models, this issue has not been fully studied especially concerning cyber-physical systems.

The work aims to develop an agent-based model for assessing the quality of the cyber-physical system, consisting of a worker performing intensive work and a robot to support the worker.

2 Literature Review

To assess the interaction between human and machine, and efficiency of production systems, modeling and various models are used [3, 4].

Some issues of human interaction with technical systems are raised within the framework of the so-called human factor, which is important to consider in the design and operation of systems. The work [5] presents an approach to integrating the human factor into the production system model. The necessity of using the Human-System Integration concept for integrating the individual abilities of workers into a cyber-physical system and improving the production process is noted. SysML models were used.

A significant place among the causes of the human factor is given to employee fatigue, which leads to a slowdown in activity and a decrease in productivity. As a result, workers may spend more time than expected on their assigned tasks, and the average time to complete an operation increases. The work [6] reviews and summarizes the literature on work-related fatigue and its impact on performance.

To combat fatigue and reduce health-related risk factors use work planning, time standards, scheduling work and rest, and shift schedules. The solution to this problem is described in some works using various approaches [6–13]. For example, in [12], a simulation model is proposed for modeling a production system, in which, in addition to the arrival of parts, breakdowns of machines, and conveyors, learning and fatigue processes are integrated. Three configurations of the production system are considered. Modeling has shown that the cellular configuration with autonomous interruptions provides operators with sufficient allowances and a productive system of work compared to other configurations. The article [13] provides a mathematical model for planning work in reactive systems, compatible with the working paradigm of Industry 4.0. A greedy randomized adaptive search procedure is used to compute the model and schedule jobs based on worker fatigue and reliability considering production goals. The described method can improve the quality of scheduling, allowing the employee to recover from fatigue due to breaks and reduce the error probability.

The issues of workplace planning are raised in some works [14–17]. For example, [16] presents an agent-based simulation tool for performance and working conditions assessment in production systems using workers' margins of manoeuver. In [17], the problem of dynamically assigning multidisciplinary workers to perform tasks, taking into account fatigue, is solved to minimize the average time to complete a task in the system and avoid overloads. The proposed methodology is based on a heuristic algorithm using multicriteria analysis.

In addition to fatigue, environmental conditions and stress are important. In [18], the influence of stress and environmental conditions on the productivity of workers was studied. An agent-based model was developed to analyze workplace stress management policies. The results of the analysis of four different stress management policies were shown, measuring the impact of different levels of adaptation on stress and productivity. The model showed that workload conditions have a significant impact on stress and productivity.

Some recent works [19–22] are devoted to the distribution of roles between a man and a robot (cobot) in production. Thus, in [22], a digital worker assistance system based on a business process model and the BPMN notation language is presented, which allows non-professionals to create processes for adaptive distribution of tasks between human workers and cobots in production. The results of the study showed an excellent assessment of the usability of the system, low indicators of task load, acceptable average task duration, and error rate.

Based on the review of the literature, the following conclusions can be drawn.

- When planning production work, it is important to consider the processes of worker fatigue, recovery, and productivity improvement. Many recent papers propose approaches based on the use of agent-based models.
- Work planning is mainly implemented based on minimizing the average time to complete tasks or achieving the required level of performance. At the same time, few works use the maximization of quality or the "price/quality" ratio of the system as a criterion for planning work [23, 24].
- Given the development of the concept of cyber-physical production systems, some recent works are devoted to the distribution of roles and responsibilities between a man and a machine (robot). However, many questions remain open.

Thus, the actual task is to build a model for planning and distributing work between a worker and a machine in a cyber-physical production system, considering worker fatigue and recovery based on maximizing the quality (price/quality ratio) of the system. In this work, an agent-based model will be developed, since this type of model is effective for modeling systems in which elements are objects with individual behavior.

3 Agent-Based Model of Worker and Robot Activity in a Cyber-Physical System

As elements of a cybernetic system, we consider two active participants: a human worker and a robot. The task of a person is to perform some intense activity, and the task of a robot is to help a person. If a robot had all the qualities that are superior to a human, then it would be used instead of a human. Therefore, initially, a person has superiority over a robot. However, with the passage of time and the development of fatigue, the qualities of a person decrease. At such times, the robot must take over the execution of tasks in the system instead of a human. After the restoration of working capacity, the person performs the work, and the robot is idle. Given the discreteness of many operations, it is worth waiting for the completion of operations by a human or a robot, and only then change the active participant.

Thus, we get two agents: a human agent and a robot agent. The human agent has the following states: performance of activities, rest, and control of the end of the robot operation. The robot agent has the following states: performance of activities, inactivity, control of the state of a person, and control of the completion of a human operator.

To describe the dynamics of the worker's performance, we will use the exponential law, which is used in models in [24, 25] and other authors (for example, [10, 11, 13]). Let us introduce the following notation:

$B(t)$ is the instantaneous level of the worker's performance at time t;

λ is the intensity of the worker's fatigue during activity;

μ is the intensity of the worker's recovery in the absence of activity;

B_0 is worker's performance at the beginning of work, i.e. at time $t = 0{:}B(t = 0) = B_0$. We will assume that upon restoration, the performance will not be able to exceed the B_0.

If at time t_0 the worker's activity begins, then the dynamics of the level of worker's working capacity can be described by the expression:

$$B(t) = B(t_0) \cdot e^{-\lambda \cdot (t - t_0)}.$$

If at t_0 time the robot starts to work and the worker is idle and resting, then the dynamics of the level of worker's performance can be described by the expression:

$$B(t) = (B_0 - B(t_0)) \cdot \left(1 - e^{-\mu \cdot (t - t_0)}\right) + B(t_0).$$

We will consider the activity of the system with a duration of $T = 9$ h (540 min), of which 1 h from 240 to 300 min falls on the worker's lunch break.

The work activity of a worker is generally characterized by some parameters: information processing time, the accuracy of work, the number of correctly solved tasks, the number of tasks solved in a timely manner, and the probability of erroneous actions. Since the worker's performance determines the efficiency of his activity, it is advisable to introduce a set of coefficients that will link the level of the worker's performance with specific parameters of his work to build a model of the quality of the system's functioning during worker activity [24]:

k_t^w is the coefficient of information processing time;
k_{ac}^w is the coefficient of work accuracy;
k_{cor}^w is the correctness coefficient of problem-solving;
k_{tl}^w is the coefficient of timeliness of problem-solving;
k_{er}^w is the coefficient of the error of actions.

The meaning of these coefficients lies in the extent to which the level of $B(t)$ performance affects the values of specific activity parameters. By this meaning of the coefficients, the following expressions can be introduced into the model to assess the change in the parameters of a worker's activity when the level of working capacity changes over time [24]:

- information processing time $t_t^w(t)$:

$$t_t^w(t) = {k_t^w}/{B(t)};$$

- accuracy of work at the moments of activity $\Delta Y^w(t)$:

$$\Delta Y^w(t) = k_{ac}^w(B_0 - B(t));$$

- the number of correctly solved problems m_{cor}^w:

$$m_{cor}^w(t) = k_{cor}^w \cdot B(t);$$

- the number of timely solved tasks n_{tl}^w:

$$n_{tl}^w(t) = k_{tl}^w \cdot B(t);$$

- probability of erroneous actions P_{er}^w:

$$P_{er}^w(t) = {k_{er}^w}/{B(t)}.$$

For the robot, we can also enter the t_t^r, ΔY^r, m_{cor}^r, m_{tl}^r, P_{er}^r indicators but they will not depend on time.

The integral assessment of the quality of the system can be determined through a set of particular indicators of the qualities of the selected parameters included in the overall assessment with a certain weight characterizing their importance in the cyber-physical system. To do this, it is necessary to convert them to a dimensionless and normalized form concerning some norm. Let's set the $a_t, a_Y, a_{cor}, a_{tl}, a_P$ weight coefficients of importance and reference values of the $t_{min}, Y_{min}, m_{max}, n_{max}, P_{min}$ selected parameters.

Then the integral assessment of the quality of the system in the case of worker activity will be defined as:

$$Q^w(t) = a_t \cdot {t_{min}}/{t_t^w(t)} + a_Y \cdot {Y_{min}}/{\Delta Y^w(t)} \\ + a_{cor} \cdot {m_{cor}^w(t)}/{m_{max}} + a_{cor} \cdot {n_{tl}^w(t)}/{n_{max}} + a_P \cdot {P_{min}}/{P_{er}^w(t)}.$$

In the case of robot activity, the integral assessment of the quality of the system will be determined as:

$$Q^r(t) = a_t \cdot {t_{min}}/{t_t^r} + a_Y \cdot {Y_{min}}/{\Delta Y^r} + a_{cor} \cdot {m_{cor}^r}/{m_{max}} + \\ a_{cor} \cdot {n_{tl}^r}/{n_{max}} + a_P \cdot {P_{min}}/{P_{er}^r}.$$

Then the integral assessment of the quality of the system activity for the entire time of operation can be found by integrating $Q^w(t)$ and $Q^r(t)$ on the intervals of work of the T_1 worker and the T_2 robot:

$$Q(T) = \int_0^{T_1} Q^w(t)dt + \int_0^{T_2} Q^r(t)dt,$$

where $T_1 + T_2 = T$.

4 Modeling the Results of the Cyber-Physical System

The $Q(T)$ value will depend on the features of the functioning of the system (i.e., the $a_t, a_Y, a_{cor}, a_{tl}, a_P$ weight coefficients of importance and reference values of the $t_{min}, Y_{min}, m_{max}, n_{max}, P_{min}$ parameters), the quality of the robot and worker (i.e., t_t^r, ΔY^r, m_{cor}^r, n_{tl}^r, P_{er}^r, t_t^w, ΔY^w, m_{cor}^w, n_{tl}^w, P_{er}^w values) and $B(t)$ worker's performance dynamics. The performance dynamics are affected not only by the λ, μ values but also by the duration of the discrete operation (T_o); the level of performance at which the robot takes over the execution of a new operation ($d1$); the level of operability, above which the execution of a new operation is again taken over by the worker ($d2$). For modeling, we will consider the integral value of the quality of the robot and the worker. Then it is possible to introduce and use the Q^r index, which is equal to the ratio of the quality of the robot's activity to the quality of the worker's activity at the initial moment.

Let's simulate the operation of the system. Let us consider the case when the intensity of recovery is much greater than the intensity of fatigue, for example, at $\lambda = 0.05 \text{ min}^{-1}$, $\mu = 0.25 \text{ min}^{-1}$.

The following conclusions can be drawn from the simulation results.

The $d2$ value in combination with the values of other indicators affects the performance of the system. However, in most cases, $d2$ greater than 0.9 reduces the overall performance of the system.

The higher the *kr* value, the better the performance of the system. However, the high value of this coefficient is associated with high costs for the implementation of the robot.

For *Tw* $\geq$ 15 min, the productivity is practically independent of $d1$ but decreases with increasing *Tw*.

As the value of $d1$ increases to some value, the performance increases and then remains constant. Decreasing the Tw value increases performance. Increasing kr also increases system performance.

For example, on Fig. 1 we present a series of curves of the influence of Tw on the $Q(T)$ performance for various kr values.

Consider the case when the intensity of recovery is equal to the intensity of fatigue, for example, at $\lambda = 0.05\ \text{min}^{-1}$, $\mu = 0.05\ \text{min}^{-1}$. Consider the effect of Tw on performance. In this case, unlike the previous one, there was an increase in performance but at low kr values, a decrease in performance is observed. The nature of the curves has changed. If in the first case there is an increase in productivity with an increase in Tw, and starting from 15 min there is a decrease, then in the second case, at small Tw values, a decrease in productivity is observed, which continues up to $Tw = 10$ min, then it does not increase significantly, and starting from 15 min slight decline (Fig. 2).

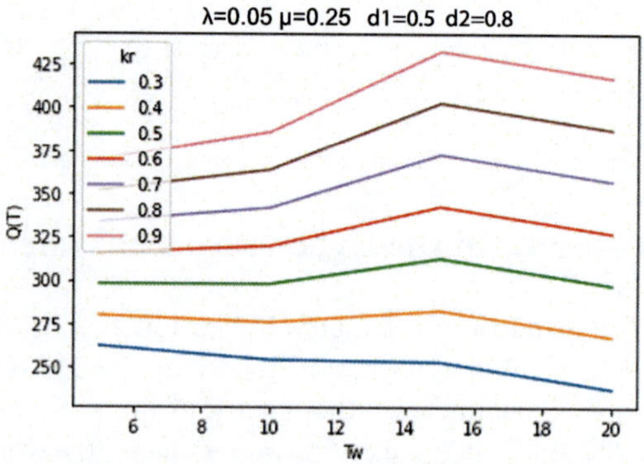

Fig. 1. A series of curves of the influence of Tw on the characteristics $Q(T)$ for different values of kr with $\lambda < \mu$

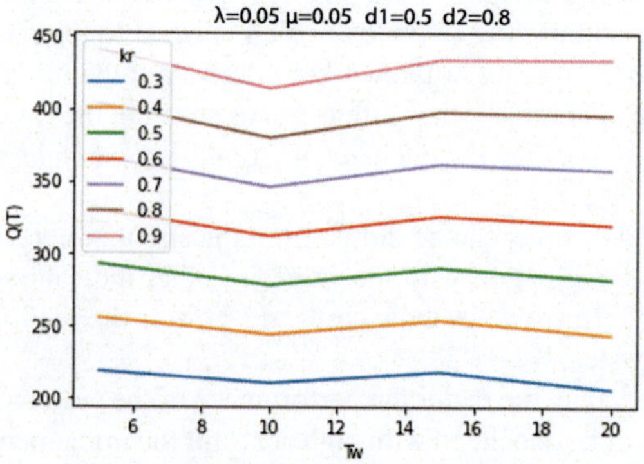

Fig. 2. A series of curves of the influence of Tw on the characteristics $Q(T)$ for different values of kr with $\lambda = \mu$

Let us consider the case when the intensity of recovery is much less than the intensity of fatigue, for example, at $\lambda = 0.25\ \text{min}^{-1}$, $\mu = 0.05\ \text{min}^{-1}$. Consider the effect of *Tw* on performance. In this case, with the growth of *Tw*, a decrease in productivity is observed. At large *kr* values of the performance increased compared to the previous case, however, at medium and low *kr* values the performance of the system greatly decreased (Fig. 3).

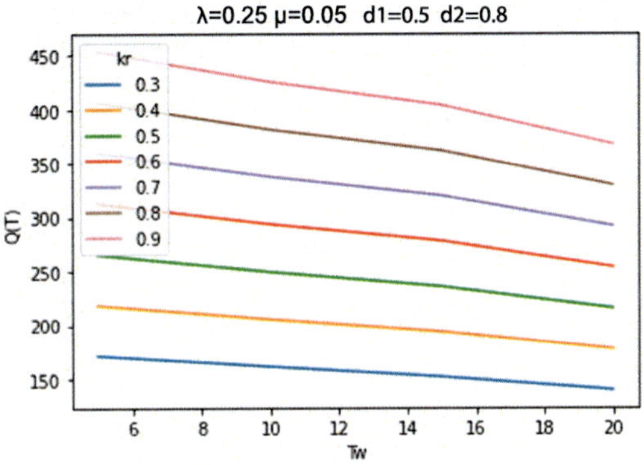

Fig. 3. A series of curves of the influence of *Tw* on the characteristics $Q(T)$ for different values of *kr* with $\lambda > \mu$

5 Conclusion

In this work, a series of curves are obtained as simulation results. Several parameters are considered, each of which affects performance. Both the influence on the performance of the system parameters of the worker and the parameters of the labor activity of the worker and the quality of the robot's activity are considered.

The simulation results show that, depending on the properties of the worker and the quality of the robot, it is possible to set the optimal values of $d1$, $d2$, and *Tw*, at which the performance of the system will be maximum.

References

1. Monostori, L.: Cyber-physical production systems: roots, expectations and R&D challenges. Procedia CIRP **17**, 9–13 (2014). https://doi.org/10.1016/J.PROCIR.2014.03.115
2. Longo, F., Nicoletti, L., Padovano, A.: Smart operators in industry 4.0: a human-centered approach to enhance operators' capabilities and competencies within the new smart factory context. Comput. Ind. Eng. **113**, 144–159 (2017). https://doi.org/10.1016/J.CIE.2017.09.016
3. Perez, J., de Looze, M.P., Bosch, T., Neumann, W.P.: Discrete event simulation as an ergonomic tool to predict workload exposures during systems design. Int. J. Ind. Ergon. **44**, 298–306 (2014). https://doi.org/10.1016/J.ERGON.2013.04.007
4. Negahban, A., Smith, J.S.: Simulation for manufacturing system design and operation: literature review and analysis. J. Manuf. Syst. **33**, 241–261 (2014). https://doi.org/10.1016/J.JMSY.2013.12.007

5. Gräßler, I., Wiechel, D., Roesmann, D.: Integrating human factors in the model based development of cyber-physical production systems. In: Procedia CIRP, pp. 518–523. Elsevier B.V., (2021)
6. Xu, S., Hall, N.G.: Fatigue, personnel scheduling and operations: review and research opportunities. Eur. J. Oper. Res. **295**, 807–822 (2021)
7. Konz, S.: Work/rest: Part II—the scientific basis (knowledge base) for the guide. Int. J. Ind. Ergon. **22**, 73–99 (1998). https://doi.org/10.1016/S0169-8141(97)00069-3
8. de la Riva, J., Garcia, A.I., Reyes, R.M., Woocay, A.: Methodology to determine time allowance by work sampling using heart rate. Procedia Manuf. **3**, 6490–6497 (2015). https://doi.org/10.1016/J.PROMFG.2015.07.934
9. El ahrache, K., Imbeau, D.: Comparison of rest allowance models for static muscular work. Int. J. Ind. Ergon. **39**, 73–80 (2009). https://doi.org/10.1016/J.ERGON.2008.10.012
10. Givi, Z.S., Jaber, M.Y., Neumann, W.P.: Modelling worker reliability with learning and fatigue. Appl. Math. Model. **39**, 5186–5199 (2015). https://doi.org/10.1016/J.APM.2015.03.038
11. Jaber, M.Y., Neumann, W.P.: Modelling worker fatigue and recovery in dual-resource constrained systems. Comput. Ind. Eng. **59**, 75–84 (2010). https://doi.org/10.1016/J.CIE.2010.03.001
12. el Mouayni, I., Etienne, A., Lux, A., et al.: A simulation-based approach for time allowances assessment during production system design with consideration of worker's fatigue, learning and reliability. Comput. Ind. Eng. **139**, 105650 (2020). https://doi.org/10.1016/j.cie.2019.01.024
13. El Mouayni, I., Demesure, G., El Haouzi, H.B., et al.: Jobs scheduling within Industry 4.0 with consideration of worker's fatigue and reliability using Greedy Randomized Adaptive Search Procedure. In: IFAC-PapersOnLine, pp. 85–90. Elsevier B.V., (2019)
14. Vinod, V., Sridharan, R.: Dynamic job-shop scheduling with sequence-dependent setup times: simulation modeling and analysis. Int. J. Adv. Manuf. Technol. **36**, 355–372 (2008). https://doi.org/10.1007/S00170-006-0836-4
15. Carnahan, B.J., Redfern, M.S., Norman, B.: Designing safe job rotation schedules using optimization and heuristic search. Ergonomics **43**, 543–560 (2010). https://doi.org/10.1080/001401300184404
16. El Mouayni, I., Etienne, A., Siadat, A., et al.: AEN-PRO: Agent-based simulation tool for performance and working conditions assessment in production systems using workers' margins of manoeuver, pp. 14236–14241. Elsevier B.V. (2017)
17. Ferjani, A., Ammar, A., Pierreval, H., Elkosantini, S.: A simulation-optimization based heuristic for the online assignment of multi-skilled workers subjected to fatigue in manufacturing systems. Comput. Ind. Eng. **112**, 663–674 (2017). https://doi.org/10.1016/j.cie.2017.02.008
18. Muñoz, S., Iglesias, C.A.: An agent based simulation system for analyzing stress regulation policies at the workplace. J. Comput. Sci. **51**, 101326 (2021). https://doi.org/10.1016/j.jocs.2021.101326
19. Ranz, F., Hummel, V., Sihn, W.: Capability-based task allocation in human-robot collaboration. Procedia Manuf. **9**, 182–189 (2017). https://doi.org/10.1016/J.PROMFG.2017.04.011
20. Schmidbauer, C., Schlund, S., Ionescu, T.B., Hader, B.: Adaptive task sharing in human-robot interaction in assembly. In: IEEE International Conference on Industrial Engineering and Engineering Management 2020, pp. 546–550 (2020). https://doi.org/10.1109/IEEM45057.2020.9309971
21. Lamon, E., de Franco, A., Peternel, L., Ajoudani, A.: A capability-aware role allocation approach to industrial assembly tasks. IEEE Robot. Autom. Lett. **4**, 3378–3385 (2019). https://doi.org/10.1109/LRA.2019.2926963

22. Schmidbauer, C., Hader, B., Schlund, S.: Evaluation of a digital worker assistance system to enable adaptive task sharing between humans and cobots in manufacturing. Procedia CIRP **104**, 38–43 (2021). https://doi.org/10.1016/j.procir.2021.11.007
23. Muślewski, L., Woropay, M., Bojar, P.: The evaluation method of human–machine–environment systems operation quality. In: Engineering Asset Management and Infrastructure Sustainability, pp. 675–691 (2012). https://doi.org/10.1007/978-0-85729-493-7_52
24. Varnavsky, A.N., Mironov, V.V.: Dependence modeling of "price/quality" of human-machine system from the ratio of the elements duration of the "work: Rest" cycle: dependence modeling of "price/quality" of human-machine system. In: 2017 6th Mediterranean Conference on Embedded Computing, MECO 2017 - Including ECYPS 2017, Proceedings (2017). https://doi.org/10.1109/MECO.2017.7977205
25. Varnavsky, A.N.: Research of simple genetic algorithm parameters for estimating duration of the elements of the "work-rest" cycle of a production worker. In: IEEE 12th International Conference on Application of Information and Communication Technologies, AICT 2018 - Proceedings (2018). https://doi.org/10.1109/ICAICT.2018.8747071

Mathematical Modeling and Research of Ways to Reduce the Load Factor of the 110/6 Kv Main Substation

V. Pupin[1], D. Safonov[1(✉)], and O. Fedorov[2]

[1] LLC "NPK Promir", 26/1, Storojevaya Street, Moscow 111020, Russia
dsafonov@npkpromir.com

[2] Nizhny Novgorod State Technical University n.a. R.E. of Alekseev, 24, Minin Street, Nizhny Novgorod 603950, Russia

Abstract. In order to meet the requirements of Rostechnadzor for unloading the supplying substation CPS-11 different technical solutions were considered, which were then implemented in practice. For this purpose, a mathematical model of the power supply scheme of the CPS-11 substation was developed. Calculations of normal modes of operation of electrical equipment CPS-11 have been made in order to determine steady-state parameters of the substation CPS-11 at changing: power of compensating devices, the power factor of synchronous motors, the use of backup synchronous motors as reactive power compensators, modes of operation of sections of the switchgears. The mathematical model takes into account the operating synchronous motors, type and mode of operation of the excitation system. Calculation studies were performed for the following cases: separate operation of 35 kV transformers, parallel operation of 35 kV transformers and during preventive maintenance on one transformer. The calculation of economic effect from the introduction of additionally installed capacitor banks was performed in this work.

Keywords: Modeling · Pumping station · Compensating devices · Optimization · Synchronous motor · Compensator · Steady-state mode · Power supply system

1 Introduction

Due to the growth of unit and total motor load capacity of enterprises, the problem of ensuring the regulatory requirements for transformer loading, ensuring the required standard power factors of electrical complexes of oil production enterprises [1–3] is now becoming more and more relevant. A sudden short-term power failure of oil production enterprises can lead to a violation of operational safety, disruption of a complex technological process, significant product rejection and other undesirable consequences.

Synchronous motors are widely used as a drive of mechanisms at pumping and compressor stations. Consumed power of such drives, as a rule, exceeds 100 kW and reaches $30 \div 50$ MW in some productions [4, 5]. In this regard, there are increased requirements for the stability of power supply systems (PSS) with synchronous motors (SM) in

A. A. Radionov and V. R. Gasiyarov (Eds.): RusAutoCon 2022, LNEE 986, pp. 226–238, 2023.
https://doi.org/10.1007/978-3-031-22311-2_23

steady-state and in transient modes [1, 6–10]. This is due to the need to ensure the uninterrupted operation of driven mechanisms and to reduce the damage from the disruption of the technological process during emergency power interruptions. For example, when electric centrifugal pumps of oil production at the cluster pumping station (CPS) are stopped, their restart in winter takes 4 ÷ 7 days.

The method of mathematical modeling with the use of computers is effective and convenient for the research of the PSS. This method makes it possible to investigate with high accuracy a wide range of problems, the solution of which by experimental means is costly or impossible at all [5–8, 11–17]. To assess the static stability of substations, from which synchronous and asynchronous motors are powered, the mathematical model must take into account the mutual influence of motors, electric network parameters, load nodes and use numerical solution methods [1, 6, 10–12, 17]. Mathematical models should reflect real electromagnetic and electromechanical processes of the SM, take into account the work of the SM excitation systems [5, 6, 18], the influence of the PSS.

Mechanisms powered by SM are technologically responsible and are subject to high requirements for reliable operation in steady-state and transient modes [6, 7, 11, 12, 19]. Besides, synchronous motors are effective means for regulation of reactive power and voltage in load nodes. In this connection, their excitation systems should also have high reliability. With the help of excitation systems the static and dynamic stability of the motors themselves and their power supply systems as a whole can be increased [5, 6, 11, 18].

The considered scheme of CPS-11 power supply is a two-transformer substation 110/35/6 CPS-11 with three-winding transformers with a capacity of 25000 kW. From four feeders through the 35 kV overhead line and through two double-winding transformers of 16000 kW capacity, the synchronous high-voltage load of CPS-11A is powered. At CPS-11A there are STM-4000 motors (4 pcs., two of them are in permanent operation) and STM-1250 (2 pcs., one of them is in permanent operation). Passport catalog parameters of electric motors are given in Tables 1 and 2. Low-voltage asynchronous motors (AM) with capacity up to 75 kW are powered from transformer substations 6/0,4 kV. According to the received data we know: one pump CNS-500 is operated 24 hours a day at CPS-11, the second pump CNS-500 is operated 19 hours a day; and a centrifugal pump CNS-180 is operated periodically by necessity.

Table 1. Catalog data for STM motors and drive mechanism (beginning).

Type of SM	S_{nom}	P_{nom}	$\cos\varphi$	η_{nom}	X_d	X_d'	T_{f0}	I_P	U_{fnom}	I_{fnom}
	MVA	*MW*	*Relative units*	*%*	*Relative units*	*Relative units*	*Sec*	*Relative units*	*kV*	*kA*
STM-1250	1,45	1,25	0,90	96,8	1,625	0	2,8	6,48	0,046	0,253
STM-4000	4,56	4,00	0,90	97,5	2,14	0	3,9	6,69	0,102	0,283

Table 2. Catalog data for STM motors and drive mechanism (continued).

Type of SM	M_P	M_{VX}	M_{max}	GD^2_{DV}	GD^2_{MX}	K_3	M_0	T_J	Z
	Relative units	*Relative units*	*Relative units*	*T•M²*	*T•M²*	*Relative units*	*Relative units*	*Sec*	*kV*
STM-1250	2,07	1,27	1,67	0,102	0,200	0,850	0,080	5,190	2,0
STM-4000	1,92	1,50	1,65	0,278	0,580	0,850	0,008	4,641	2,0

2 Modeling of the PSS of CPS-11

The used model [3] of the industrial power supply system (Fig. 1) makes it possible to represent the PSS of any structure (by type and number of elements), configuration and state.

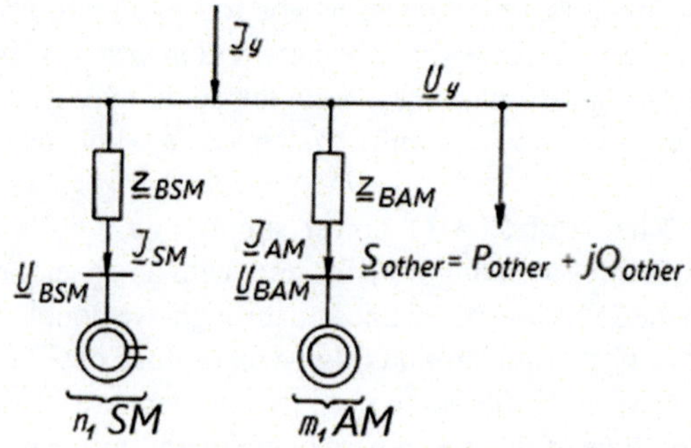

Fig. 1. Substitution scheme of the industrial load node.

Calculation of steady state parameters of electric motors and load nodes is proposed to perform on the basis of nodes voltage equations [3]:

- for the first level (from the electrical system to the load nodes, Fig. 2):

$$U_y = E_c - Z_y \cdot J_y \tag{1}$$

- for the second level (from the load nodes to the terminals of the sections, SM and AM):

$$\underline{J_y} = \underline{J_{other}} + \sum \underline{J_{SM}} + \sum \underline{J_{AM}} + \underline{J_{BK}} \tag{2}$$

$$\underline{U_{BSM}} = \underline{U_y} - Z_{BSM} \cdot \underline{J_{SM}} \tag{3}$$

$$\underline{U_{BAM}} = \underline{U_y} - Z_{BAM} \cdot \underline{J_{AM}} \tag{4}$$

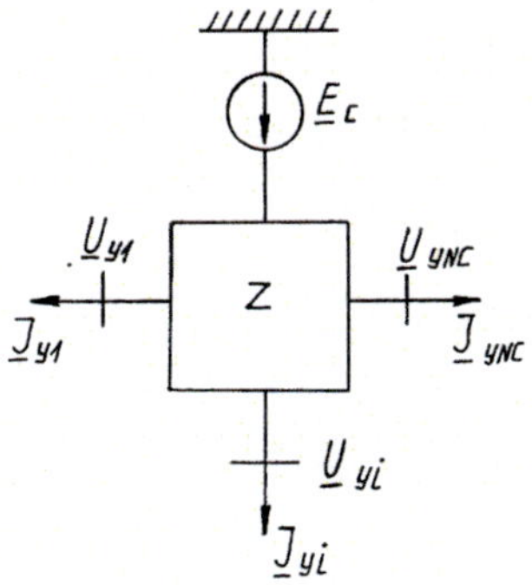

Fig. 2. Substitution scheme of the first level of the industrial power supply system.

- for the third level (for synchronous and asynchronous motors):

$$\underline{J_{SM}} = \frac{P_{SM} - jQ_{SM}}{|U_{SM}|^2} \cdot \underline{U_{SM}} \tag{5}$$

$$\underline{J_{AM}} = \frac{P_{AM} - jQ_{AM}}{|U_{AM}|^2} \cdot \underline{U_{AM}} \tag{6}$$

As the initial data for the calculation of the steady-state mode of the PSS are taken: U_{ON} – initial (nominal) voltage on the first section in the initial mode; $K_{Z,i}$ and $K_{ZA,j}$ – load factors of the *i*-th synchronous and *j*-th asynchronous motors ($i = 1, N_{SM}; j = 1, N_{AM}$); $cos\varphi_i$ – power factor of the *i*-th SM in the initial mode ($i = 1, N_{CD}$); $P_{other,i}$ and $Q_{other,j}$ – active and reactive power of other load of the *i*-th node at nominal voltage ($i = 1, N_C$); $Q_{BC,j}$ – power of capacitor banks connected to the *j*-th node ($j = 1, N_C$).

Calculation of the steady state mode is carried out by iterative method. The condition of the process termination is fulfilment of the inequality:

$$\left|\underline{U_{y1}} - \underline{U_{OH}}\right| < \varepsilon \tag{7}$$

where U_{y1} is the voltage at the first load node; ε - required accuracy of voltage calculations (0.001).

In accordance with the adopted model [3] of electric motors, the active resistance of the stator windings of SM and AM is related to the resistances $\underline{Z_{BSM}}$ and $\underline{Z_{BAM}}$, characterizing the electrical distance of the motors from the‘ switchgear section (SG). Thus, voltages $\underline{U_{BSM}}$ and $\underline{U_{BAM}}$ Eqs. (3) and (4) are actually applied not to motor leads but behind an active resistance of stator winding.

All the initial data are the widespread catalogue data of the equipment in the reference literature [4, 5]. Part of the parameters determining the system mode (load factors of motors, power of other loads, etc.) is set by the user [6, 7]. All the switchgear section parameters are given to $S_B = 25$ MVA, and the parameters of SM, AM are given to the rated full power of the corresponding motor S_{nom} [6, 11, 14]. The substitution scheme of the substation CPS-11 was developed on the basis of the presented normal circuits of electrical connections of substations 6 kV No. 1,2,3,4, substations 6/0,4 kV No. 1,3,4,5;

the circuit diagram of networks 35 kV and substation 35/6 kV and is shown on Fig. 3. The scheme contains 49 branches of the supply network and the connected load; 6 synchronous motors, 15 asynchronous electric motors of 6 kV and 20 nodes of the load.

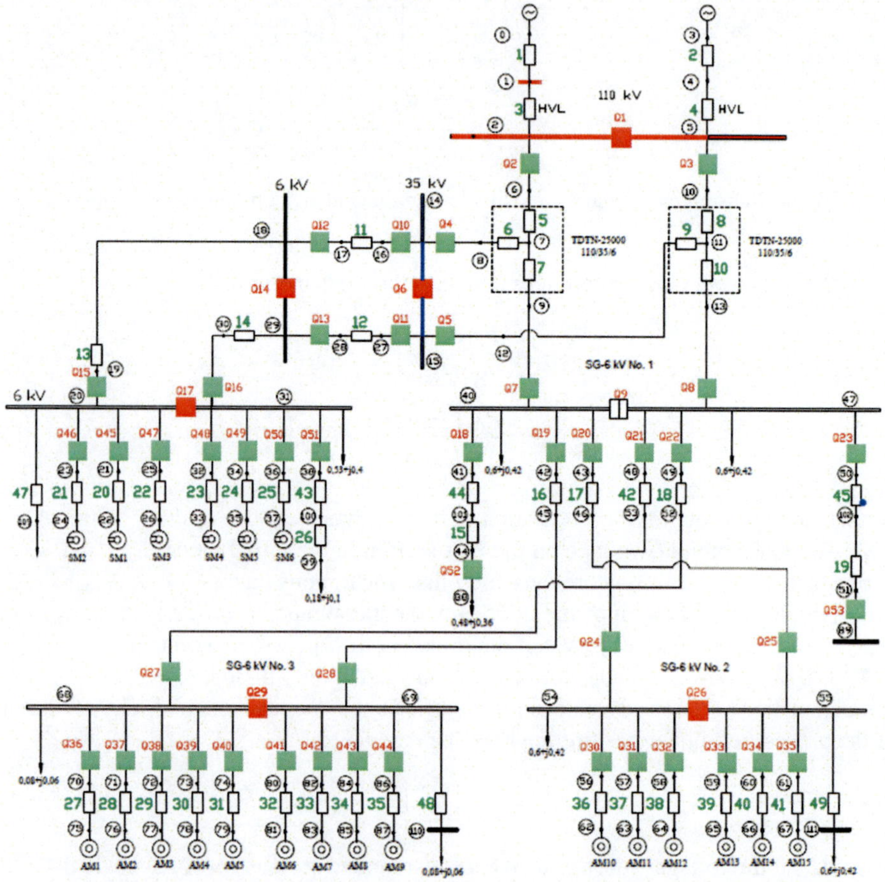

Fig. 3. Mathematical model of the substation substitution scheme CPS-11.

3 Calculation and Experimental Studies of Steady-State Modes of CPS-11 Electrical Equipment

For the developed model of CPS-11 electrical equipment, we calculated the initial steady-state mode, the results of which are summarized in Tables 3, 4, 5, and 6 separately for sections, SM, AM and branches of the considered scheme of CPS-11.

All parameters, specified in Tables 3, 4, and 5, are given in relative units, and in Table 6, the average effective averages of the specified parameters are given.

Table 3. Parameters of the initial operating mode of the SM.

Type and number of the SM on the substitution diagram	Active power, relative units	Reactive power, relative units	Current, relative units	Voltage, relative units	δ, °	E_q, relative units'
2. STM-4000 (SM2)	0.7459	−0.0749	0.7196	1.009	1.113	1.858
5. STM-1250 (SM5)	0.6143	−0.0389	0.5774	1.045	0.994	1.591
6. STM-4000 (SM6)	0.6098	−0.0000	0.5722	1.045	1.019	1.519

Table 4. Parameters of the initial operating mode of the AM.

Type and number of the AM on the substitution diagram	Active power, relative units	Reactive power, relative units	Current, relative units	Voltage, relative units	ω, relative units
1. VAM2-450LB4	0.7202	0.4209	0.8042	1.024	0.989
3. VAM2-560 M4	0.7306	0.3979	0.8013	1.025	0.993
4. VAM2-315 M4	0.7054	0.4653	0.8156	1.023	0.990
5. VAM2-560LA4	0.7321	0.3972	0.8030	1.024	0.993
8. VAM2-560 M4	0.7306	0.3927	0.8128	0.999	0.993
9. VAM2-560 M4	0.7306	0.3927	0.8128	0.999	0.993
10.VAM2-560 S4	0.7267	0.3770	0.7835	1.032	0.985
14. VAM2-315 M4	0.7054	0.4407	0.8103	1.006	0.990
15. VAM2-560 S4	0.7267	0.3646	0.7915	1.007	0.989

Table 5. Parameters of the sections of the considered model.

Switchgear and sections	Power		Section current, relative units	Section voltage, relative units	$E_C \cdot U_{sec}$, radians	φ, radians'
	Active, relative units	*Reactive, relative units*				
1. Electrical system	0.000	−0.029	0.028	1.049	0.003	−1.571
2. HVL-110 kV Beloz-Flame No. 1	0.000	−0.028	0.027	1.024	0.019	−1.569
3. HVL-110 kV Beloz-Flame No. 2	0.000	−0.028	0.027	1.031	0.017	−1.570
4. SG 35/6 kV feed. No. 3	0.380	0.357	0.503	1.036	0.137	0.755
5. SG 35/6 kV feed. No. 1	0.260	0.244	0.332	1.073	0.111	0.755
6. SG-6 kV No. 1 2 sec.	0.023	0.016	0.028	1.019	0.097	0.603
7. SG-6 kV No. 1 1 sec.	0.024	0.017	0.028	1.044	0.081	0.610
8. SG-6 kV No. 4 2 sec.	0.020	0.014	0.024	1.012	0.160	0.626
9. SG-6 kV No. 4 1 sec.	0.147	−0.006	0.140	1.049	0.133	−0.038
10. SG-6 kV No. 2 2 sec.	0.041	0.025	0.046	1.042	0.081	0.556
11. SG-6 kV No. 2 1 sec.	0.048	0.029	0.056	1.017	0.096	0.548
12. SG-6 kV No. 3 2 sec.	0.074	0.042	0.082	1.034	0.079	0.517
13. SG-6 kV No. 3 1 sec.	0.046	0.025	0.051	1.012	0.096	0.501
14. SG-6/0.4 kV No. 4 1 sec.	0.007	0.004	0.008	1.046	0.141	0.497
15. SG-6/0.4 kV No. 1 1 sec.	0.018	0.018	0.022	1.046	0.103	0.633
16. SG-6/0.4 kV No. 1 2 sec.	0.018	0.013	0.022	1.021	0.118	0.626
17. STM-4000 6 kV No. 2	0.132	−0.013	0.131	1.013	0.161	−0.101

According to the calculations of the initial steady-state mode of the CPS-11 electric equipment operation it was found (branches 6 and 9 in Table 6. Further the data on the branches $1 \div 5$ and $15 \div 44$ are not given) that T-2 and T-1 transformers at 35 kV are loaded with 23,64 MW. Power losses in the power supply system of KNS-11 for the pre-defined loads are as follows: total active power losses $\Delta P_{total} = 0.619$ MW, total reactive power losses in the industrial power grid $\Delta P_{mains} = 0.534$ MW, active power losses in the SM $\Delta P_{SM} = 0.030$ MW, active power losses in the AM $\Delta P_{AM} = 0.055$ MW, total power losses $\Delta Q_{total} = 6.147$ Mvar, reactive power losses in the industrial power grid $\Delta Q_{mains} = 6.145$ Mvar.

Table 6. Parameters of the mode of branches in steady state mode.

Branch number	Branch name	Branch current, kA	Branch voltage, kV		Active power, MW		Reactive power, Mvar	
			Beginning	*End*	*Beginning*	*End*	*Beginning*	*End*
1	Electrical system No. 1	0.108	115.77	115.43	16.99	16.98	13.39	13.30
2	Electrical system No. 2	0.086	115.77	115.52	14.47	14.46	9.44	9.38
3	HVL-110 kV Beloz-Flame No. 1	0.110	115.43	112.69	16.98	16.79	14.02	13.43
4	HVL-110 kV Beloz-Flame No. 2	0.088	115.52	113.46	14.46	14.34	10.10	9.72
5	Transformer No. 1 HV side	0.112	112.69	105.67	16.79	16.73	14.12	11.96
6	Common point – transformer No. 1 LV1 side	0.091	105.67	102.48	13.34	13.30	9.85	9.04
7	Common point – transformer No. 1 LV2 side	0.022	105.67	105.65	3.40	3.39	2.11	2.11
8	Transformer No. 2 HV side	0.090	113.46	108.27	14.34	14.30	10.42	9.03
9	Common point – transformer No. 2 LV1 side	0.065	108.27	106.17	10.36	10.34	6.58	6.15
10	Common point – transformer No. 2 LV2 side	0.025	108.27	108.25	3.94	3.94	2.45	2.46
11	From HVL-35 kV to SG-6 kV No. 4 2 sec.	0.061	36.24	36.24	3.81	3.81	0.12	0.12
12	From HVL-35 kV to SG-6 kV No. 4 1 sec.	0.059	37.55	37.55	3.85	3.85	0.04	0.04
13	Transformer No. 1 16 MVA	0.061	36.24	36.20	3.81	3.81	0.12	0.03
14	Transformer No. 2 16 MVA	0.059	37.55	37.52	3.85	3.85	0.04	−0.04

4 Evaluation of Various Compensating Measures for the Purpose of Unloading the Supply Substation

4.1 Maximum Use of the Compensating Capacity of the Working SM

Due to the fact that in the initial operating mode of the synchronous motors of the switchgear No. 4 CPS-11 power factor is equal to $cos\varphi = 1$, then by changing the excitation mode of the SM, we can solve the problem of unloading the supplying substation [6, 11, 19]. At the same time it should be done in such a way that: 1) the stator winding current of SM not to exceed the rated current; 2) and the excitation winding current not to exceed the rated current. In the considered model, the power factor of all working SM: for the second SM $cos\varphi = 0.80$, for the fifth and sixth SM $cos\varphi = 0.70$.

Due to this, the total load of the 35 kV windings of the three-winding transformers of the CPS-11 substation decreased and was: $P_{total} = 23.76$ MW, $Q_{total} = 9.68$ Mvar, $S_{total} = 25.656$ MVA (instead of $P_{total} = 23.64$ MW, $Q_{total} = 15.19$ Mvar, $S_{total} = 28.10$ MVA in the initial mode).

To check the permissibility of this mode, we calculated the mode of parallel operation of three-winding transformers of KNS-11 substation on the 35 kV side, when the circuit breaker Q6 is on in the substitution diagram (Fig. 3). As a result, we got: $P_1 = 11{,}88$ MW, $Q_1 = 4{,}84$ MVA, $S_1 = 12{,}828$ MVA, which is close to the permitted load of the windings, equal to 12,5 MVA.

4.2 Using the Compensating Capacity of Capacitor Batteries not Switched on at 6 kV, on the Winding Side of 35 kV Transformers of CPS-11 Substation

In spite of the fact that capacitor batteries of oil producing companies are often switched off due to overvoltages and harmonic components, we checked the possibility of using the compensating ability of capacitor batteries not switched on at 6 kV, which is recommended in the sources [9, 11, 15, 20]. According to the initial data, capacity of the capacitor banks at the CPS-11 substation is equal to: for feeder No. 1 – $S_{BK} = 3200$ kvar, for feeder No. 3 – $S_{BK} = 1800$ kvar. We have increased the maximum capacity of capacitors and calculated this steady-state mode. The total load of the windings of three-winding transformers of the 35 kV CPS-11 substation has decreased and made up: $P_{total} = 23.8$ MW, $Q_{total} = 4.54$ Mvar, $S_{total} = 24.23$ MVA. During parallel operation of the three-winding transformers of the CPS-11 substation on the 35 kV voltage side, the load on the 35 kV winding of one transformer became less than the permitted load [6] and was: $P_1 = 11.9$ MW, $Q_1 = 2.27$ Mvar, $S_1 = 12.115$ MVA.

4.3 Use of Redundant Synchronous Motors as Synchronous Compensators

For large pump station units, synchronous compensators are often used for additional reactive power compensation on the 35 kV input side of three-winding transformers. This solution is more practical in operational terms, as it will eliminate the use of capacitor banks at 6 kV for feeders No. 1 and No. 3 (which are often taken out of operation). We have simulated the specified mode of operation and determined that the total load of the windings of three-winding transformers of the 35 kV CPS-11 substation decreased less than in the cases considered above and was: $P_{total} = 24.01$ MW, $Q_{total} = 6.62$ Mvar, $S_{total} = 24.906$ MVA. This method of unloading the supplying substation is ineffective. When the three-winding transformers of the CPS-11 substation work in parallel on the 35 kV voltage side, the load on the 35 kV winding of one transformer will be: $P_I = 12.005$ MW, $Q_I = 3.31$ Mvar, $S_I = 12.453$ MVA, which is less than the allowable winding load of 12.5 MVA.

4.4 Parallel Operation of TDTN-25000/110/35/6 Transformers on 35 kV Voltage Side

Parallel operation of transformers TDTN-25000/110/35/6 of the substation CPS-11 on the 35 kV side is interesting because it provides a uniform loading of the 35 kV windings of these transformers. To check the permissibility of this mode of operation we calculated three-phase short circuit currents, which were equal to 2.607 kA for a short circuit in the node No. 14 of the equivalent circuit (Fig. 3) and 2.47 kA for a short circuit in the node No. 15 of the equivalent circuit.

At parallel operation of transformers TDTN-25000/110/35/6 of the substation CPS-11 on the 35 kV side for the mode with the use of redundant synchronous motors as synchronous compensators, the three-phase short circuit current was: $I_{SC} = 5.08$ kA, which is less than 5.1 kA (permissible tripping current of 35 kV circuit breakers). Parallel operation of transformers TDTN-25000/110/35/6 of the substation CPS-11 on the 35 kV side is acceptable.

5 Conclusions

As technical solutions to reduce the load of 35 kV windings to the permitted 12,5 MVA measures have been considered: a) maximum use of compensating ability of the working SM; b) use of compensating ability of not included capacitor batteries at voltage 6 kV on the side of 35 kV winding of transformers of CPS-11 substation; c) use of backup synchronous motors as synchronous compensators; d) parallel operation of TDTN-25000/110/35/6 transformers on the 35 kV voltage side. The results obtained are summarized and summarized in Table 7.

Table 7. Consolidated results of technical measures to reduce 35 kV winding load of CPS-11 transformers to 12.5 MVA.

CPS-11 operating mode	Load T-1/T-2 on the 35 kV side of the winding		35 kV winding load factor as a fraction of 12.5 MVA, relative units	Active power losses, MW*'
	Active, MW	*Reactive, Mvar*		
Separate operation of transformers				
Initial steady state mode (separate operation of transformers)	13.3/10.34	9.04/6.15	1.289/1.000	0.619 (2.48%)
At the maximum use of the compensating capacity of the SM	13.37/10.39	6.95/2.73	1.254/0.859	0.559 (2.24%)

(continued)

Table 7. (*continued*)

CPS-11 operating mode	Load T-1/T-2 on the 35 kV side of the winding		35 kV winding load factor as a fraction of 12.5 MVA, relative units	Active power losses, MW*'
	Active, MW	*Reactive, Mvar*		
Use of compensating capacity of capacitor batteries not switched on at 6 kV on the winding side of 35 kV transformers of CPS-11 substation and maximum use of compensating capacity of working SM	13.39/10.41	5.23/-0.69	1.15/0.835	0.516 (2.06%)
When using redundant synchronous motors as synchronous compensators	13.62/10.39	3.40/3.22	1.123/0.870	0.542 (2.17%)
Parallel operation of transformers on U_{nom} = 35 kV side				
Initial steady state mode (separate operation of transformers)	11.82/11.82	7.59/7.59	1.12/1.12	0.609 (2.44%)
At the maximum use of the compensating capacity of the SM	11.88/11.88	4.84/4.84	1.02/1.02	0.535 (2.14%)
Use of compensating capacity of capacitor batteries not switched on at 6 kV on the winding side of 35 kV transformers of CPS-11 substation and maximum use of compensating capacity of working SM	11.9/11.9	2.22/2.22	0.96/0.96	0.516 (2.06%)
When using redundant synchronous motors as synchronous compensators	12.0/12.0	3.13/3.13	1.00/1.00	0.532 (2.13%)
Operation with one transformer				
Initial steady state mode (separate operation of transformers)	24.35	17.66	1.525	1.409 (5.64%)
At the maximum use of the compensating capacity of the SM	24.35	11.54	1.41	1.131 (4.52%)

(*continued*)

Table 7. (*continued*)

CPS-11 operating mode	Load T-1/T-2 on the 35 kV side of the winding		35 kV winding load factor as a fraction of 12.5 MVA, relative units	Active power losses, MW*'
	Active, MW	*Reactive, Mvar*		
Use of compensating capacity of capacitor batteries not switched on at 6 kV on the winding side of 35 kV transformers of CPS-11 substation and maximum use of compensating capacity of working SM	24.19	6.95	1.34	0.976 (3.90%)
When using redundant synchronous motors as synchronous compensators	25.46	9.21	1.38	1.075 (4.30%)

* Losses are given in relation to the full transformer capacity

The main conclusions are as follows:

1) As the main measure to unload the supplying substation CPS-11 calculations determined the use of the compensating capacity of the 6 kV capacitor banks not switched on and the maximum use of the compensating capacity of the working SM.
2) The installation and inclusion of capacitor banks at the input of the transformer T-1 CPS-11 will be $tg\varphi = 0{,}24$ and $cos\varphi = 0{,}972$, and at the input of the transformer T-2 – will be $tg\varphi = 0{,}26$ and $cos\varphi = 0{,}968$, which is less than the regulatory value $tg\varphi = 0{,}4$ and $cos\varphi = 0{,}93$ for the considered networks.
3) Parallel operation of transformers TDTN-25000/110/35/6 on the 35 kV side of the substation CPS-11 provides uniform loading of the 35 kV windings of these transformers and is permissible, since three-phase short circuit currents on the 35 kV bars of the substation CPS-11 in this mode is $I_{SC} = 5.08$ kA, which is less than the allowable tripping current of circuit breakers 35 kV.
4) The economic effect [19, 21] from the introduction of additionally installed capacitor batteries will be:

$$E_{KB} = \Delta W \cdot C_{EE} \cdot T_{year} = (0.976 - 0.472) \cdot 1000 \cdot 2.0 \cdot 8760 = 8830000$$

5) In the post-emergency mode (at operation of one three-winding transformer) using the compensating capacity of non-involved capacitor banks on the 6 kV side of 35 kV windings of transformers of CPS-11 substation and the maximum use of compensating capacity of operating SM, the load of T-2 transformer on the 110 kV side of the winding is: $P_5 = 31{,}55$ MW, $Q_5 = 11{,}22$ Mvar, $S_5 = 33{,}486$ MVA, which corresponds to the transformer load factor is 1,34 with allowable in the mode of continuous emergency overloads 1,5 (according to GOST 14209–85 for transformers).

References

1. Gurevich, Yu.E, Libova, L.E., Khachatryan, E.A.: Load Stability of Electrical Systems, p. 208. Energoizdat (1981)
2. GOST 14209-85: Power Oil Transformers for General Purposes, p. 38. FGUP Standartinform (2009)
3. Christensen, P.J., Graf, W.H., Yeung, T.W.: Refinery Power Failures: Causes, Costs and Solutions, pp. 1–6. Hydrocarbon Publishing Company (2013). www.digitalrefining.com/article/1000773
4. Syromyatnikov, I.A.: Operating Modes of Asynchronous and Synchronous Motors, p. 528. Gosenergoizdat (1963)
5. Hindmarsh, J., Renfrew, A.: Electrical Machines and Drives, 3rd edn. p. 336. Elsevier Science (1996)
6. Gamazin, S.I., Pupin, V.M.: Computer calculation methods for starting powerful synchronous motors. Ind. Power Eng. **10**, 38–42 (1983)
7. Ershov, M.S., Egorov, A.V., Trifonov, A.A.: Stability of Industrial Electrical Systems, p. 319. Publishing House Nedra (2010)
8. Meleshkin, G.A., Merkuryev, G.V.: Sustainability of Energy Systems. Monograph, vol. 1, p. 369. Center for Power Engineering Training (2006)
9. Merkuryev, G.V., Shargin, Y.M.: Sustainability of Energy Systems. Monograph, vol. 2, p. 300. Center for Power Engineering Training (2006)
10. Rausand, M., Hoyland, A.: System Reliability Theory: Models, Statistical Methods and Application. Wiley Interscience (2004)
11. Venikov, V.A., Stroev, V.A.: Electrical Systems and Networks, p. 512. Higher School (1998)
12. Idelchik, V.I.: Calculation of Steady-State Modes of Electrical Systems, p. 192. Energia, Moscow (1977)
13. Anderson, G.O.: Development of a mathematical model for reactive power transmission costs in electric power systems. SAIEE Afr. Res. J. **99**(3), 87–93 (2008)
14. Abur, A., Exposito, A.G.: Power System State Estimation. Theory and Implementation. Marcel Dekker (2004)
15. Kwasinski, A.: Quantitative model and metrics of electrical grids' resilience evaluated at a power distribution level. Energies **9**(2), 1–27 (2016). https://doi.org/10.3390/en9020093
16. Luzyanin, I., Petrochenkov, A.: Practical aspects of software developing for the system of structural and functional analysis of power supply systems in oil companies. In: Proceedings of the 5th International Conference on Applied Innovations in IT, vol. 5, pp. 65–69 (2017). https://doi.org/10.13142/KT10005.31. WOS: 000402660300009
17. Mariesa, L.: Crow Computational Methods for Electric Power Systems, 2nd edn., p. 304. CRC Press (2010)
18. Shchedrin, V.A., Kapitonov, O.K.: Influence of excitation multiplicity of synchronous generators on operation of load node of commensurate power. Vestnik (Herald) Chuvash Univ. **3**, 121–126 (2012)
19. Gruntovich, N.V., Kapanskiy, A.A., Pupin, V.M., Safonov, D.O., Fedorov, O.V.: Influence of running engines on the residual stresses of the nodes of complex loading stations. Bull. P.O. Sukhoi Gomel State Tech. Univ. **2**(85), 78–90 (2021)
20. Gurevich, Y., Ilyushin, P.V.: Features of Calculations of Modes in Power Districts with Distributed Generation: A Monograph, p. 278. Nizhny Novgorod Institute of Management - Branch of the Russian Presidential Academy of National Economy and Public Administration (2018)
21. Gruntovich, N.V., Kapansky, A.A.: Forecasting the energy efficiency of technological systems of water supply and wastewater disposal under the implementation of energy saving measures. Energ. Effi. **1**, 20–24 (2016)

Digital Observer of Elastic Torque of Rolling Stand Two-Mass System

B. M. Loginov[1], V. R. Khramshin[2], O. A. Gasiyarova[3], A. Y. Semitko[1], and V. R. Gasiyarov[4](✉)

[1] South Ural State University (National Research University), 76, Lenin Ave., Chelyabinsk 454080, Russia
[2] Nosov Magnitogorsk State Technical University, 38, Lenin Ave., Magnitogorsk 455000, Russia
[3] Uchtekh-Profi LLC, 83, Lenin Ave., Chelyabinsk 454080, Russia
[4] Moscow Polytechnic University, 38, Bolshaya Semenovskaya Str., Moscow 107023, Russia
gasiyarovvr@gmail.com

Abstract. The relevance of monitoring elastic torques in mechatronic systems of the rolling mill stand is defined by high dynamic loads at the roll biting. This causes torsional oscillations of the torsion shafts (spindles) and, consequently, the fatigue failure of mechanical joints. This problem is particularly relevant for plate and wide-strip hot rolling mills, the stands of which are designed with individual roll drives. To limit mechanical overloads, first of all, the elastic torque should be continuously monitored. This has led to developing observers of elastic torques on spindles. The paper describes a digital observer developed for a plate rolling mill 5000. Unlike known analogs, it is simply configured and has no speed differentiation. The elastic torque calculation (recovery) is based on the processing of data arrays imported from the stand controller into the observer model implemented with the help of the Matlab-Simulink software. The authors provide an example for the emergency recovery of an elastic torque. The paper further explores the structure of a simplified diagnostic data collection system.

Keywords: Rolling mill · Mechatronic system · Elastic torque · Continuous monitoring · Digital observer · Structure · Model · Experiments · Implementation

1 Introduction

An analysis of the rolling mill stand's mechanical part failures shows that they are mostly caused by one-time or accumulated torque overload cycles. Rolling stands operate under recurrent shock loads caused by the roll bite of a workpiece. Therefore, the use of transient process parameters for diagnostic purposes is advisable. A workpiece shall be interpreted here as an intermediate product between the blank and the finished product.

The elastic torque in the drive line is highly informative. However, elastic torque direct measurement systems based on physical sensors are complex and have a low service life in the aggressive environment of a hot rolling mill. This is confirmed by the

A. A. Radionov and V. R. Gasiyarov (Eds.): RusAutoCon 2022, LNEE 986, pp. 239–249, 2023.
https://doi.org/10.1007/978-3-031-22311-2_24

operating history of the elastic deformation measurement system, previously mounted on the mill 5000 [1]. Such short service life is caused by harsh operating conditions and scheduled spindle replacements complicating the reinstallation of instruments. The benefit of this system implementation is the ability to check the adequacy of the signals sent by the spindle elastic torque observer developed (the observer parameters are given below).

An alternative option to direct elastic torque measurement is recovering it with the observers that are, in fact, digital shadows of a physical system [2]. According to the definition provided in [3], "*a state observer is a model connected in parallel to the controlled object, continuously receiving data on changes in the control action and the controlled parameter*." Of relevance is the development of observers for the parameters of two- and three-mass mechatronic systems facilitating the shaft torque and the gravitational mass speed recovery by the drive parameters. Papers [4–6] can be singled out in terms of the problem consideration.

2 Problem Statement

The study is exemplified by the horizontal stand drive of the mill 5000 (Fig. 1a). Slabs are rolled in a reverse mode with accelerations and decelerations in each pass.

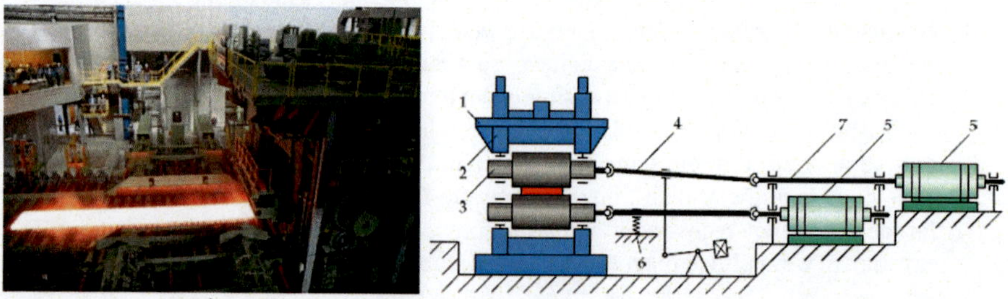

a) b)

Fig. 1. Photo of a 5000 mill stand (a) and working line scheme for the rolling mill's main drive with independent roll drives (b): 1 - roll stand; 2 - housing; 3 - working rolls; 4 - universal spindles; 5 - motors; 6 - spindle balance; 7 - countershaft.

The workpiece entry into the stand is accompanied by a shock load change. Figure 1b shows a simplified kinematic diagram of the stand. The torque is transmitted from the motors to the rolls by universal bar spindles. Dynamic loads at biting cause elastic deformations (twisting) of the spindle and damped oscillations. This predetermines the accumulation of fatigue loads and the destruction of mechanical parts. The resulting accidents elongate downtime and entail significant costs for equipment replacement and restoration.

The kinematic diagram shows that the motor-roll system can be represented as a two-mass mechatronic one with an elastic coupling and a mechanical transmission gap (Fig. 2a). Figure 2b shows the block diagram (model) of the stand drive with a double-loop speed control system. Such a representation of the rolling stand's main line is justified in [7]. The same paper considers a technique for experimentally defining the two-mass system parameters. The model parameters calculated according to this technique are given in Table 1 [8].

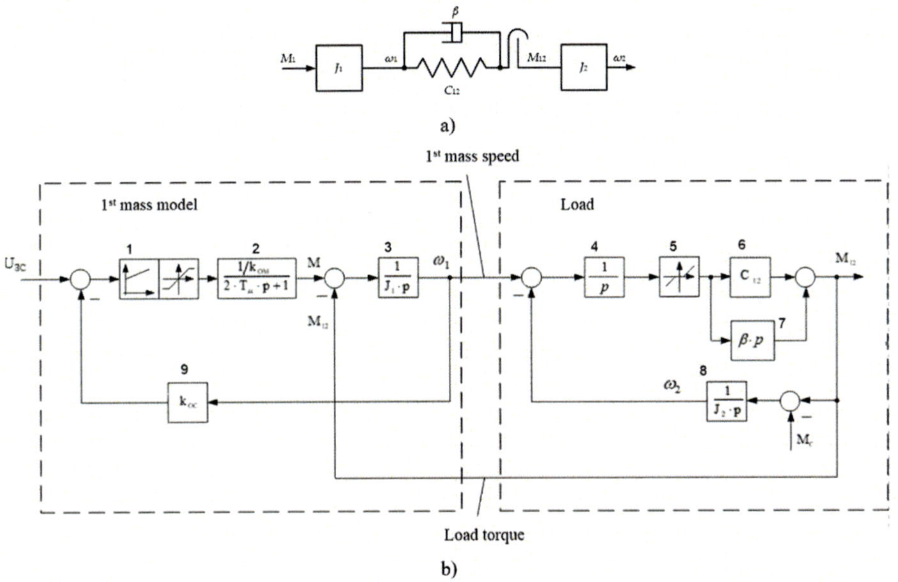

Fig. 2. Kinematic diagram of a two-mass system (a) and model structure explaining the observer development (b).

The diagram legend: T_{μ} – uncompensated time constant; J_1, J_2 – the 1st and 2nd mass inertia torques; C_{12} – elastic coupling rigidity; β – natural damping (such as viscous friction) ratio; M_1 – motor torque; M_{12} – spindle elastic torque; ω_1, ω_2 – motor and roll speeds (1st and 2nd mass, respectively).

The author's publications [8–10] consider developing an elastic torque digital observer for a two-mass system. The observer of the rolling stand's spindle elastic torque and roll speed has been considered. The results of its pilot testing on the mill 5000 are provided. Further, the authors present the results of studies conducted for the automated system, developed on the observer's basis to control elastic torque. The system drawback is the elastic torque recovery by differentiating the signal of the difference between the measured motor speed and the calculated second mass speed of the two-mass system. As is known, differentiation reduces noise immunity and negatively affects the recovered signal reliability. The strive to eliminate this procedure and simplify the elastic torque recovery has led to the development of a simpler digital observer discussed below.

Table 1. Parameter of the two-mass system model, Mill 5000 [8].

Parameter	Symbol	Dimensions	Value
Torque of inertia of the first moving mass (the motor)	J_1	kg·m^2	125,000
Torque of inertia of the second moving mass (the roll)	J_2	kg·m^2	114,571
Elastic coupling rigidity	c_{12}	N·m/rad	5,934,842
Eigenfrequency of elastic oscillations	ω_{12}	rad/s	9.96
Electric drive acceleration	ε_0	rad/s^2	1–3
Transmission gap	δ	rad	0.017–0.051 (1°–3°)
Mean elastic torque	M_{12}	MN·m	1.9
Damping ratio	β	-	2.817
Attenuation decrement	ξ	-	0.172
Speed controller gain	k_{sc}	-	19.5
Speed controller time constant	T_{sc}	s	0.0041

3 The Text Body

3.1 The Observer Structure

For the structure in Fig. 2b, the following differential equation system is true, justified in [8, 10]:

$$\frac{dM_1}{dt} = -\frac{1}{T_m}M_1 + \frac{1}{T_m K_{oM}}M_{1ref}$$

$$\frac{d\omega_1}{dt} = \frac{1}{J_1}M_1 - \frac{1}{J_1}M_{12}$$

$$\frac{dM_{12}}{dt} = C_{12}\omega_1 - C_{12}\omega_2 + \frac{\beta}{J_1}M_1 - \beta\frac{J_1 + J_2}{J_1 * J_2}M_{12} + \frac{\beta}{J_2}M_c$$

$$\frac{d\omega_2}{dt} = \frac{1}{J_2}M_{12} - \frac{1}{J_2}M_c$$

The development considers that the friction torque is insignificant, and module 7 can therefore be removed from the scheme. It is assumed that for this structure, the load is a "black box". It affects the 1st mass model through the spindle elastic torque. The idea is that under these conditions, an elastic torque observer can be implemented as that of a single-mass system load. For this purpose, a scheme is proposed without calculating the speed derivatives (Fig. 3a). Figure 3b shows the observer structure, implemented in Matlab-Simulink. It is built according to the reduced differential equation system.

To calculate the load torque, a special speed mismatch compensation controller is used - module 4 (autotuning controller). It is a conventional PI controller with a quite a response to minimize calculation errors. The procedure for autotuning of controllers in the microcontroller is described in [11, 12]; the controller parameters are given in Table 2. It calculates the first mass load torque and strives to zero out the difference between the measured and calculated speeds.

In the model, the dynamic torque is calculated as a difference between the total motor torque by the stand controller and the load torque recovered by the observer. Then the dynamic torque is integrated with the $(1/J_1)$ ratio to calculate the 1st mass speed. Thus, in the observer structure, differentiation is replaced by integration, which improves the signal's noise immunity. This is an advantage over known technical solutions. The second advantage is a simple configuration. Compared to the development in [8], the lack of the second mass speed signal recovery is a drawback. Such a signal may be required when developing a closed control system for elastic torque. However, this study does not pose such a problem; the speed signal can be extracted in further studies.

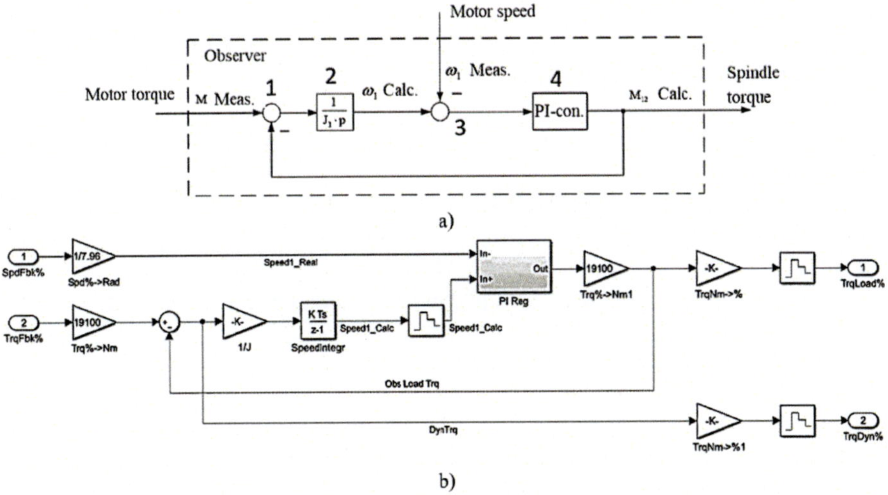

Fig. 3. The observer scheme (a) and its implementation in Matlab-Simulink (b).

The observer has been configured, where the algorithm tried-and-true in Matlab-Simulink is imported into the industrial controller software. As a result, the observer is virtually commissioned. For the mill drives under study, this approach is acceptable. However, to generalize the results and recommendations for using the observer at other facilities, providing a technique for calculating the autotuning controller parameters is advisable. This may be the subject of a separate study.

3.2 Adequacy Check for Elastic Torque Calculation

When estimating the adequacy of the recovered values provided by the observer, an approach has been proposed, according to which the motor speed and torque signals are imported into the observer's model (Fig. 3b). They can be measured in advance and stored as arrays or fed online to the observer's inputs. Pre-recorded signals, imported

Table 2. The elastic torque observer parameters.

Parameter	Unit of measure	Value
Motor's rated speed	rad/s	7.96
Motor's rated torque	N·m	1,910,000
Motor's torque of inertia	$N{\cdot}m^2$	1,250,000
Autotuning controller's P-part ratio	–	400
Autotuning controller's I-part ratio	–	1,000

into Matlab from the IBA PDA information system, are used below. Further, the authors propose a method consisting of the following actions:

- Data stored as arrays are imported into Matlab, where they are fed to the discrete model input.
- Processes are simulated, previously recorded on oscillograms.
- The oscillograms are compared with the calculated dependencies by their superimposition or by comparing coordinates at characteristic points.
- To assess reliability, statistical result processing techniques can be applied.

Figure 4 shows oscillograms of the drive sequential deceleration and acceleration. This mode is the most informative one in terms of estimating the observer's adequacy since the object non-linear properties manifest themselves, determined by gaps in the spindle joints. In the deceleration and acceleration mode, the gap, respectively, opens and closes. Figure 4b shows similar oscillograms recorded when rolling a single billet. The motor torque M_{mot} (window 2) reaches the limit of 240% of the rated value immediately after biting.

In both figures, dependencies between the spindle torques $M_{\mathrm{sp(meas)}}$ measured by the PDA system and $M_{\mathrm{sp(rec)}}$ recovered by the observer completely coincide. Curve $M_{\mathrm{sp(rec)}}$ of the observer-recovered torque is on the background of that of the torque $M_{\mathrm{sp(meas)}}$ read by the physical measurement system, so it becomes invisible in black-and-white printing. This allows concluding that the signal is recovered with absolute accuracy, despite the impact of the non-linearity due to the gap.

The oscillograms provided allow asserting that, having a properly configured autotuning controller, the proposed principle of the measured signal processing facilitates obtaining the recovered signals that are approximated to the physical ones as much as possible. In this case, the object non-linear properties are automatically considered. A comparative analysis of processes using mathematical methods is not required since the transient curves coincide.

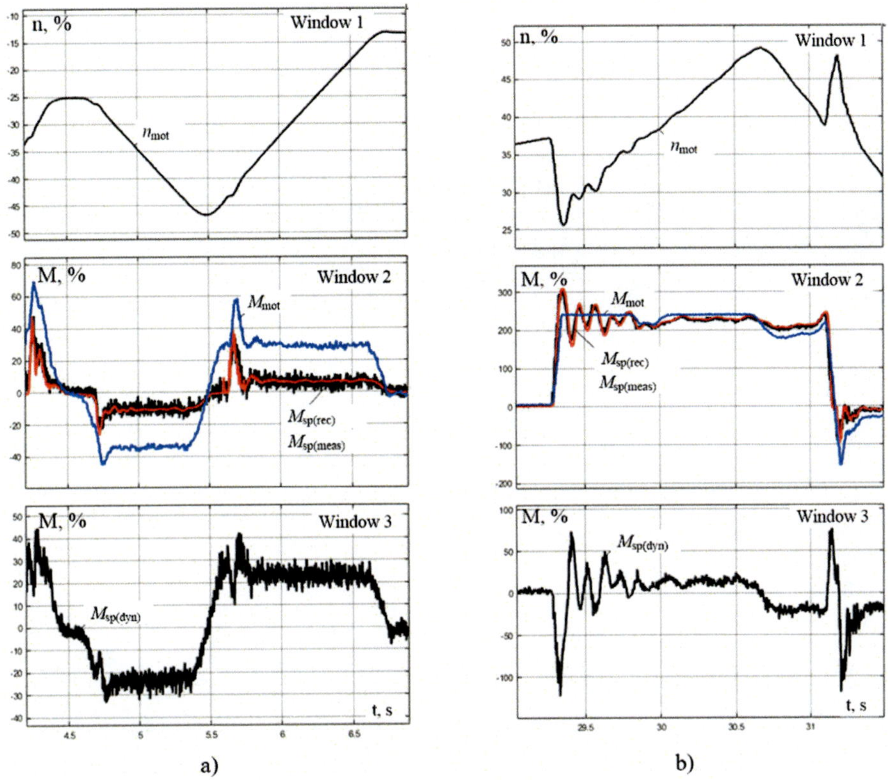

Fig. 4. Oscillograms recorded at sequential acceleration and deceleration of the drive (a) and in the single workpiece biting mode (b): n_{mot} – the motor speed, % of the rated value; M_{mot} – the motor torque, $M_{sp(meas)}$ – the sensor-read spindle torque, $M_{sp(rec)}$ – torque from the observer; $M_{sp(dyn)}$ – dynamic torque from the observer.

The conclusions drawn have been confirmed by the results of studying various stand drive modes. The dynamic torques arising in an emergency have been analyzed, including those accompanied by equipment failures. Note that the literature sources have no information on such studies. As an example, one emergency is analyzed below.

4 Experimental Studies

Dynamic loads at the roll failure have been studied. Figure 5 shows the PDA oscillograms. The speeds and torques of the rolls' top (TMD) and bottom (BMD) main drives have been recorded. The metal was bitten at the time instant t1. The bottom roll broke at the time instant t2 with the entire load applied to the top spindle (window 2) since the broken one (window 3) was not loaded. Then the TMD speed damped oscillations occur (window 1) relative to the zero value.

The BMD torque (window 3) decreases to zero, but the motor rotates for another three seconds (window 1) until an emergency shutdown by the operator. This situation

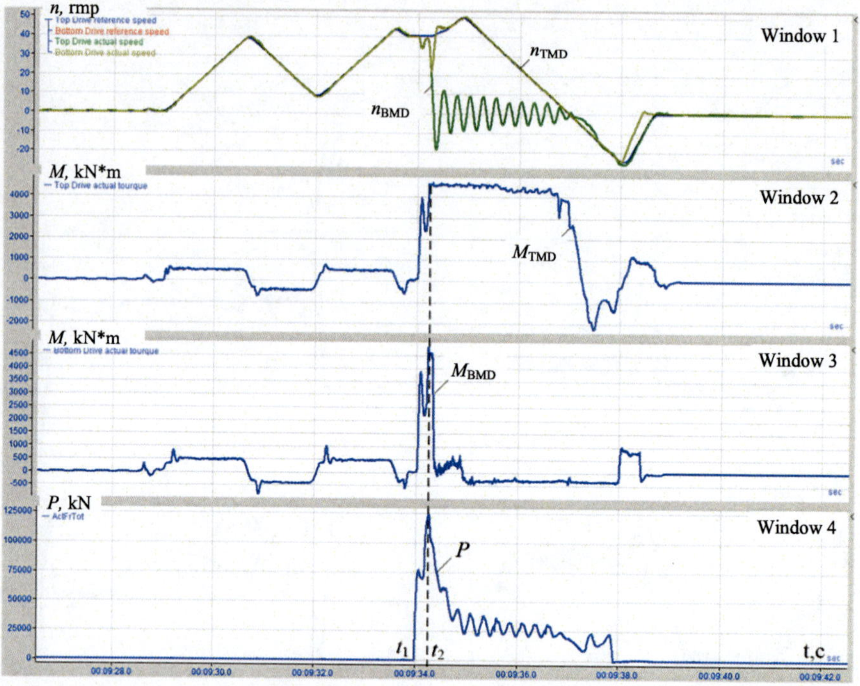

Fig. 5. Speeds and torques oscillograms at a roll breakage: window 1 – set and actual motor speeds, rpm; window 2 – TMD motor torque, kN·m; window 3 – BMD motor torque, kN·m; window 4 – rolling force, kN.

is extremely dangerous since the spindle rotates, being fixed only on the motor side. The uncontrolled large mass rotation can be devastating.

To estimate the elastic torque, the signals recorded during the accident were exported from the PDA to Matlab and processed there using the software implementing the observer (Fig. 3b). Figure 6 shows the recovered speed and torque oscillograms. It is seen that the top spindle with an intact roll (window 2) experiences a sixfold overload relative to the rated value ($M_{SUmax} = 600\%$) at the motor workload $M_{ST} = 240\%$, i.e., exceeds M_{ST} by 2.5 times. The bottom spindle elastic torque amplitude (window 2) exceeds the rated load and the workload 5-fold (M_{SLmax} reaches ~ 500%) and 2-fold, respectively.

The key emergency prevention problems are the timely fixation of the accident onset and generating a signal for a quick drive shutdown. The operator should also be urgently informed of such situations, which will provide his/her timely response to the developing accident.

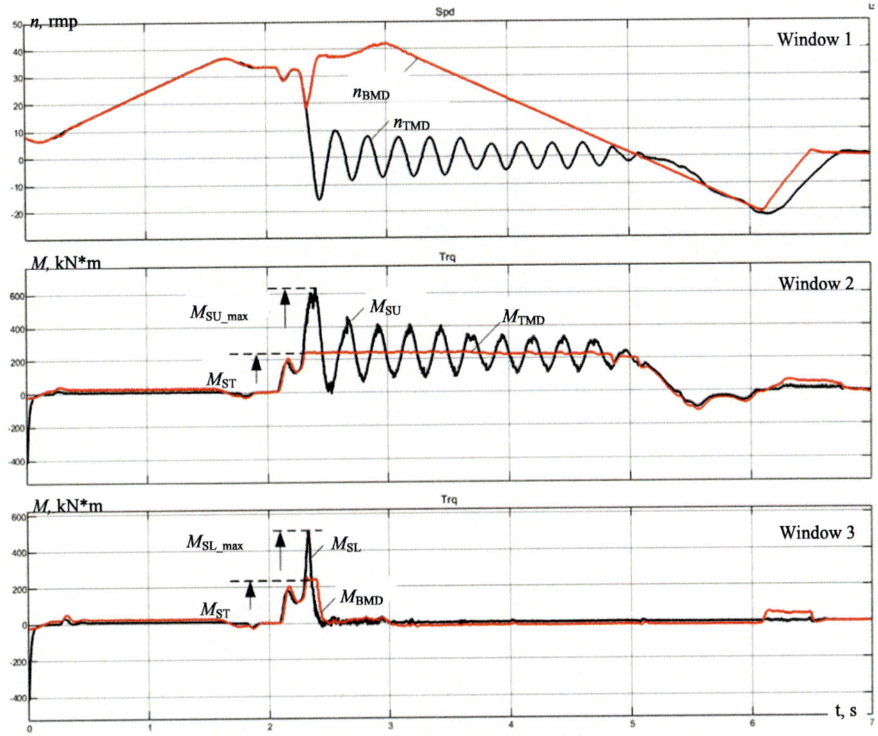

Fig. 6. Speeds and torques oscillograms at a roll breakage, recovered using an observer: window 1 – actual TMD and BMD speeds, rpm; window 2 – motor and TMD spindle torques, %; window 3 - the same dependencies for BMD.

5 Discussion of the Results and Conclusions

The developed observer has been currently implemented in the pilot operation mode at mill 5000. A simple system for collecting and processing data (Fig. 7), implemented in the stand controller, has been proposed. It is an analog of the system for monitoring angular gaps in spindles, considered in [13]. Compared to contact measuring systems, the observer developed has the following advantages:

- It is simple and highly reliable,
- It is maintenance-free,
- It has virtually no value since it is a piece of software.

Since the drive load (Fig. 2b) is represented as a black box, it can be used in both single-mass and two- and three-mass systems. The controller's virtual configuring principle does not change in these cases.

The received spindle elastic torque signals are not intended for use in drive control systems. Their main purpose is to provide elastic torque data in dynamic modes. The observer also allows monitoring of pre-accident situations to prevent accidents, and if occur, analyze their causes.

The spindle dynamic torque data should be stored in a special database while the amplitude values multifold exceeding the set limits should be displayed in a special window on the operator's screen. This allows for avoiding successive dynamic shocks causing fatigue failures. To use the observer opportunities in their entirety, it is advisable to develop and implement the following at the mill:

- A spindle overload monitoring system ensuring recording and calculation of torque overloads exceeding the set limits.
- A technique for calculating the spindle life based on the overload calculation results and their amplitude estimates.
- Ways to limit dynamic loads.

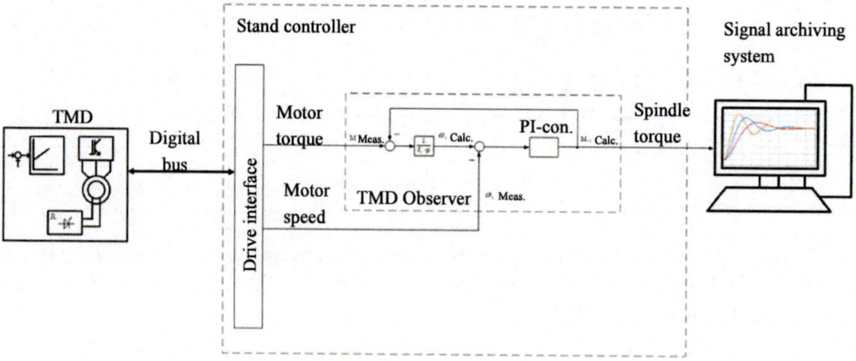

Fig. 7. The structure of the system for collecting the spindle torque data.

The solution to the listed problems will allow transforming the developed observer into the spindle technical condition online monitoring system. On its basis, a set of developments aimed at preventing accidents and mitigating their consequences will be implemented.

References

1. Radionov, A.A., Gasiyarov, V.R., Tverskoi, M.M., Khramshin, V.R., Loginov, B.M.: Implementation of telemetric on-line monitoring system of elastic torque of rolling mill line of shafting. In: 2nd International Ural Conference on Measurements (UralCon) , pp. 450–455. IEEE (2017). https://doi.org/10.1109/URALCON.2017.8120750
2. Radionov, A.A., Karandaev, A.S., Loginov, B.M., Gasiyarova, O.A.: Conceptual areas of creating digital twins of electrical engineering systems of rolling units. News High. Educ. Inst. Electromech. **1**(64), 54–68 (2021). https://doi.org/10.17213/0136-3360-2021-1-54-68
3. Kalachev, Y.: State Observers in a Vector Electric Drive. EFO, Moscow (2015)

4. Szabat, K., Orlowska-Kowalska, T., Dybkowski, M.: Indirect adaptive control of induction motor drive system with an elastic coupling. IEEE Trans. Industr. Electron. **10**(56), 4038–4042 (2009). https://doi.org/10.1109/TIE.2009.2022514
5. Szabat, K., Orlowska-Kowalska, T.: Control of the drive system with stiff and elastic couplings using adaptive neuro-fuzzy approach. IEEE Trans. Industr. Electron. **1**(54), 228–240 (2007). https://doi.org/10.1109/TIE.2006.888787
6. Muszynski, R., Deskur, J.: Damping of torsional vibrations in high-dynamic industrial drives. IEEE Trans. Industr. Electron. **2**(57), 544–552 (2010). https://doi.org/10.1109/TIE.2009.2036034
7. Karandaev, A.S., Radionov, A.A., Loginov, B.M., Gasiyarova, O.A., Gartlib, E.A., Khramshin, V.R.: Experimental definition of the rolling mill two-mass mechatronic system parameters. News High. Educ. Inst. Electromech. **3**(64), 24–35 (2021). https://doi.org/10.17213/0136-3360-2021-3-24-35
8. Radionov, A.A., Karandaev, A.S., Gasiyarov, V.R., Loginov, B.M., Gartlib, E.A.: Development of an automatic elastic torque control system based on a two-mass electric drive coordinate observer. Machines **9**, 305 (2021). https://doi.org/10.3390/machines9120305
9. Radionov, A.A., Gasiyarov, V.R., Gartlib, E.A., Loginov, B.M., Odintsov, K.E.: Controlling the rolling mill mechatronic system coordinates based on the elastic torque observer. Bull. SUSU. Energy Ser. **2**(21), 115–129 (2021). https://doi.org/10.14529/power210212
10. Radionov, A.A., Gasiyarov, R.V., Loginov, B.M., Gartlib, E.A., Gasiyarova, O.A.: Developing an elastic torque digital observer for the rolling mill stand mechatronic system. Electr. Eng. Syst. Complex. **2**(51), 19–29 (2021). https://doi.org/10.18503/2311-8318-2021-2(51)-19-29
11. Hang, C.C., Åström, K.J., Ho, W.K.: Refinements of the Ziegler-Nichols tuning formula. IEE Proc. D Control Theor. Appl. **138**(2), 111 (1991). https://doi.org/10.1049/ip-d.1991.0015
12. Ziegler, J.G., Nichols, N.B.: Optimum settings for automatic controllers. Trans. ASME **65**, 759–765 (1945)
13. Gasiyarova, O.A., Karandaev, A.S., Erdakov, I.N., Loginov, B.M., Khramshin, V.R.: Developing digital observer of angular gaps in rolling stand mechatronic system. Machines **10**, 141 (2022). https://doi.org/10.3390/machines10020141

Control Theory

Definition of the Nested Metagraph

S. V. Astanin(✉) and N. K. Zhukovskaya

Taganrog Branch Russian New University, 47, St. Alexandrovskaya, Taganrog 347949, Russia
astser@mail.ru

Abstract. The metagraph is a graphical construct that has proven to be very useful for graphical visualizations and quantitative analysis. This advantage determines the most promising area for their application: the problems of big data analysis, provided that this data can be described using a graph representation. The article proposes an approach to modeling hierarchical systems based on nested metagraphs. Category theory is used to eliminate ambiguity in the interpretation of concepts. Formalized descriptions of static and dynamic nested metagraphs, methods for defining them, as well as basic operations are given. The basis of the nested metagraph concepts is a monoid, that is considered as an information object, which is characterized by an internal form, internal and external contents. Thus, a monoid is a generalization of a graph structure and is interpreted as a vertex of a generalized graph.

Keywords: Nested metagraph · Metagraph monoid · Metagraph adjacency matrix · Operations on nested metagraphs

1 Introduction

In 1994, A. Basu proposed a new graphical-theoretic structure, called a metagraph, and designed to solve, first of all, the problems of modeling, analysis and design of decision support systems (DSS) [1]. It was assumed that the metagraph would solve the issues of generalization, presentation and processing of big data by reducing the volume of processed data without losing the original information.

In our opinion, the loss of interest in metagraphs is determined by the absence of strict theoretical provisions, similar to those that were formulated in the classical graph theory. Moreover, the concept of a meta-vertex introduced by A. Basu is not meaningfully defined and contradicts this theory. So, for different researchers, a metavertex can be considered both as a graph vertex containing a set of other vertices [2], and as a hyperedge, which, for unknown reasons, can become a vertex [3].

The existing problems, from the semantic representation and processing of big data to the modeling of enterprise management systems, require new solutions related to the generalization of data and knowledge. One of the approaches to such a generalization was the representation of knowledge based on a nested metagraph [4]. When defining such a concept, the concepts of ordinary graphs, metagraphs and hypergraphs were used. At the same time, the metavertex was interpreted as a hyperedge, which led to ambiguity in the definitions of the term's «vertex» and «edge». In addition, the definition of an

A. A. Radionov and V. R. Gasiyarov (Eds.): RusAutoCon 2022, LNEE 986, pp. 253–261, 2023.
https://doi.org/10.1007/978-3-031-22311-2_25

n-dimensional nested metagraph turned out to be a rather cumbersome construction that complicates the description. To eliminate these shortcomings, this paper defines a nested metagraph as a category. Different fields of science use different definitions of the category. So, in philosophy, a category is understood as a concept that reflects the most general properties and connections of the phenomena of the material world. In mathematics, the category defines the algebraic properties of sets of morphisms of the same type of mathematical objects (sets, topological spaces, groups, etc.) [5]. In applied applications, the meaningful meaning of the concepts used in the definition of a nested metagraph, including the concept of «category», as a mathematical object, is important.

1.1 Basic Definitions

Let there be a set of objects $F = \{F_1, \ ..., \ F_n\}$, each of which is a graph, and a set of morphisms $f = \{f_1, \ ..., \ f_m\}$.

Definition 1. A static nested metagraph (SNM) of the 1-st level is a monoid C_1, consisting of a single object (directed graph) $F_1 = (V_1, \ E_1)$, an associative composition law (*) and a unit morphism α, where $V_1 = v_{1,1}, \ v_{1,2}, \ ..., \ v_{1,k}$ - set of graphs vertices; $E_1 = \{v_i, \ v_j\}$ - family of pairs of elements from V_1 forming arcs (edges); $\alpha : \ C_1 \rightarrow C_1$.

Since the monoid is represented by a single object in the form of a graph $F_1 = (V_1, E_1)$, the operations of union and intersection of graphs can be used as an associative law of composition. For our case $F_1 = (V_1, E_1) \cup F_1 = (V_1, E_1)$, $F_1(V_1, E_1) \cup F_1(V_1, E_1)$.

A vertex of the graph F_1 is called elementary if it is not a monoid. Obviously, for a statically nested metagraph of the 1-st level, all vertices of the graph F_1 are elementary, and the graph F_1 is an ordinary directed or undirected graph.

Category C_1 will be interpreted as an information object, which is characterized by an internal form, internal and external content. Signs, traits or properties of an object that characterize its single (private) manifestation will be called the internal content of the object. The internal content is structurally organized, i.e., there is a way of interrelation of signs, which acts as an internal form of the object and determines the general manifestation of the object. In addition to the internal content and internal form, an object may have an external content, i.e., a set of properties (structurally organized or not) that characterize the internal form of an object.

In our case, the internal content is determined by the set of designated vertices and edges (arcs). The internal form is formed by a directed (undirected) graph F. The external content semantically characterizes the category C_1.

The foregoing allows us to consider the monoid C_1 as an attribute vertex.

According to [6], the metagraph $S = (X, \ XM, E, \ EM)$, where X is a generating set; XM – set of metavertices; E is the set of edges defined on the set X; EM is the set of meta-edges defined on the set $XM \cup X$. In other words, the generating set of the nested metagraph is the set of vertices of the graph F, the metavertices are monoids, and the only metaedge is the identity morphism of the category C_1.

The category C_1 is a generalization of the structure, what is its fundamental difference from a hyperedge, which is a relation on the set of vertices of a hypergraph. Such a generalization can also be considered as a vertex of a generalized graph or metagraph.

An illustration of such a metagraph can serve as a representation of the states of complex systems. To describe the states of such systems, as a rule, a significant number of variables and their measurements are required. To reduce the space of descriptions, an approach based on the use of linguistic variables of the theory of fuzzy sets is often used. In particular, this approach is applicable when presenting the states of a business process. Formally, the current state S of a business process with dependent components is as follows: $S = (Y,\ \mu_S(v_i))$, where.

$Y = \{y_1,\ y_2,\ \ldots,\ y_n\}$ – is the set of linguistic variables (components) of the business process, with $v_i = \{T_i^1,\ T_i^2,\ \ldots\}$, $i = 1,\ 2,\ \ldots,\ n$; $\{\mathrm{T}_\mathrm{i}^1,\ \mathrm{T}_\mathrm{i}^2,\ \ldots\}$ – term-set of fuzzy variables that determine the values (properties) of linguistic variables; $\mu_S(v_i)/v_i = \{< \mu_S(T_i)\} > /T_i$ – fuzzy business process situation with membership functions of fuzzy values of linguistic variables.

Let the state of the business process be described by three dependent features: y_1 is the stock of resources, y_2 is the motivation of the performers, y_3 is the risk of failure. Also, $y_1 = \{T_i^1,\ T_i^2\} = \{min,\ \mathrm{max}\}$, $y_2 = \{T_2^1,\ T_2^2\} = \{low,\ \mathrm{high}\}$, $y_3 = \{T_3^1,\ T_3^2\} = \{low,\ \mathrm{high}\}$. Then, the current state S_t of the business process can be displayed by a nested metagraph of the 1-st level (Fig. 1).

Definition 2. A static nested metagraph of the 2-nd level is a monoid C_2 consisting of a single object $F_1 = (V_1,\ E_1)$ and a unit morphism, where $V_1 = \{v_{1,1},\ v_{1,2},\ \ldots,\ v_{1,k}\}$ - set of graphs vertices; $E_1 = \{(v_i,\ v_j)\}$ - family of pairs of elements from V forming edges; $\alpha : C_2 \to C_2$, and $v_{1,1},\ v_{1,2},\ \ldots,\ v_{1,k}$ is also monoids $v_{1,1} = (V_{2,i},\ E_{2,i})$, with unit morphisms $f = \{f_1,\ \ldots,\ f_k\}$.

For a statistical nested metagraph of the 2-nd level, all vertices of the categories $v_{1,\mathrm{i}}$ are elementary.

Definition 3. An n-th level static nested metagraph is a monoid C_n consisting of a single object $F_1 = (V_1,\ E_1)$ and a unit morphism, where $V_1 = \{v_{1,1},\ v_{1,2},\ v_{1,k}\}$ - set of graphs vertices; $E_1 = \{(v_i,\ v_j)\}$ - family of pairs of elements from V forming edges; $\alpha : C_n \to C_n$; $v_{1,1},\ v_{1,2},\ \ldots,\ v_{1,k}$ is monoids of directed graphs $v_{1,i} = (V_{2,i},\ E_{2,i})$, with unit morphisms $f = \{f_1,\ \ldots,\ f_k\}$, $V_{2,i}\ldots V_{n-1,r}$, $r = 1,\ 2,\ \ldots,\ p$, in turn, are monoids of directed graphs with corresponding unit morphisms, and $V_{n-1,r}$ is monoids of directed graphs $F_{n,1} = (V_{n,l},\ E_{n,1})$, $l = 1,\ 2,\ \ldots,\ u$ with elementary vertices V_n.

Thus, at each level, except for the nth level, there is a set of monoids generalizing the corresponding directed graphs.

The n-th level SNM can be considered as a stratified object with n strata, the first (uppermost) stratum being represented by the monoid C_n.

The second stratum, (n-1)-level, is represented by a set of monoids, the number of which coincides with the number of vertices of the graph V_1, etc., up to the n-th stratum (the first level), whose graph vertices are elementary.

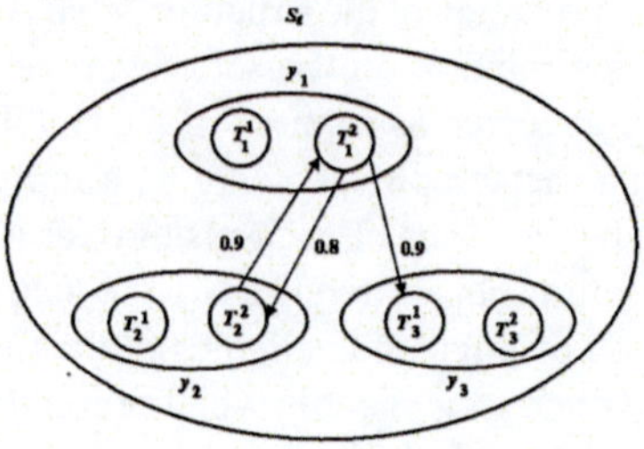

Fig. 1. Description of the current state of a business process with a 1-st level static nested metagraph.

Definition 4. A dynamic nested metagraph (DNM) of the *n*-th level is a stratified situational network, the initial vertices of which are the monoids of the *n*-th level of the CNM, and the edges are the control actions (morphisms) that transform the monoids at the time *t*-1 into monoids at time *t*.

In other words, the DNM of the nth level is represented by a set of categories of monoids $S_1, \ldots, S_n$, the initial vertices of which are the monoids $C_1, \{C_2\}, \ldots, \{C_n\}$, while morphisms are control actions that transform their form (structure) and content. It is assumed that only the first level, as the initial vertex, is represented by a single monoid C_1 with one object $F_1 = (V_1, E_1)$. The lower levels have sets of initial monoids $\{C_2\}, \ldots, \{C_n\}$, for each of which the category $S_{i,j}$ can be constructed, where i is the number of levels, and j is the number of monoids of the *i*-th level. So, formally, the DNM of the 1st level can be defined as follows: $S_1 = (X, *)$, where X_1 is the set of monoids, and $C_1 \in X_1$, and «*» is a morphism of monoids.

An example of a DNM of the 1-st level can be a representation of the dynamics of a business process based on its description by means of linguistic variables (Fig. 2) [7].

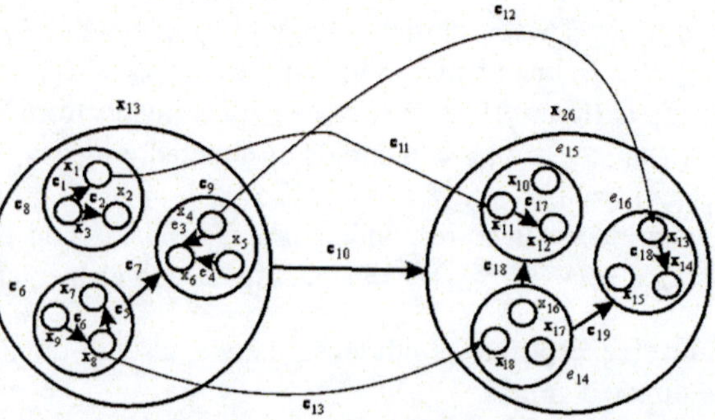

Fig. 2. Fragment of the DNM of the 1st level on the example of the dynamics of a business process.

Here the vertices $C_1(t)$ and $C_1(t+1)$ are monoids, and the monoid $C_1(t+1)$ is the result of the compound action of e_{10} on the monoid $C_1(t)$. In turn, the impact e_{10} includes control decisions e_{11}-e_{13}.

From the standpoint of category theory, the action e_{10} is an arrow (morphism) that transforms the monoid $C_1(t)$ into the monoid $C_1(t+1)$.

1.2 Matrix Method for Specifying Nested Metagraphs and Operations on Nested Metagraphs

The use of CNM and DNM monoids as graph vertices greatly simplifies their representation, and reduces to various ways of specifying ordinary directed graphs. In fact, the graphical and analytical methods for specifying nested metagraphs have been considered above. Let us dwell in more detail on the matrix method, on the example of constructing the SNM adjacency matrix of the 2-nd level (Fig. 3).

Let C_n, $C_{n-1,\,k}$, ..., $C_{2,l}$, $C_{1,p}$ C_n, be the CNM monoids of the *n*-th level corresponding to levels *n*, *n*-1, ..., 1, and let x_i $(i = \overline{1,\ r})$ be elementary vertices (generating set) of monoid graphs $C_{1,p}$.

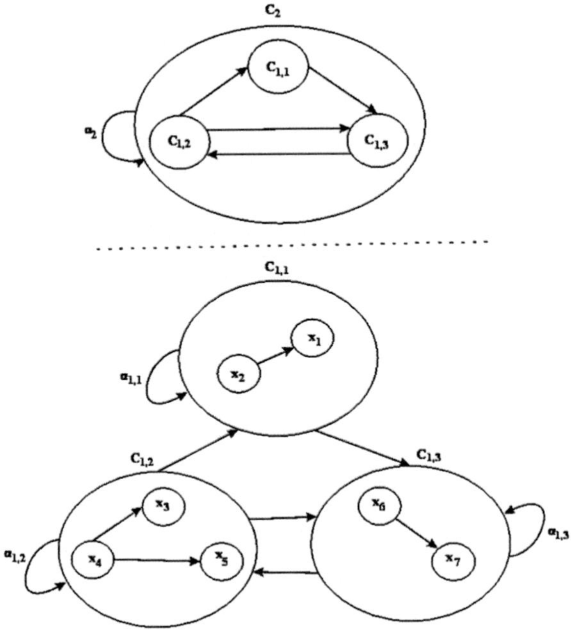

Fig. 3. Second level static nested metagraph.

Definition 5. The adjacency matrix of the *n*-th level CNM is a set of square matrices built according to the hierarchical principle: $x_i \times x_i$, $C_{1,p} \times C_{1,p}$, ..., $C_{n-1,k} \times C_{n-1,k}$. At the intersection of the rows and columns of each matrix, the values are set to 1 or 0, depending on whether there is an edge between a pair of vertices of the corresponding graphs.

The adjacency matrix of the nested metagraph C_2 (Fig. 3) has the following form (Tabl. 1).

The incidence matrix can be constructed in a similar way.

Let us consider the operations of union and intersection for the case of SNM.

Definition 6. The union of the SNM $C_{n,1}$ and $C_{n,2}$, denoted as $C_{n,1} \cup C_{n,2}$, represents such a SNM C_n, such that, if for a particular i-th level, there are coinciding monoids, and some vertices of the graphs of these monoids also coincide, then the graphs are combined according to the rules of graph theory.

Example. Two SNM $C_{2,1}$ (Fig. 3) and $C_{2.2}$ (Fig. 4) are given. The resulting SNM $C_2 = C_{2,1} \cup C_{2,2}$ is shown in Fig. 5 (Table 1).

Table 1. Setting the 2-nd level SNM by the adjacency matrix.

$C_{1,1}$	1			0		1				
$C_{1,2}$	1			1		1				
$C_{1,3}$	0			0		1				
	$C_{1,1}$	x_1	x_2	$C_{1,2}$	x_3	x_4	x_5	$C_{1,3}$	x_6	x_7
	x_1	0	0	x_3	0	0	1	x_6	0	1
	x_2	1	0	x_4	1	0	1	x_7	0	0
				x_5	0	0	0			

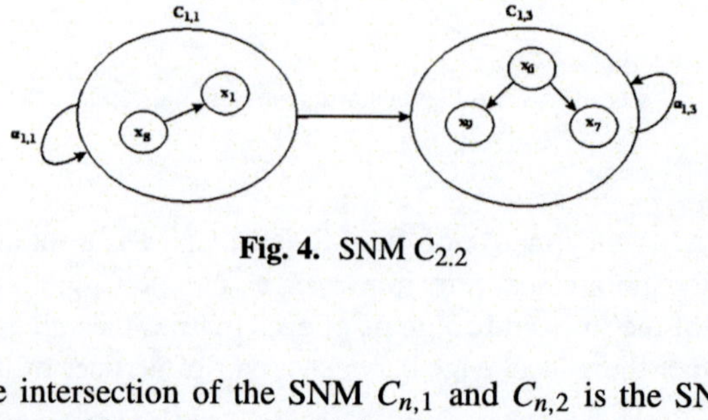

Fig. 4. SNM $C_{2.2}$

Definition 7. The intersection of the SNM $C_{n,1}$ and $C_{n,2}$ is the SNM $C_n = C_{n,1} \cap C_{n,2}$ the set of monoid graphs of which $F_{1n,1} = (V_{1n,1}, E_{1n,1}) \cap F_{1n,2} = (V_{1n,2}, E_{1n}), \ldots, F_{n,1} = (V_{n,1}, E_{1n}) \cap F_{n,2} = (V_{n,2}, E_{n,2})$.

Example. Two SNMs $C_{2,1}$ (Fig. 3) and $C_{2,2}$ (Fig. 4) are given. The result of their intersection is shown in Fig. 6.

The unary operations on the SVM include the operations of deleting a vertex and deleting an edge (arc).

Consider unary operations on the SNM.

Definition 8 SNM C_n^* is called a submetagraph of SNM C_n (denoted $C_n^* \subseteq C_m^*$) if $V_{n,1}^* \subseteq V_{n,1}, E_{n,1}^* \subseteq E_{n,1}$.

Definition 9. If $C_{n-1,k}$, ..., $C_{2,l}$, $C_{1,p}$ are SNM monoids C_n corresponding to levels, *n*-1, ..., *1*, and x_i ($i = \overline{1, r}$) are elementary vertices (generating set) of graphs monoids $C_{1,p}$ then C_{n-x_i} or $C_n - C_{n-1,k}$ is a generated submetagraph of the SNM C_n, i.e. (C_n-x_i) or (C_n- $C_{n\text{-}1,k}$) is the SNM C_n^* obtained after removing the vertex x_i or $C_{n-1,k}$ from C_n and all edges (arcs) incident to this vertex.

Note that deleting the monoid $C_{n-1,k}$ means deleting the graph corresponding to it. This also means deleting all vertices of the given graph at all lower levels.

Definition 10. Let $\{E_{n,1}\}$ be the set of edges (arcs) of C_n. If $E_{n,h} \in \{E_{n,1}\}$, then $C_n - E_{n,h}$ is a comptograph of the SNM C_n obtained after deleting the arc $E_{n,h}$ from C_n.

Unlike deleting a vertex, deleting an edge (arc) concern only the current level C_n.

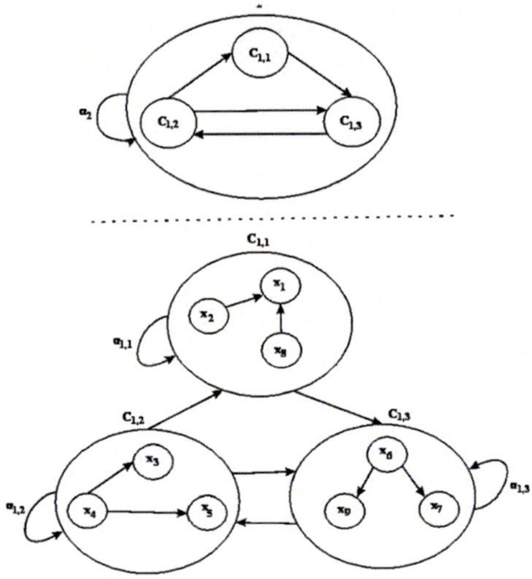

Fig. 5. Resulting CNM C2.

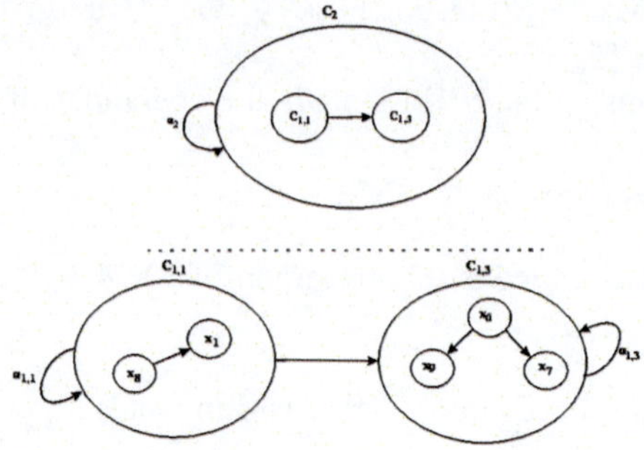

Fig. 6. The result of the operation of crossing SNM $C_{2.1}$ and $C_{2.2}$

2 Conclusion

The categorical approach to the representation of a metagraph makes it possible to overcome the ambiguity of the definitions of the term's «vertex» and «edge», as well as to simplify its description, using the methods of setting and operations on the CCM of classical graph theory.

SNM is a hierarchical structure that allows the description of complex objects with different levels of detail. DNM is a situational network that models possible processes of changes in initial situations given by SNM monoids under the influence of homomorphic morphisms.

Today, knowledge representation is one of the most important tasks in the field of artificial intelligence and software development. This, in particular, is due to the fact that modern software is becoming more and more complex, combining elements of AI, Internet technologies and other latest technologies. The use of a planar graph model and a hypergraph does not allow the implementation of the emergence principle when modeling complex systems. It seems that it is possible to implement this principle on the basis of the nested metagraph model.

Further research may be related to the development of the theory of nested metagraphs, following the example of classical graph theory, as well as to modeling reasoning based on SNM and DNM.

References

1. Basu, A., Blanning, R.: Metagraphs: a tool for modeling decision support systems. Manage. Sci. **40**(12), 1579–1600 (1994)
2. Samokhvalov EN, Revunkov GI, Gapanyuk YuE (2015) Use of metagraphs to describe semantics and pragmatics of information systems. N.E. Bauman. Series "Instrument making" 1(100). https://cyberleninka.ru/article/n/ispolzovanie-metagrafov-dlya-opisaniya-semantiki-i-pragmatiki-informatsionnyh-sistem
3. Blyumin, S.L., Prinkov, A.S.: Development of a matrix representation of generalized graph structures in problems of description and analysis of big data. Comp. nanotechnol **2**, 9–15 (2018)

4. Astanin SV, Dragnysh NV and Zhukovskaya NK (2012) Nested Metagraphs as Models of Complex Objects, IVD 4–2. https://cyberleninka.ru/article/n/vlozhennye-metagrafy-kak-modeli-slozhnyh-obektov
5. Kurosh, A.G., Livshits, A.K., Shulgeifer, E.G.: Fundamentals of category theory. Uspekhi Mat. Nauk **6**(96), 3–52 (1960)
6. Basu A, Blanning R (2007) Metagraphs and Their Applications. Springer Science+Business Media, LLC, p 174
7. Astanin, S.V., Zhukovskaya, N.K.: Management of business processes based on their modeling by fuzzy situational networks. UBS **37**, 145–163 (2012)

Modeling and Study of Maintenance Service System

G. V. Redreev(✉), V. D. Chervenchuk, and P. V. Kiyko

FSBEI HE Omsk SAU, 1 Institutskaya Square, 1, Omsk 644008, Russia
gv.redreev@omgau.org

Abstract. To ensure the high efficiency of production activities, it is necessary to understand the processes taking place in social and technical systems. This issue is solved quite simply for systems that produce material products, and the theoretical representation of these processes is known. However, this issue remains unresolved for systems that provide intangible product services. One of the examples of such social and technical systems are maintenance service systems. The representation of a maintenance and repair system that has initially formed more than 30 years ago has been developed into a maintenance service information system, which includes four interrelated and mutually agreed elements. The degree of consistency was expressed through the logistic function, which allowed to factor in certain nondeterminism of processes in the information system. At the same time, the question immediately arose about assessing the qualification of maintenance service performers, which determines the system's operation quality. The qualification of performers significantly depends on the amount of available knowledge. In addition to the knowledge acquired by performers during training, production activities, and studying production-related literature, large amounts of information about equipment state change are now becoming available, as well as tools for obtaining knowledge from this information. The natural development logic predetermined the need to create an information model of a maintenance service system, which includes five elements. A methodological device, the system of five interacting elements was developed by Russian philosophers after reevaluating certain provisions of Chinese philosophy. We have analytically described the interactions of five elements of the information model of the maintenance service system. An interpretation is presented of the obtained model analysis results.

Keywords: Component · Formatting · Style · Styling · Insert (key words)

1 Introduction

Advancements in the production activity occur with the development of theoretical concepts for this activity. Previously presented as a maintenance and repair system, the maintenance service system was standardized in the past [1]. In addition to the serviced tractor fleet, it included maintenance and repair workers, maintenance and repair facilities, and related documentation (see Fig. 1).

A. A. Radionov and V. R. Gasiyarov (Eds.): RusAutoCon 2022, LNEE 986, pp. 262–268, 2023.
https://doi.org/10.1007/978-3-031-22311-2_26

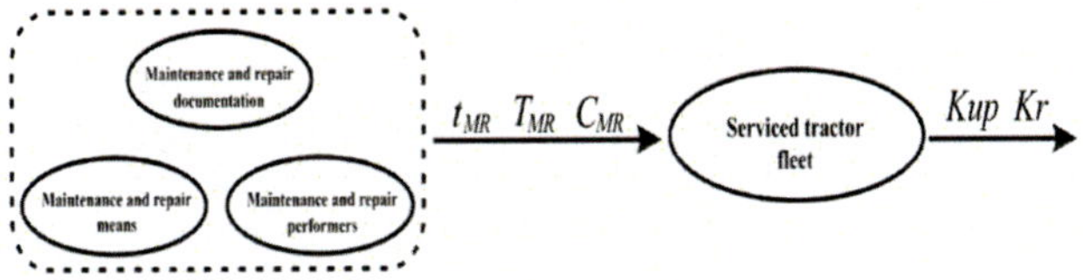

Fig. 1. Model of the maintenance and repair system [1]

The system performance was evaluated by the time spent on maintenance and repair, as well as maintenance and repair complexity and cost. Evaluation of tractor fleet performance was carried out by the uptime percentage and operational readiness.

The purpose of the maintenance and repair service activity turned out to be unrelated to the serviced tractor fleet operation results. This approach was justified for agricultural enterprises with their own maintenance and repair service. In modern conditions, with the existing limited staff, when maintenance service is carried out by service organizations, it is necessary to consider the satisfaction of customer expectations when evaluating the performance of maintenance service systems.

We have developed an information model of a maintenance service system that satisfies the above conditions [2]. This system includes the goal of service and maintenance activities in the form of a maintenance service quality indicator being part of the uptime coefficient of the serviced tractor fleet (Fig. 2).

Fig. 2. Relations of maintenance service information system elements.

Being a significant step in the development of theoretical concepts of maintenance service systems, this model nevertheless does not address the possibility of using the available knowledge of maintenance service performers, but also the knowledge obtained by processing data from sensors of continuous monitoring of the technical condition of components and assemblies of modern agricultural machinery.

2 Research Methods

Methodological definitions of modern Russian philosophers and methods of mathematical modeling of processes in social and technical systems were selected as research methods [3].

Taking into account the need to use databases of technical condition monitoring devices and specialized software, a maintenance service system model was developed, which includes five elements (Fig. 3) [4, 5]:

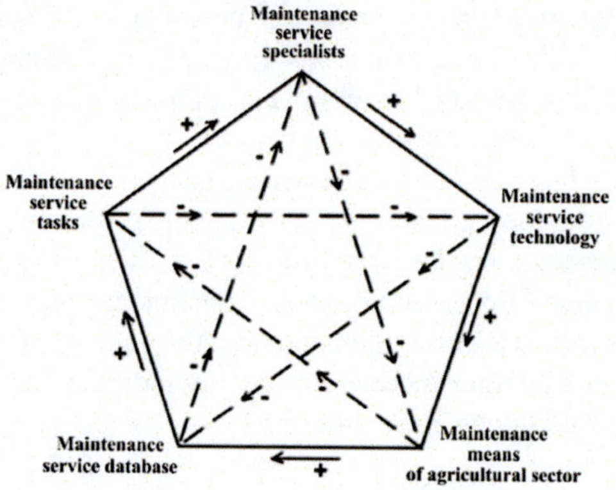

Fig. 3. Maintenance service system model (+) – positive effects, (-) – negative effects.

According to the classical representation of a pentagram, when moving clockwise along the solid lines of the pentagon, each element has a positive effect on the next element on the path. Traveling along the dashed lines of the "star", each element has a negative (limiting) effect on the next element on the path.

To move from a system model to an information system model, formalize the representation of an information model of a maintenance service system, the system's elements and their supposed mutual effects are shown below [6–8] (Fig. 4).

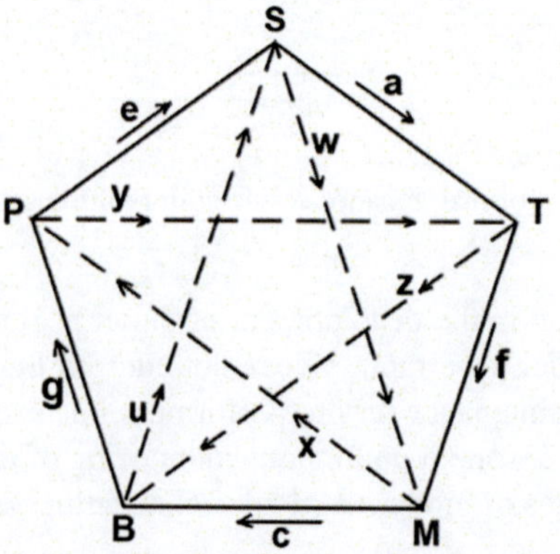

Fig. 4. Maintenance service system information model.

Where:

S—information (knowledge) available to maintenance service specialists;

T—information about the applied maintenance service technologies;

M—information about the serviced agricultural machinery;

B—information in databases with statistics on maintenance, failures and repairs;

P—information about maintenance service tasks;

a, f, c, g, e—coefficients corresponding to the successive positive effects of the system's elements on each other; obviously (a, f, c, g, e) ≥ 1.

w, x, y, z, u—coefficients corresponding to the successive negative (limiting) effect of the system's elements on each other; obviously $0 \leq$ (w, x, y, z, u) ≤ 1.

Let us assume that changes to the system's elements depend on the positive and negative effects of neighboring elements [9–12]. Let us compose the following system of homogeneous differential equations:

$$\left.\begin{array}{l} \frac{dS}{d\tau} = eP - uB; \\ \frac{dT}{d\tau} = aS - yP; \\ \frac{dM}{d\tau} = fT - wS; \\ \frac{dB}{d\tau} = cM - zT; \\ \frac{dP}{d\tau} = gB - xM \end{array}\right\} \quad (1)$$

where τ is time.

Let us introduce the initial conditions:

$$S(0) = s_0;\ T(0) = t_0;\ M(0) = m_0;\ B(0) = b_0;\ P(0) = p_0$$

Let us assume that the knowledge of specialists and information about agricultural machinery improve linearly.

$$S(\tau) = s_0 + s_1\tau \text{ and } M(\tau) = m_0 + m_1\tau \quad (2)$$

Under the accepted conditions: $0 < s_1 << s_0$ and $0 < m_1 << m_0$.

Then from the 1st and 3rd equations of system (1) we obtain

$$s_1 = eP - uB \quad (3)$$

$$m_1 = fT - wS \quad (4)$$

From (2) and (4) equations of system (1) we obtain

$$T(\tau) = \frac{m_1 + ws_0}{f} + \frac{ws_1}{f}\tau \quad (5)$$

From (5) and the 2nd equation of system (1) we find $P(\tau)$:

$$\frac{dT}{d\tau} = \frac{ws_1}{f} = aS - yP = as_0 + as_1\tau - yP(\tau)$$

where

$$P(\tau) = \left(\frac{as_0}{y} - \frac{ws_1}{fy}\right) + \frac{as_1}{y}\tau \quad (6)$$

And finally, from (3) we find $B(\tau)$. The resulting form is as follows:

$$B(\tau) = \frac{eP - s_1}{u} = \frac{1}{u}\left(\frac{aes_0}{y} - \frac{ews_1}{fy} - s_1\right) + \frac{aes_1}{uy}\tau. \quad (7)$$

Thus, the mathematical model of the maintenance service information system will have the following form:

$$\left.\begin{array}{c} S(\tau) = s_0 + s_1\tau \\ T(\tau) = \frac{m_1 + ws_0}{f} + \frac{ws_1}{f}\tau \\ M(\tau) = m_0 + m_1\tau \\ B(\tau) = \frac{1}{u}\left(\frac{aes_0}{y} - \frac{ews_1}{fy} - s_1\right) + \frac{aes_1}{uy}\tau \\ P(\tau) = \left(\frac{as_0}{y} - \frac{ws_1}{fy}\right) + \frac{as_1}{y}\tau \end{array}\right\} \quad (8)$$

with initial data values and constraints:

$$\left.\begin{array}{c} S(0) = s_0. \\ M(0) = m_0 \\ (a, f, e) \geq 1 \\ 0 \leq (w, y, u) \leq 1 \end{array}\right\} \quad (9)$$

The obtained dependencies reflect the extent of interconnectedness between the maintenance service information system elements. The system's efficiency can be controlled using the values of the coefficients that reflect positive and negative mutual effects of elements [13–15].

3 Research Results

We have made calculations using the analytical expressions of the system of Eqs. (8).

We have accepted $s_0 = 100$ information units; $m_0 = 100$; initial values of coefficients (a, f, e) = 1.1; (w, y, u) = 0.1.

- The values of T, B and P were calculated with a change in time τ, at a simultaneous change in the values of w, u, y.
- A graphic representation of the calculation results is shown in Figs. 5, 6, and 7.

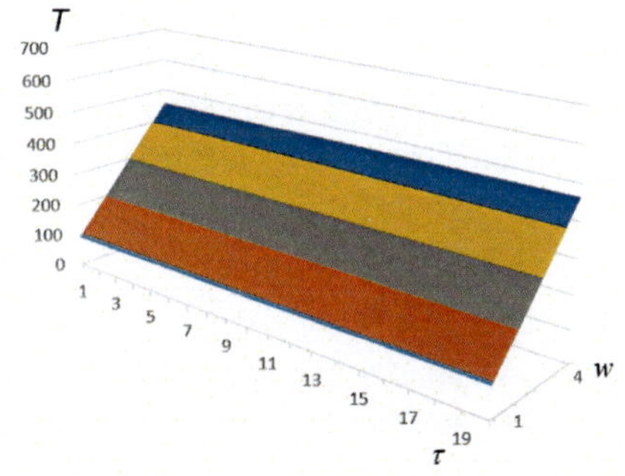

Fig. 5. Dependence of T on τ with changes in w.

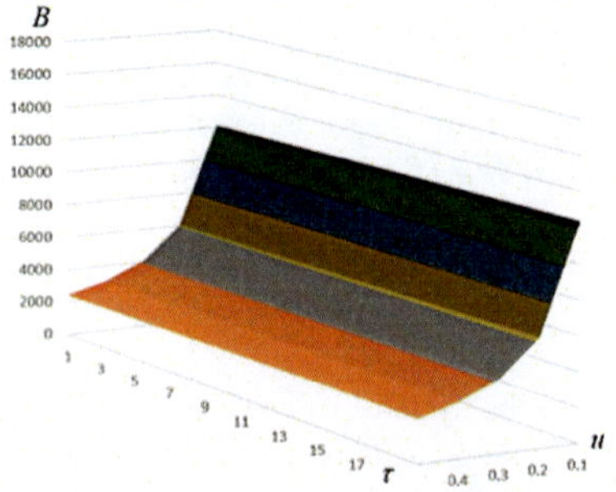

Fig. 6. Dependence of B on τ with changes in u.

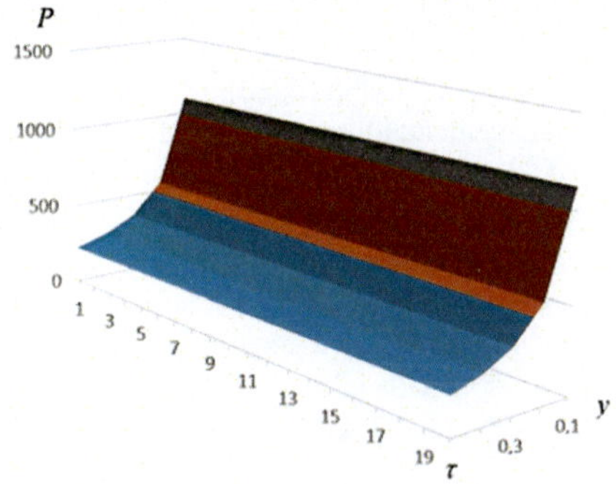

Fig. 7. Dependence of P on τ with changes in y.

4 Conclusions

The analysis of the mathematical model of the maintenance service information system led to a conclusion about the amount of information contained in the maintenance service information system elements. Preliminary calculations using the values of s0 = 100; m0 = 100; a, f, e = 1.1; w, y, u = 0.1 allowed to obtain the current values of the conditional information volumes Si = 100, Ti = 100, Mi = 100, Bi = 12100, Pi =

1100. These preliminary results allow evaluating the ratio of the information volumes of the information system elements, as well as the prospects for using our model for the theoretical description of maintenance service system processes.

References

1. GOST 18322–78 (ST SEV 5151–85): Equipment maintenance and repair system. Terms and Definitions, p. 12. Publishing House of Standards, Moscow (1991)
2. Redreyev, G.V.: Mathematical model of the process of ensuring the maintenance service quality. J. Omsk State Agrarian Univ. **3**(31), 75–85 (2018)
3. Cover, T.M., Thomas, J.A.: Elements of Information Theory. John Wiley & Sons (1991)
4. Teslinova, A.G., Razumov, V.I.: Categorical and systematic methodology for training scientists: textbook. Introductory Page, p. 277. Omsk State University, Omsk (2004)
5. Redreyev, G.V.: Ensuring the performance of a heterogeneous tractor fleet with the maintenance service system for agricultural sector enterprises. Abstract of thesis for the degree of Doctor of Technical Sciences. SFNTSA RAS, Novosibirsk (2020)
6. Chudinov, I.L., Osipova, V.V.: Information Systems and Technologies: Textbook, p. 145. Publishing House of Tomsk Polytechnic University, Tomsk (2013)
7. Pushkar, O., Sibilyev, K.: Information Systems and Technologies: Textbook, p. 264. Publishing House of KhNUE, Kh (2012)
8. Steven, A.: Defining information systems as work systems: implications for the IS field. In: Business Analytics and Information Systems, p. 22 (2008). http://repository.usfca.edu/at/22
9. Lin, A.Z.: Using information value, information gain and gain ratio for detecting two-way interaction effect. In: SAS SUGI Proceedings: Breakout (SUGI 31 and Later), Loan Depot, Lake Forest, CA (2020)
10. Ananthanarayanan, R., et al.: Photon: fault-tolerant and scalable joining of continuous data streams. In: SIGMOD '13: Proceedings of the 2013 International Conference on Management of Data, New York, NY, USA, pp. 577–588 (2013)
11. Information Storage and Management: Storing, Managing, and Protecting Digital Information in Classic, Virtualized, and Cloud Environments. In: Somasundaram, G., Alok, S. (ed.). 2nd edn. EMC Corporation Published by John. Wiley & Sons, Inc., Indianapolis, Indiana (2012)
12. Slaev, V.A., Chunovkina, A.G., Mironovsky, L.A.: Metrology and Theory of Measurement. Walter de Gruyter GmbH, Berlin/Boston (2013)
13. Epstein, M.J., Rejc, A.: Evaluating Performance in Information Technology. The Society of Management Accountants of Canada (CMA Canada), the American Institute of Certified Public Accountants, Inc. (AICPA) and The Chartered Institute of Management Accountants (CIMA) (2005)
14. Sculley, D., et al.: Hidden Technical Debt in Machine Learning Systems. In: NIPS Conference (2015). https://neurips.cc/Conferences/2015
15. Lu, H.K., Lin, P.C., Lo, C.H., Wu. M.Y.: A review of information system evaluation methods. In: International Conference on Software and Computer Applications (ICSCA 2012), vol. 41. IACSIT Press, Singapore (2012)

Machine Learning, Big Data, Internet of Things

Multi-factor Authentication in Innovative Business Systems of Industrial Clusters

A. Zaenchkovski[1], A. Lazarev[1(✉)], and S. Masyutin[2]

[1] National Research University "Moscow Power Engineering Institute" in Smolensk, 1, Energy Passage, Smolensk 214013, Russia
no@sbmpei.ru

[2] Russian Academy of Economy and State Service of the Russian President, 82, Vernadskogo Prospekt Moscow 117571, Russia

Abstract. Development of information and communication technologies and the Internet significantly modifies tools and approaches to innovation process implementation. At current stage, the importance of creating continuous mutually conditioning interaction of individual business systems of industrial clusters is emphasised. Rapid progress in innovation infrastructure development, ensuring the implementation of scientific and industrial cooperation, affects almost all stages of this process. Ensuring the data transmission security in such systems is implemented using methods and models of symmetric and asymmetric cryptography, which are currently used in almost all spheres of human activity, including banking, e-mail protection and corporate data. However, the critical vulnerabilities presence in their operation leaves the developing new methods and algorithms problem for ensuring security in the implementation of information and communication exchange between individual industrial entities in clusters still relevant. To solve it within the framework of this study, a multi factor authentication method was developed based on a dynamic model of increasing the target function using a deep learning neural network model. The developed method of multi-factor authentication is based on dynamic patterns prediction acting as keys for two-way automatic authentication. Among the main advantages of this algorithm are increased security due to the use of a key pair on the client and server side, as well as the absence of the need to re-enter the authenticated sequence. An important implemented system feature is the support for the separation of information flows into parallel data protection keys for isolated transmission and processing of information and increased security.

Keywords: Information security · Multi-factor authentication · Innovation · Neural networks · Fuzzy logic

1 Introduction

Currently, increasing number implementation of functions and even entire innovation processes stages are moving into the digital space. The formation of a sustainable clusters innovation environment to realise the cumulative effect of them is currently being carried

A. A. Radionov and V. R. Gasiyarov (Eds.): RusAutoCon 2022, LNEE 986, pp. 271–281, 2023.
https://doi.org/10.1007/978-3-031-22311-2_27

out by reconfiguring value chains, in conjunction with a personal informal connexions system. The network nature of industrial and economic relations, the globalisation trends contribute to active introduction of "open" practises in innovation [1]. The creation and dissemination of the innovative results activity is now a single complex process, structural elements of which should be interconnected, in which the information and communication technologies development and Internet play an essential role [2]. Such an ecosystem makes it possible to reduce costs, facilitate interaction, realising the possibility of rapid resources exchange, abilities and accumulation of experience, but at the same time they require increased attention to implementation security of these processes in information space [3]. Most of approaches to information security area based on the use of applied data encryption tools, which in turn are divided into symmetric (one-sided) and asymmetric (two-sided) encryption methods. These methods are also used to provide two-factor/multi-factor identity verification, consisting in the presence of several stages of digital fingerprint validation. Development of information technologies, including methods of "penetration" leads to the exploits (vulnerabilities) emergence in encryption methods (algorithms AES, Triple DES (TDES), RSA, etc.) [4]. For example, AES algorithm when using a small 128-bit encryption key size is subject to possible vulnerabilities such as Brute-Force, MITM [5, 6]. Triple DES algorithm, due to the limitations in the encryption block, is subject to the CVE-2016-2183 operation, and the presence of an incorrect configuration in RSA encryption opens up opportunities for private keys theft and leakage of confidential data [7]. In addition, existing cryptography and data protection methods represented by application software have static methods of key exchange, which in turn can be a bottleneck when selecting a key or using phishing methods to obtain the required data.

Various software solutions for two-factor authentication also have a number of advantages. For example, the google-authenticator-libpam package in conjunction with Google 2FA in bash allows to access the root terminal after entering the confirmation code from a registered device by modifying the sudo file (see Fig. 1) [8].

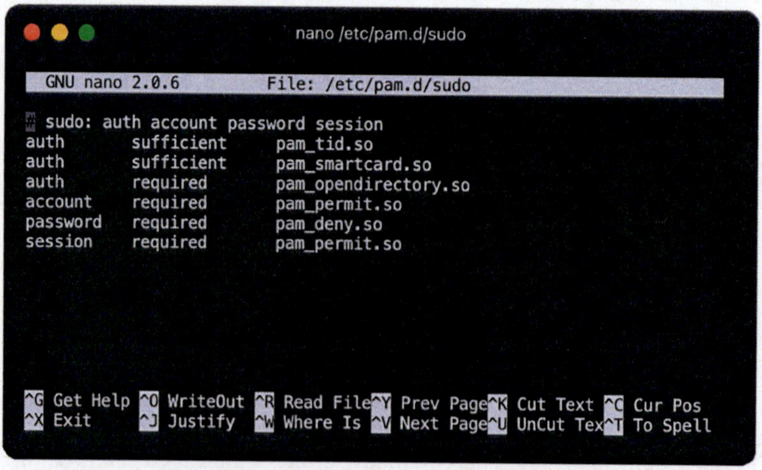

Fig. 1. Modification of the sudo file to support 2FA.

The use of multi-factor authentication systems also presents opportunities in solving OTP (One-Time Password) tasks, as they allow for identity verification in system and application software solutions. For example, OTP tokens allow authentication using USB keys, and ESET Secure Authentication/GateKeeper solutions allow to bind two-factor verification to Active Directory services, RADIUS server and other services using the appropriate API [9, 10]. However, even taking into account the described systems capabilities to ensure their security and some special functions of the solutions described above, problems of automating the input of the generated key, storing backup copies of temporary PIN codes to restore access, as well as the loss of physical USB tokens remain relevant.

So, based on the above, a multi-factor authentication method was developed, the basis of which is the neural prediction module of a dynamically increasing function on deep learning neural networks. The basic function used to determine the dynamic interval of key action is a static variable when predicting a time series, and as the main neural model is the trained Elman neural network, which is based on the deep learning neural networks. To ensure the primary segmentation process of two devices, a temporary static key is used, linked to the client ID, and generated as an output QR code. Also, the advantage of proposed solution is the implemented interface for iOS and Android platforms. The result of using this methodology is a significant increase in security by multi factor access verification methods field.

2 Main Algorithm

Taking into account the fact that in implemented model it is proposed to use multi-factor authentication methods for their testing in various operating systems, it is assumed the support for both operating families of Microsoft Windows, and systems based on the Linux kernel and OS X (Darwin). The following software system packages were used to operate the system: unixtime (for dynamic generation), qrencode (for generating a QR key for connexion initialization), as well as libraries of neural network data processing pytorch, pandas, which are Python programming language part [11]. Based on the fact that the system under development consists of several digital fingerprint verification procedures, system forming was focused on creating and processing a hierarchical set of modules (see Fig. 2):

- "Primary training of the neural model" – module performs primary training of model on the presented identification patterns set;
- "Generating the device's private key" – this module includes the process of generating a private key for a client-server topology;
- "Initialization of the primary connexion" – the module determines the available identifiers in the system, as well as the generation of the primary function (fingerprint) for synchronisation with the client;
- "Prediction of open fingerprint" – this process is carried out by neural network prediction of open fingerprint for the client/server;
- "Decision-making based on fuzzy logic" – the module makes decisions based on the output sample of the neural network model to accept or reject further verification of the client/server;

- "Backup of accumulated data" – the module saves the fingerprints used during operation and the state of neural network model for resuming work in cases of platform hardware failures.

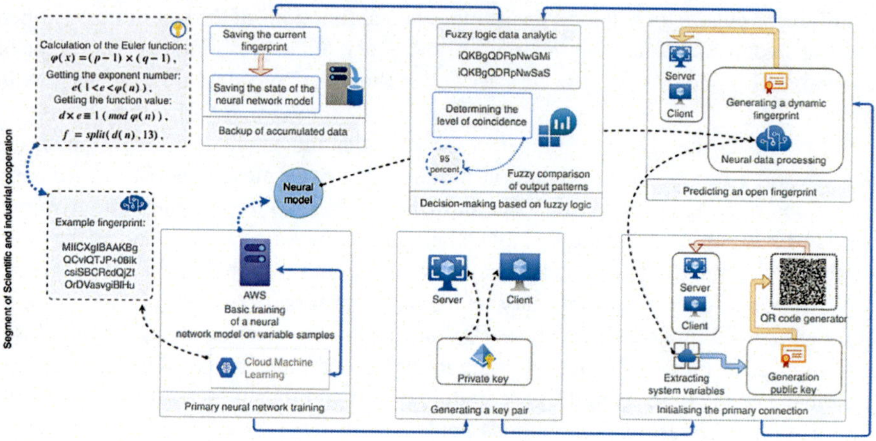

Fig. 2. Algorithm for generating encryption keys.

2.1 Primary Neural Model Training and Private Key Generation

At the presented system initial stage, the starting model training module of the predicted parameters is carried out. As can be seen from Fig. 2, trained patterns example is a set of digital prints generated with a modification of standardised RSA cryptography algorithm – at the initial stage, the value of the *Euler* function is calculated from the product modulus of 1024 bit numbers (1). Then a simple number is written using an exponent according to the condition of (2).

$$x = \left[p_{random_1} \times q_{random_2}\right],$$
$$\varphi(x) = (p-1) \times (q-1), \tag{1}$$

were x is the product of two numbers from the random function; φ is the Euler function; p is the first number from the random function; q is the second number from the random function.

$$e(1 < e < \varphi(n)), \tag{2}$$

were e is the exponent (reciprocal with the *Euler* number); φ *(n)* is the output value of the Euler function.

At the next stage of algorithm, the inverse-multiplicative exponent number is calculated modulo the *Euler* function value (3). The output value is the function *f*, which splits the value of the function into 13 characters.

$$\begin{aligned} d \times e &\equiv 1(mod\varphi(n)), \\ f &= split(d(n), 13), \end{aligned} \tag{3}$$

were f is the output function of the key for the training sample; d is a multiplicative number; *d (n)* is the generated full key; *split* is the separation function.

As the main logic for predicting digital prints, the Elman neural network (Elman RNN) is used, which is represented by a three-layer network with direct communication, and the only input for the objective function f of (3) [12, 13]. Among features of this neural network, one can distinguish the contextual neural units use in each neuron of the hidden layer (see Fig. 3).

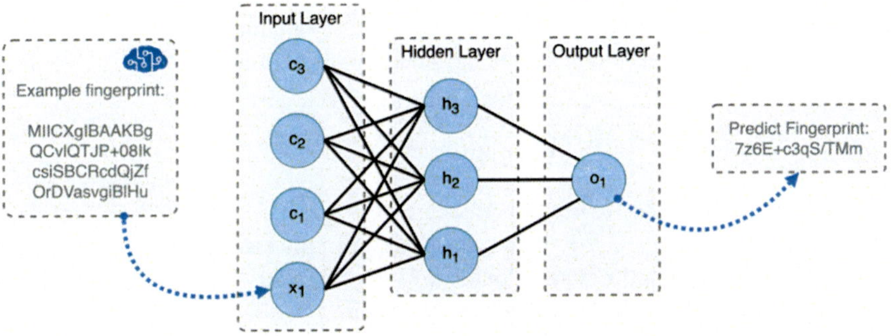

Fig. 3. The process of generating a sequence using Elman NN.

Software implementation of Elman RNN is based on the use of the PyTorch library [14, 15]. The size of input layer is total value of 6 contextual neurons and one input neuron (functions input, hidden, output size) [16, 17].

The value of the "seq_length" parameter is equal to value of input sequence length, i.e., 13. Two matrices with normal distribution are used as weight matrices. Next, the variable is initialised to store the neurons of the context, which combines the history of the input sequence by combining the input vector with the variable and the subsequent scalar product of variable and second matrix of weights. Also, in the process of work, the root-mean-square error (MSE) is calculated during cyclic passage, as a result of which, the output vector y_t is constructed according to the following (4), and the hidden vector h_t according to the (5):

$$y_t = \sigma_y\left(W_y h_t + b_h\right), \tag{4}$$

were y_t is the output vector of the neural network; σ_y is the activation function; h_t is the hidden vector of the neural network; *W, U, b* are the matrices of weights/parameters.

$$h_t = \sigma_h(W_h x_t + U_h h_{t-1} + b_h), \tag{5}$$

were h_t is the hidden vector of the neural network; σ_h is the activation function; x_t is the input vector; *W, U, b* are the matrices of weights / parameters.

At the final stage, after training, the prediction's function is executed to obtain the value at the next step (the predicted value based on the input sample). This function, interpreted as "predictions.append(pred.data.numpy().ravel()[0])" allows to get the predicted value of the private key, which is identical for server and the client [18, 19].

2.2 The Process of Initialising the Primary Key for Connexion

The initialization primary connexion process is based on the use of a partial private key generation function and a joint dynamic variable generation process. This process relies on the use of current time interval dependence on the starting point of Coordinated Universal Time (UTC) standard. The implementation of described approach makes it possible to use independent local variables from individual subjects two physical devices of the scientific and industrial cluster to initialise communication. As can be seen from (6), the process of generating a QR code is reduced to obtaining a hash function of a dynamic key, the duration of which is determined by the parameter T.

$$QR_{key} = MD5\left(\left[\frac{T_1 - T_0}{T}\right] \times round\left(static_{var}\right)\right), \tag{6}$$

were QR_{key} is the output key to be converted into a QR code; T_1 is the current time value from the time package; T_0 is the static time offset from the starting point in UTC; T is the static time offset from the starting point in UTC; *staticvar* – is the static variable of the server identifier of scientific and industrial cluster.

Thick, as a result of using this function, a temporary digital fingerprint is generated to initialise the connexion. Peculiarity of which is its 100% match on the client and server.

An alternative primary key initialization function is the use of static variables extracted from the server hardware features. In this case, the server ID is extracted by using application software that allows to get the processor ID (7). Among other things, to increase security, it is also possible to use the dynamic variable of the terminal process identifier obtained using (8).

$$QR_{key} = MD5(VendorID), \tag{7}$$

were QR_{key} is output key to be transferred and authorised on the client; MD5 is the function of obtaining the hash function of the processor offset value (*VendorID*); *VendorID* is the value of the processor offset (hardware-dependent variable).

$$QR_{key} = MD5(\$\$), \tag{8}$$

were QR_{key} is output key to be transferred and authorised on the client; *MD5* the obtaining function of hash function of the value; *$$* is the value of the terminal identifier.

Using algorithms and models presented above together with the qrencode package allows to obtain a code for further identification of the device on server.

2.3 Prediction of Open Fingerprint and Decision-Making

As can be seen from the diagram in Fig. 2, the process of predicting open fingerprint is based on using a trained model of neural network test dataset followed by glueing system variables, which in this case act as time-dependent parameters. So, the final sequence (public key) storage variable is the result of the following function representation:

```
def (pred_offset, static_offset, t):
time_offset = date +%s // 60
return pred_offset+/static_offset+/time_offset | cut -c2–13
```

As can be seen from the listing above, the software implementation involves the use of function that returns the hash value of the parameter predicted offset sum by the neural network, a static identifier variable and a time offset [20, 21]. Using this approach allows to solve the problem of unauthorised replacement of fingerprint and the possibility of unauthorised generation / substitution of them [22, 24].

Managing process based on fuzzy logic is used to make decisions on access verification according to the public key. The need for this module is due to possible discrepancies in the predicted values of the neural network on different hardware platforms. The software implementation of this module using the Fuzzy Wuzzy library allows to use the process functions for selective comparison of input values. For example, using the extract function allows to perform a character-by-character comparison with the subsequent output of the result as a percentage, as shown in Fig. 4.

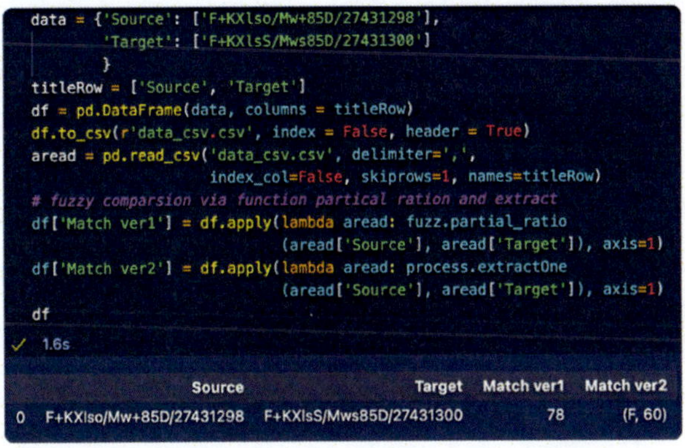

```
data = {'Source': ['F+KXlso/Mw+85D/27431298'],
        'Target': ['F+KXlsS/Mws85D/27431300']
       }
titleRow = ['Source', 'Target']
df = pd.DataFrame(data, columns = titleRow)
df.to_csv(r'data_csv.csv', index = False, header = True)
aread = pd.read_csv('data_csv.csv', delimiter=',',
                    index_col=False, skiprows=1, names=titleRow)
# fuzzy comparsion via function partical ration and extract
df['Match ver1'] = df.apply(lambda aread: fuzz.partial_ratio
                            (aread['Source'], aread['Target']), axis=1)
df['Match ver2'] = df.apply(lambda aread: process.extractOne
                            (aread['Source'], aread['Target']), axis=1)
df
✓ 1.6s
```

	Source	Target	Match ver1	Match ver2
0	F+KXlso/Mw+85D/27431298	F+KXlsS/Mws85D/27431300	78	(F, 60)

Fig. 4. Decision-making process for access verification.

As can be seen from Fig. 4, the verification process of two lines is reduced to verification using functions "partial_ratio" and "extractOne", where discrepancies in percentage are detected.

Thus, even in the case of a symbolic comparison, where the result of 78% is visible, which is formed due to a partial discrepancy in the key and the verification time [25]. It can be said that authentication is possible even under the condition (validate > = 70%) and this option will provide a high level of security.

2.4 Backup of Accumulated Data

Backup of accumulated data is a two-layer saving the current fingerprint process and the state of neural network model. As can be seen from the algorithm in Fig. 5, the module allows to save the last authenticated fingerprint on the server or client after a certain time interval.

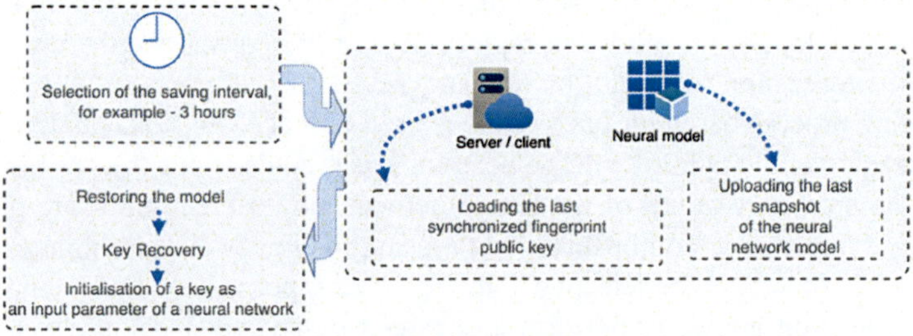

Fig. 5. The process of backup and accumulated data recovery.

The restoration of the neural network for subsequent work is implemented on "hot start" principles [26]. According to this, can save the current model at any time with the appropriate accuracy and loss coefficients.

3 Results and Discussions

Thus, it can be said that the result of this multi-factor authentication system development will significantly increase information and communication interaction security in scientific and industrial clusters. Providing structured possible variations forecasting of the public key both on the client and on the server in the developed proposal is solved by implementing a neural network forecasting module. The interface of interaction with the user from a mobile application is essential for visualising work of the developed system and coordinating its functioning (see Fig. 6) [27–29].

As can be seen in Fig. 6, the user is provided with the automated authorization possibility of the client with server, as well as the functions of controlling output of QR code, neural network model, selecting the interval of the public authentication key and changing the server private key [30].

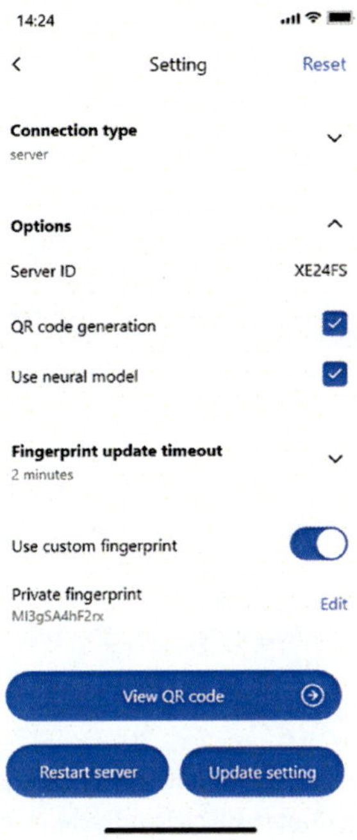

Fig. 6. Developed user interaction interface.

4 Conclusion

The algorithmic-programmatic approach developed in this study to the implementation of information interaction between individual industrial enterprices in clusters will allow to minimise existing systems disadvantages of organising interaction between devices during cooperation, based on the automated method of multi-factor authentication. Due to the use of existing symmetric and asymmetric encryption methodologies, an algorithm was developed both for the initial initialization of connexion by means of QR code, and subsequent verification of clients on the server due to two-way system for generating cryptographic open fingerprints.

The implemented system uniqueness is due to the combined use of neural network forecasting methods of time series based on deep learning neural networks with the use of algorithms for dynamic system variable calculation. An important aspect in this system implementation is the use of a fuzzy logic system, which made it possible to perform an analytical open fingerprint comparison with subsequent output of the result as a percentage.

Acknowledgments. The research was supported by RSF (project No. 22-21-00487).

References

1. Kirillova, E.A., Zaenchkovski, A.E.: Model of integration of information logistics technologies for managing elements and processes at various levels of functioning of the innovation

environment with the allocation of logistics optimization contours. J. Legal Econ. Res. **2**, 166–170 (2019)
2. Dli, M.I., Zaenchkovski, A.E., Kakatunova, T.V.: Prerequisites for using controlling to manage industrial clusters. Manag. Econ. Syst. Electr. Sci. J. **128**, 14–17 (2019)
3. Kalayda, S.A.: Model of creating an economic ecosystem in the framework of economic convergence under the influence of Digitalization. J. Appl. Inform. **16**, 28–42 (2021). https://doi.org/10.37791/2687-0649-2021-16-6-28-42
4. Dooley, J.: History of Cryptography and Cryptanalysis Codes, Ciphers, and their Algorithms. Springer, Cham (2018)
5. Rangisetti, A.K., Dwivedi, R., Singh, P.: Denial of ARP spoofing in SDN and NFV enabled cloud-fog-edge platforms. Clust. Comput. **24**(4), 3147–3172 (2021). https://doi.org/10.1007/s10586-021-03328-x
6. Almon, L., Krause, A.M., Fietze, O., Hollick, M.: Desynchronization and MITM attacks against neighbor awareness networking using opennan. In: Proceedings of the 19th ACM International Symposium on Mobility Management and Wireless Access, pp. 97–105 (2021). https://doi.org/10.1145/3479241.3486689
7. Travedi, B.: Cryptography and Network Security: Demystifying the ideas of Network Security, Cryptographic Algorithms, Wireless Security, IP Security, System Security, and Email Security. BPB Publications, S.l. (2021)
8. Pappu, S., Kangane, D., Shah, V., Mandwiwala, J.: Ai-assisted risk based two factor authentication method (AIA-RB-2FA). In: 2021 International Conference on Innovative Computing, Intelligent Communication and Smart Electrical Systems (ICSES), pp. 1–5 (2021). https://doi.org/10.1109/icses52305.2021.9633937
9. Malikovich, K.M., Turakulovich, K.Z., Tileubayevna, A.J.: A method of efficient OTP generation using pseudorandom number generators. In: 2019 International Conference on Information Science and Communications Technologies (ICISCT), pp. 1–4 (2019). https://doi.org/10.1109/icisct47635.2019.9011825
10. Boonkrong, S.: Authentication and Access Control: Practical Cryptography Methods and Tools, 1st edn. Apress, Berkeley (2021)
11. Romano, F., Kruger, H.: Learn Python Programming: An In-Depth Introduction to the Fundamentals of Python. Packt Publishing, Birmingham (2021)
12. Feng, C., et al.: Compact design of on-chip Elman optical recurrent neural network. In: Conference on Lasers and Electro-Optics, pp. 1–2 (2020). https://doi.org/10.1364/cleo_at.2020.jth2b.8
13. Ab Aziz, M.F., et al.: Integrating Elman recurrent neural network with particle swarm optimization algorithms for an improved hybrid training of multidisciplinary datasets. Expert Syst. Appl. **183**, 115441 (2021). https://doi.org/10.1016/j.eswa.2021.115441
14. Jha, A.R., Pillai, G.: Mastering Pytorch: Build Powerful Neural Network Architectures Using Advanced PYTORCH 1.x Features. Packt Publishing, Birmingham (2021)
15. Loy, J.: Neural Network Projects with Python: The Ultimate Guide to Using Python to Explore the True Power of Neural Networks Through Six Projects. Packt Publishing Ltd, Birmingham (2019)
16. Hu, G., Zhang, Z., Chen, J., et al.: Elman Neural Networks combined with extended Kalman filters for data-driven dynamic data reconciliation in Nonlinear Dynamic Process Systems. Ind. Eng. Chem. Res. **60**, 15219–15235 (2021). https://doi.org/10.1021/acs.iecr.1c02916
17. Suri, M., Raj, N., Sireesha, K.: Prediction of EV arrivals at battery swapping station using hybrid neural network models. J. Phys: Conf. Ser. **2062**, 012021 (2021). https://doi.org/10.1088/1742-6596/2062/1/012021
18. Jain, A., et al.: Performance characterization of DNN training using tensorflow and pytorch on modern clusters. In: 2019 IEEE International Conference on Cluster Computing (CLUSTER), pp. 1–11 (2019). https://doi.org/10.1109/cluster.2019.8891042

19. Park, A.: Python Machine Learning: A Complete Guide for Beginners on Machine Learning and Deep Learning with Python. Independently Published, New York (2019)
20. Chan, J., Chung, R., Huang, J.: Python API Development Fundamentals: Develop a Full-Stack Web Application with Python and Flask. Packt Publishing Ltd., Birmingham (2019)
21. Romano, F.: Learn Python Programming: The No-Nonsense, Beginner's Guide to Programming, Data Science, and Web Development with Python 3.7. Packt, Birmingham, Mumbai (2018)
22. Li, X., Chen, T., Cheng, Q., Ma, J.: An efficient and authenticated key establishment scheme based on fog computing for healthcare system. Front. Comp. Sci. **16**(4), 1–12 (2022). https://doi.org/10.1007/s11704-021-0537-z
23. Bibak, K., Kapron, B.M., Srinivasan, V.: Authentication of variable length messages in quantum key distribution. EPJ Quant. Technol. (2022). https://doi.org/10.1140/epjqt/s40507-022-00127-0
24. Andola, N., Prakash, S., Yadav, V.K., et al.: A secure searchable encryption scheme for cloud using hash-based indexing. J. Comput. Syst. Sci. **126**, 119–137 (2022). https://doi.org/10.1016/j.jcss.2021.12.004
25. Kamide, N.: Sequential fuzzy description logic: reasoning for fuzzy knowledge bases with sequential information. In: 2020 IEEE 50th International Symposium on Multiple-Valued Logic (ISMVL), pp. 218–223 (2020). https://doi.org/10.1109/ismvl49045.2020.000-2
26. Nanjappan, M., Natesan, G., Krishnadoss, P.: An adaptive neuro-fuzzy inference system and black widow optimization approach for optimal resource utilization and task scheduling in a cloud environment. Wireless Pers. Commun. **121**(3), 1891–1916 (2021). https://doi.org/10.1007/s11277-021-08744-1
27. Frain, B.: Responsive Web Design with HTML5 and CSS: Develop Future-Proof Responsive Websites Using the Latest HTML5 and CSS Techniques. Packt Publishing, Birmingham, UK (2020)
28. Romano, F.: Learn Web Development with Python: Get Hands-on with Python Programming and Django Web Development. Packt Publishing, Birmingham, UK (2018)
29. Grinberg, M.: Flask Web Development: Developing Web Applications with Python. O'Reilly Media, Beijing (2018)
30. Northwood, C.: The Full Stack Developer Your Essential Guide to the Everyday Skills Expected of a Modern Full-Stack Web Developer. Apress, New York (2019)

Optimal Energy Consumption Scheduling for Enterprises with Local Energy Sources

O. Yu. Maryasin(✉) and A. I. Lukashov

Yaroslavl State Technical University, 88, Moskovsky Prospect, Yaroslavl 150023, Russia
maryasinou@ystu.ru

Abstract. The paper reviews the optimization of energy consumption of an enterprise using a centralized power grid along with local generators and energy storage systems. To solve this problem, the authors suggest an optimal power consumption scheduling algorithm with a moving scheduling horizon. The advantage of this algorithm is the use of more precise power output values of local power generation and storage systems at every stage to reduce the probability of discrepancies between the planned and actual energy consumption profiles. This paper presents an evaluation of savings due to the use of different power supply variants. The results obtained can be used for the dynamic control of the enterprise's energy consumption and the selection of local power generation and storage system parameters to facilitate the economic and technical requirements of the company operations.

Keywords: Energy optimization · Optimal consumption scheduling · Renewable energy sources · Smart Grid · Forecasting · Linear programming

1 Introduction

The spreading of local power generators like photoelectric panels, wind turbines, diesel and gas generators resulted in the emergence of distributed Smart Grids (SG) that form local power ecosystems where electricity consumers are also its producers [1–3]. Currently, many companies purchase local generators, including those based on renewable energy resources (RER), and energy storage systems to cut the electricity costs, reduce greenhouse gas emissions, and improve the reliability of power supply to company consumers. In the nearest future, we expect a boost in the number of distributed generators and energy storage systems at enterprises. Thus, we can talk about the gradual transformation of the energy networks of enterprises into SG.

The enterprises of the Russian Federation that pay for electricity based on price categories 3–6 can save on electricity costs through energy consumption control considering the following aspects: changes in hourly market electricity prices; the power payment during peak load hours.

In the first case, electricity costs can be reduced by cutting down the consumption when the prices are high and increasing the loads when they become lower. In the second case, it is necessary to reduce the energy consumption during the peak load hours

A. A. Radionov and V. R. Gasiyarov (Eds.): RusAutoCon 2022, LNEE 986, pp. 282–293, 2023.
https://doi.org/10.1007/978-3-031-22311-2_28

as much as possible. It is also necessary to consider the limitations on the minimum and maximum energy consumption that are determined by the main area of the company activities. These limitations can include limitations associated with the production equipment operation, production safety, health risks, etc. Thus, the task of optimal energy consumption scheduling (OECS) of the enterprise can be set, taking into account the market electricity prices, peak load hours, and other consumption limitations.

The OECS problems are relevant for open-loop systems when the values of the optimized parameters are determined in advance for the entire scheduling period based on the forecasts for the key system parameters. These problems are often solved in demand side management (DSM) systems and SGs [4–6]. More specifically, they include the energy consumption scheduling problems for household appliances that have recently become popular [7, 8].

Another approach that is widely used in practice for the optimization of energy consumption is the optimal energy consumption control (OECC). The OECC problems are relevant for closed-loop systems when the values of the optimized parameters are determined based on the current state of the system. An example of such a problem is the optimal energy consumption control for a building [9, 10]. Model predictive control (MPC) is one of the most popular optimal energy consumption control methods [11, 12].

In this paper, the authors continue the research they previously published in [13]. Compared to the OECS problem set out in [13], the problem solved in this work has several significant complements. One of the complements allows for the use of the central grid, along with local power generation and storage facilities, as energy sources. This is linked to the previously mentioned trend of transforming the energy networks of enterprises into SGs. Another complement stipulates solving the problem using the moving scheduling horizon employed in MPC algorithms. This allows for the dynamic changing of the enterprise's energy consumption profile, taking into account the actual power output of local generation and storage devices. Yet another important complement is the possibility of accounting for the uncertainties occurring as a result of peak load hours forecasting.

2 The OECS Problem

It is assumed that the enterprise may consume electricity both from the central grid and from local energy sources. Power generators can be divided into RER-based and those using non-renewable or fossil energy sources (FES). RER power can be consumed directly or through distributed energy storage (DES) systems.

Assume x_h is the power consumption of the enterprise from the central grid at h, $h = t+\Delta h,\ldots, t+H$, $p_{g,h}$ is the power output of generator g, $g = 1,\ldots,G$ at h, $G = R+F$ is the total number of generators at the enterprise, R is the number of RER generators, F is the number of FES generators, $d_{s,h}$ is the power output of DES device s, $s = 1,\ldots,S$ at h, t is the start time, H is the scheduling horizon. The time step Δh is deemed equal to one hour. Each of the DES device s, $s = 1,\ldots,S$ is connected to RER generator g, $g = 1,\ldots,R$. The optimality criterion for the optimal energy consumption scheduling problem looks as follows:

$$E_{OECS} = E_1 + E_2 + \ldots + E_h + \ldots + E_H, \tag{1}$$

where

$$E_h = c_h x_1 + c_g p_{1,h} + \ldots + c_g p_{G,h} + \ldots + c_s d_{1,h} + \ldots + c_s d_{S,h}, \tag{2}$$

c_h is the forecast market electricity price at h, $h = t+\Delta h,\ldots, t+H$, c_g is the price for the energy from generator g, c_s is the price for the energy from DES device s.

Denote the average daily power consumption required by the enterprise to perform its operations as X_d. Then the following condition is true:

$$(x_1 + Re_1 + De_1) + \ldots + (x_h + Re_h + De_h) + \ldots + (x_H + Re_H + De_H) = X_d, \tag{3}$$

where

$$Re_h = p_{1,h} + \ldots + p_{G,h}, \tag{4}$$

$$De_h = d_{1,h} + \ldots + d_{S,h}. \tag{5}$$

If facility i does not have any renewable energy sources or storage facilities, the respective variables $p_{s,h}$ and $d_{s,h}$ are equated to zero.

Let us define the lower and upper limits for per-hour power consumption, as well as generator and DES device power output:

$$x_{h,l} \leq x_h \leq x_{h,u}, \quad h = t + \Delta h, \ldots, t + H, \tag{6}$$

$$p_{g,h,l} \leq p_{g,h} \leq p_{g,h,u}, \quad g = 1, \ldots, G, \quad h = t + \Delta h, \ldots t + H, \tag{7}$$

$$d_{s,h,l} \leq d_{s,h} \leq d_{s,h,u}, \quad s = 1, \ldots, S, \quad h = t + \Delta h, \ldots t + H, \tag{8}$$

$$x_h + Re_h + De_h \leq P_u, \quad h = t + \Delta h, \ldots, t + H, \tag{9}$$

where $x_{h,l}$, $x_{h,u}$ are the lower and upper limits for per-hour power consumption from the central grid, $p_{g,h,l}$, $p_{g,h,u}$ are the lower and upper limits for per-hour power consumption from generators, $d_{s,h,l}$, $d_{s,h,u}$ are the lower and upper limits for per-hour power consumption from the DES device, P_u is the maximum power consumption of the enterprise's electrical equipment.

The limits of per-hour power consumption are determined by the enterprise's ability to reduce or increase its power consumption, as well as the capacities of the local power generation and storage devices. The upper limit value for per-hour power consumption $x_{h,u}$ depends on the overlapping of the current hour and the peak load interval h_{pp}, which is unknown for the current month but can be assessed through forecasting. During the peak load interval h_{pp}, it is necessary to reduce the power consumption to the acceptable minimum X_{pp}

$$x_{h,u} = X_{pp}, \quad i \in h_{pp}. \tag{10}$$

Peak load interval limits

$$h_l(\theta) \leq h_{pp} \leq h_u(\theta), \tag{11}$$

where $h_l(\theta)$, $h_u(\theta)$ are lower and upper limits, θ is the uncertainty associated with peak load hour forecasting. Condition (11) can be rewritten using probabilities

$$P\{h_l(\theta) \leq h_{pp} \leq h_u(\theta)\} \geq \alpha, \tag{12}$$

where α is the set probability.

Thus, the OECS problem will include the minimization of criterion (1) while fulfilling conditions (3)–(10) and probabilistic limiting (12). The desired variables for this problem are $x_h, p_{g,h}, d_{s,h}$. With $G = 0, S = 0, H = 24$ and without any consideration of uncertainty θ, problem (1),(3)–(10),(12) becomes equivalent to the problem described in [13].

3 The OECS Problem with Moving Horizon

The solving algorithm for the optimal energy consumption scheduling problem with a Moving Horizon (OECSMG) includes the following key steps:

1. Forecasting the market electricity prices c_h and peak load hours h_{pp} for the next month. These forecasts can be obtained using the application developed previously by the authors and described in [14]. Forecasting the power output of RER generators $p_{g,h}$, $g = 1,\ldots,R$ during the interval $[t+\Delta\ h,\ t+H]$. This forecasting is equivalent to the calculation of the following dependency

$$p_{g,h+1} = f(p_{g,h}, \ldots, p_{g,h-l}), \tag{13}$$

   where f is a non-linear function in the general case, l is the number reflecting the number of power output values during the previous periods used in the forecast. The power output of DES devices $d_{s,h}$ during the interval $[t+\Delta\ h,\ t+H]$ can be determined using the following formula [15]

$$d_{s,h+1} = d_{s,h}(1 - \sigma) + (p_{g,h} - x_h/\eta_{inv})\eta_b, \tag{14}$$

   where σ is the self-discharge rate factor of the battery, η_{inv} is the inverter efficiency factor, η_b is the battery efficiency factor.

   The obtained values of $p_{g,h}$ and $d_{s,h}$ are used to set $p_{g,h,u}$ and $d_{s,h,u}$ under conditions (7) and (8) respectively. The lower $x_{h,l}$ and upper $x_{h,u}$ per-hour power consumption limits are set. The probability value α is set.
2. Solving the OECS problem for the scheduling horizon H and determining the optimal values of variables x_h, $p_{g,h}$ and $d_{s,h}$ during $h = t+\Delta\ h,\ldots,\ t+H$. The solution of this problem is the optimal profile of energy consumption using the central grid, RER generators, and DES devices for the scheduling horizon H. The optimal values of variables x_h, $p_{g,h}$ and $d_{s,h}$ at the current time are used to set the loads of production and power equipment.
3. Shifting the scheduling horizon one step forward. Determining the actual values of variables x_h, $p_{g,h}$ and $d_{s,h}$. The obtained actual power output values of RER generators, along with the previously accumulated data, are used for the forecasting on new values of $p_{g,h+1}$ for another step forward. In this case, forecasting is performed

over the small interval of Δ h. Rather than the horizon H, this can significantly improve the accuracy of forecasting. Based on the current values of variables x_h, $p_{g,h}$ and $d_{s,h}$, the values $d_{s,h+1}$ are calculated by formula (14) at the next time $h+1$. The new values of $p_{g,h+1}$ and $d_{s,h+1}$ are used to set the values of $p_{g,h+1,u}$ and $d_{s,h+1,u}$ under conditions (7) and (8).

4. Going to step 2 of the algorithm and repeating the procedure from that step.

Determining the optimal values of $p_{g,h}$ and $d_{s,h}$ during step 2 instead of their calculation using formulae (13) and (14) is used to prevent the inefficient use of the energy from local generators and storage devices of the enterprise's SG. The use of a moving horizon makes it possible to add feedback elements to the OECS problem and thereby bring the OECS problem closer to the OECC problem.

The presented scheme of the OECSMG algorithm, although it resembles the scheme used in MPC algorithms, differs from it in that the data obtained at the current step are used to adjust the constraint parameters in the OECS problem at the next step. For MPC algorithms, as a rule, the values of the constraint parameters do not change over the forecasting horizon.

4 Case Study

4.1 Input Data

As an example, let's take a small industrial company with a round-the-clock work schedule and with a production equipment capacity not exceeding 120 kW. The main source of power supply for the enterprise is the central power grid. To reduce the electricity costs, let us review the possibility of using local power generation and storage devices at the enterprise. To do that, we explore the following power supply variants for the enterprise. Variant 1 features the central grid complemented with an array of photovoltaic (PV) modules, providing a total of 40 kW extra power. Variant 2, apart from the power sources mentioned in Variant 1, includes a 40 kW storage device connected to the grid. Variant 3 comprises the central power grid, a 40kW wind turbine and a 40kW grid storage device. Variant 4 differs from Variant 3 because instead of the grid storage device, it uses a 40 kW diesel generator.

To solve the OECSMG problem, the information on the market electricity prices was used, which is published monthly on the website of the energy provider for the region where the enterprise is located [16]. For certainty, we used the electricity price data for June 2021. As for the specific date on which we solved the OECSMG problem, we used 18.06.2021.

The full electricity price for the energy consumers belonging to price categories 3–6 includes the market price component, generating power fees, power transmission fees (only for price categories 4 and 6), sales margin of energy provider, as well as the payments for other services. During the peak load hours, the electricity price may exceed tenfold compared with the price at other hours. This is due to the payment for generating power, the amount of which depends on the power consumption during peak load hours.

Apart from the forecasting of the market electricity prices, solving the OECSMG problem requires the forecast peak load hours for the desired date. As shown in [17], the precision of peak load hours forecasting for a one-hour interval is about 50% of cases during the month, and for a two-hour interval it is about 70%. Therefore, we can say that the probability of $\alpha = 0.5$ corresponds to the one-hour interval h_{pp}, and the probability $\alpha = 0.7$ corresponds to the two-hour interval. The actual data on the peak load hours are available on the Trade System Administrator website [18]. The peak load hour for 18.06.2021 was at 2 p.m.

The solution to the OECSMG problem also requires the RER generator power forecasts. RER generator power can be obtained from the experimental data or calculated using the well-known formulas using forecasts of the necessary climate parameters [15, 19]. There are ready-made libraries that can help determine the power of RER generators based on the open weather data in various regions all over the world [20]. The PV panel and wind turbine power data for June 2021 used in the work for forecasting purposes are shown in Fig. 1.

The price of electric power produced by RER generators was assumed 4 rubles per kW, which is comparable with the market electricity price. The price of diesel fuel was assumed to be 55 rubles per liter.

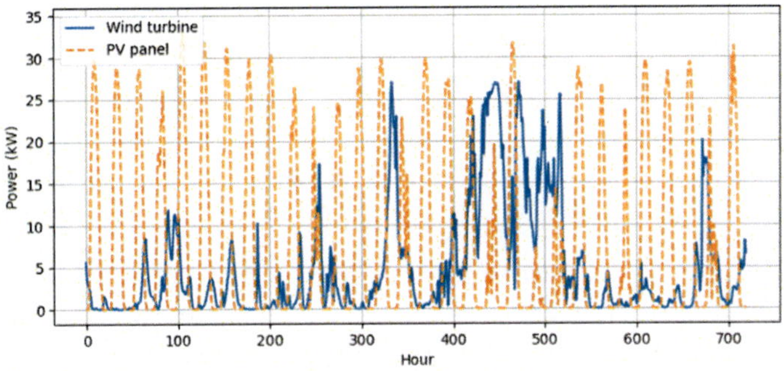

Fig. 1. The PV panel and wind turbine power charts for the June 2021.

To forecast the temporary values of RER generator power over the scheduling horizon H, we used a Long Short Term Memory Neural Network (LSTM NN). These neural networks are well-suited for the time-series forecasting [21]. In this paper, we did not aim to achieve high forecasting precision for RER generator power. The forecasts obtained were only used to illustrate the operation of the proposed OECSMG algorithm.

4.2 Optimization Results

The authors conducted many numerical experiments to solve the OECS and OECSMG problems in question for enterprise power supply variants 1–4. Since the OECS is linear and expression (12) is not involved in the optimization, the linear programming method is used at every step of the OECSMG algorithm to solve the OECS problem. An example of using the linear programming method to solve the energy consumption optimization problem is presented in [22].

For illustration purposes, we will first review the solutions for the OECS problem at the initial time, for the one-hour peak load interval, then for the two-hour interval, and finally we will compare the results of solving the OECS and OECSMG problems. The optimal power consumption profile charts for the central grid (Net Power) and the PV panel (PV power), obtained while solving the OECS problem for variant 1, as well as that of the total power consumption from all the sources (Sum Power) on 18.06.2021 under one-hour peak load interval, are shown in Fig. 2. For comparison, the same figure shows the power consumption chart (Base Power), obtained from [13]. Figure 2 shows that, compared to the basic variant from [13], the optimal total power consumption profile provides non-zero power consumption during the peak load hours.

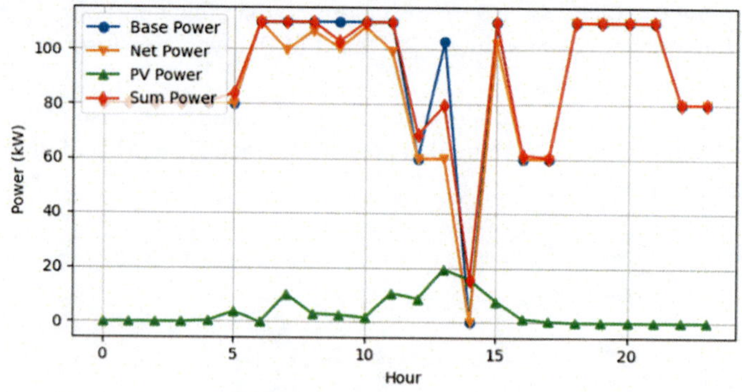

Fig. 2. Charts for variant 1.

The optimal power consumption profile charts for the central grid, PV panel, and grid storage device (Battery Power), as well as the total power consumption for variant 2 on 18.06.2021 for the one-hour peak load interval, are shown in Fig. 3. The use of the electric power from the storage device allows for the significant increase in power consumption during the peak load period, which leads to the equalization of the profile. The operation of grid storage devices must be arranged so that the storage systems were fully charged at the beginning of the peak load interval.

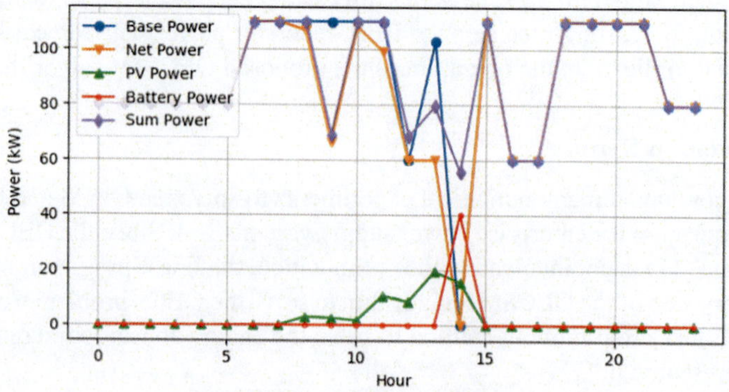

Fig. 3. Charts for variant 2.

The optimal power consumption profile charts for the central grid, wind turbine (WT Power), and grid storage system, as well as the total power consumption for variant 3 on 18.06.2021 for the one-hour peak load interval, are shown in Fig. 4. Since the wind turbine produced more power on 18.06.2021 than the PV panel, the consumption from the central power grid in variant 3 was lower than in the previous variant.

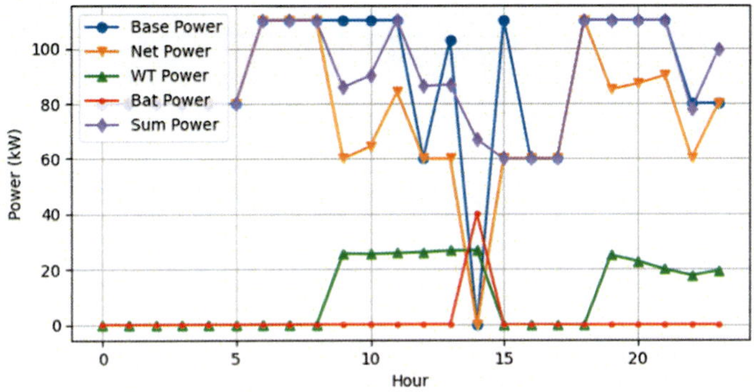

Fig. 4. Charts for variant 3.

The optimal power consumption profile charts for the central grid, wind turbine, and diesel generator (DG Power), as well as the total power consumption for variant 4 on 18.06.2021 for the one-hour peak load interval, are shown in Fig. 5. This variant facilitates the most stable total power consumption profile compared to the previous variants.

All the variants discussed can assure non-zero power consumption during the peak load hour. Variants 2–4 can also facilitate a more even power consumption profile. Profile drops during daytime can be attributed to, first, a higher electricity price during this time, and, second, to zero power consumption from the central grid during the peak load hour. To equalize the profile, it is necessary to increase the power output of the RER generators and storage devices used. For variants with grid storage devices, it is necessary to remember that most of the energy produced by RER generators is used for the charging of storage devices. With fully charged grid storage devices, RER generator power may be used for production purposes.

We must note that the power consumption profile obtained for variant 4 is the most convenient for the enterprises from the practical standpoint as it does not feature power consumption surges and is the smoothest one. Therefore, this profile can be easily implemented at the enterprise with minimum costs.

For the two-hour peak load interval, all of the power consumption variants reviewed, except for variant 4, cannot compensate for the reduction in power consumption over the entire interval h_{pp}. Consider the solution of the OECS problem for variant 2. The optimal power consumption profile charts for the central grid, PV panel, and storage system, as well as the total power consumption for variant 2 on 18.06.2021 for the two-hour interval, are shown in Fig. 6.

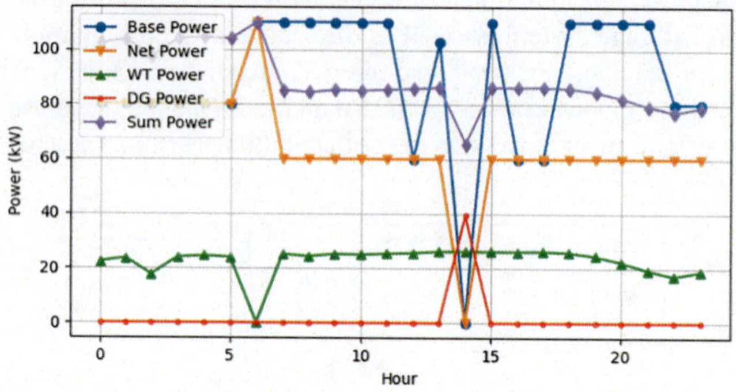

Fig. 5. Charts for variant 4.

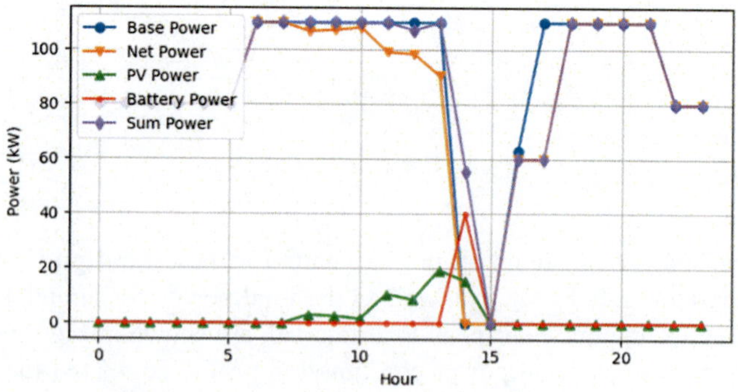

Fig. 6. Charts for variant 2, two-hour interval.

For variant 2, the capacity of the storage system is only enough to compensate for the reduction in power consumption during the first hour of the peak load interval. A similar situation can be observed in variant 1 and 3, but the compensation through the PV panel power for variant 1 is much lower, and the compensation through the use of the wind power and grid storage in variant 3 is a little higher. In terms of profile stability, variant 4 proved to be the best one for the two-hour interval as well. For this variant, the little reduction in power consumption at 2 p.m. observed in the total power consumption chart in Fig. 5 is prolonged for another hour. At other times, the total power consumption chart does not change significantly.

Let's compare the solution results of the OECS and OECSMG problems for one of the presented variants. Figure 7 shows the optimal total power consumption charts obtained while solving the OECS and full OECSMG problems for 18.06.2021 for the one-hour interval and for variant 2.

The analysis of Fig. 7 shows that the use of the OECSMG algorithm instead of OECS may have a significant impact on the power consumption profile. However, this does not provide any benefits in terms of energy savings. The same conclusion can be made for other variants as well. The key advantage of using the OECSMG algorithm is the use of actual power output values of RER generators and DES devices at every step of the algorithm. Thus, the profile obtained while solving the OECSMG program is closer to the actual one. It can significantly reduce the probability of significant discrepancies between the planned and actual power consumption profiles, which may result in the violations of enterprise production program.

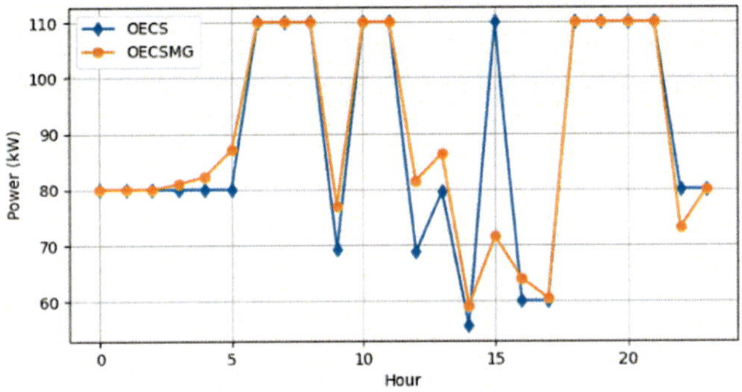

Fig. 7. The power consumption charts for the OECS and OECSMG problems for variant 2.

4.3 Energy Savings Assessment

The electricity costs for the enterprises being energy consumers 3–6 price categories include the energy costs and the power payment. The power payment is calculated as a product of the weighted average market price of power and the average energy consumption of the enterprise during the peak load hours for the current month. For the considered case, it amounted to 70,698.73 rubles in June 2021. If all forecasts of peak load hours were absolutely precise and during these hours it would be possible to reduce the energy consumption of the enterprise during the month to zero, the savings obtained from the power payment would equal the previously mentioned sum or 3366.61 rubles per day.

Table 1 shows the daily savings obtained through solving the OECSMG problem for various power supply variants at the enterprise with one- and two-hour peak load intervals. The savings were calculated relative to the basic variant amount, i.e. 12,730.81 rubles. Besides, Table 1 shows the total savings obtained by summing up savings due to energy payment and the power payment. The power payment values were used considering the probability of the peak load hour falling into the predicted time interval.

The analysis of the Table 1 data shows that variant 3 with central power grid complemented with a wind turbine and a storage device produces the most savings. Variant 2 using a PV panel instead of a wind turbine yields a rather similar result. Variant 4 produced less savings compared to variants 2 and 3 because of the diesel fuel costs. Variant 1 proved to be the least efficient for a one-hour interval, even though with the two-hour interval it produced more savings that variant 4.

Table 1. Savings for various power supply variants

Variant #	Savings for 1h interval, rub.	Total savings for 1h interval, rub.	Savings for 2h interval, rub.	Total savings for 2h interval, rub.
1	3355.25	5038.55	4044.87	6401.5
2	4209.19	5892.49	4208.79	6565.42
3	4250.87	5934.17	4252.67	6609.30
4	3747.45	5430.75	3374.17	5730.8

5 Conclusions

The paper reviews the optimization of energy consumption of an enterprise using not only a centralized power grid, but also local generators and storage devices. To solve this problem, authors suggested the OECSMG algorithm. It differs from the OECS algorithm because, after solving the OECS problem, the scheduling horizon moves one step forward, and the actual power output of the local generation and storage devices are used to calculate the new values of these variables during the next step. The presented scheme of the OECSMG algorithm may resemble the scheme used in MPC algorithms but in it, the data obtained during the current step are used for the adjustment of the constraint parameters in the OECS problem during the next step. The key advantage of the OECSMG algorithm is the use of actual power output values of local generators and storage devices at every stage to reduce the probability of significant discrepancies between the planned and actual power consumption profiles.

The paper compares various variants for the power supply of the enterprise, on the basis of which the following main conclusions can be drawn. To get more savings when paying for electricity, it is necessary to use RER generators together with grid storage devices. It can also help equalize the power consumption profile at daytime. To prevent power consumption surges, it is necessary to increase the power output of the RER generators and storage devices used. The best solution to stabilize the profile is the combined use of RER and FES generators. It is the only solution assuring the stable profile for the two-hour peak load interval. However, the use of FES generators may reduce the amount of savings due to the fuel costs.

The results obtained can be used for the dynamic control of the enterprise's power consumption and the selection of local power generation and storage system parameters to facilitate the economic and technical needs requirements of the company operations.

References

1. Ahmad, A., et al.: An optimized home energy management system with integrated renewable energy and storage resources. Energies **10**(549), 2–35 (2017)
2. Dileep, G.: A survey on smart grid technologies and applications. Renew. Energy **146**, 2589–2625 (2020)
3. Lamnatou, C., Chemisana, D., Cristofari, C.: Smart grids and smart technologies in relation to photovoltaics, storage systems, buildings and the environment. Renew. Energy **185**, 1376–1391 (2022)

4. Maharjan, I.K.: Demand Side Management: Load Management, Load Profiling, Load Shifting, Residential and Industrial Consumer, Energy Audit, Reliability, Urban, Semi-Urban and Rural Setting, p. 116. LAP Lambert Academic Publishing (2010)
5. Catherine, R.L.H., Soundarrajan, A.: A literature review on demand side and energy efficient management. Int. J. Curr. Trends Eng. & Res. **2**, 131–137 (2016)
6. Shah, A.S., Nasir, H., Fayaz, M., Lajis, A., Shah, A.: A review on energy consumption optimization techniques in IoT based smart building environments. Information **10**, 1–34 (2019)
7. Bradac, Z., Kaczmarczyk, V., Fiedler, P.: Optimal scheduling of domestic appliances via MILP. Energies **8**, 217–232 (2014)
8. Yahia, Z., Pradhan, A.: Optimal load scheduling of household appliances considering consumer preferences: an experimental analysis. Energy **163**, 15–26 (2018)
9. Kim, W., Jeon, Y., Kim, Y.: Simulation-based optimization of an integrated daylighting and HVAC system using the design of experiments method. Appl. Energy **162**, 666–674 (2016)
10. Ilbeigi, M., Ghomeishi, M., Dehghanbanadaki, A.: Prediction and optimization of energy consumption in an office building using artificial neural network and a genetic algorithm. Sustain. Cities Soc. **61**, 1–15 (2020)
11. Afram, A., Janabi-Sharifi, F.: Theory and applications of HVAC control systems – A review of model predictive control (MPC). Build. Environ. **72**, 343–355 (2014)
12. Ascione, F., Bianco, N., De Stasio, C., Mauro, G.M., Vanoli, G.P.: Simulation-based model predictive control by the multi-objective optimization of building energy performance and thermal comfort. Energy Build. **111**, 131–144 (2016)
13. Maryasin, O., Lukashov, A.I.: Optimizing the daily energy consumption of an enterprise. Lecture Notes in Electrical Engineering **857**, 370–382 (2022)
14. Maryasin, O.Yu., Lukashov, A.I.: Developing a digital model of an electricity consumer using deep learning. In: 2nd International Conference on Control Systems, Mathematical Modeling, Automation and Energy Efficiency, pp. 624–629 (2020)
15. Deshmukh, M.K., Deshmukh, S.S.: Modeling of hybrid renewable energy systems. Renew. Sustain. Energy Rev. **12**, 235–249 (2008)
16. Electricity prices TNS energo Yaroslavl, 2021. https://yar.tns-e.ru/legal-entities/prices/
17. Maryasin, O.Yu., Lukashov, A.I.: Analyzing and forecasting peak load hours. In: International Conference on Industrial Engineering, Applications and Manufacturing, pp. 25–30 (2021)
18. Peak load hours of ATS, 2021. https://www.atsenergo.ru/results/market/calcfacthour/
19. Hemeida, A.M., et al.: Optimum design of hybrid wind/PV energy system for remote area. Ain Shams Eng. J. **11**, 11–23 (2020)
20. GitHub – oemof/feedinlib, 2021. https://github.com/oemof/feedinlib
21. Chollet, F.: Deep Learning with Python, p. 384. Manning Publications Co (2018)
22. Lee, J.Y., Choi, S.G.: Linear programming based hourly peak load shaving method at home area. In: 16th International Conference on Advanced Communication Technology, pp. 310–313 (2014)

Methods for Improving the Efficiency of Neural Network Decision-Making

L. A. Lyutikova(✉)

Institute of Applied Mathematics and Automation, KBSC RAS (IPMA KBSC RAS), 89a, Shortanova St., Nalchik 360002, Russia
lylarisa@yandex.ru

Abstract. This The article proposes a method for creating neural network correctors based on a logical analysis of the initial data. In the study area, the relationship between features and objects is presented as a connection by precedents; a sigma-pi neural network was used to process the data. It is shown that the results of the neural network are only a part of the solutions that can be obtained as a result of a complete logical analysis. Therefore, to obtain a more complete solution, the neural network is proposed to be supplemented with logical correctors. The implementation of which is carried out in two stages. At the first stage, a classifier is built in the form of disjunctions of possible solutions, and the classifier can be built only on the basis of the structure of the neural network, the data may be unknown. The second stage proposes the implementation of this classifier in the form of a logical neural network. The use of such a hybrid approach can significantly affect the quality of solutions.

Keywords: Logic classifier · Neural network · ΣΠ - neuron · Decision rules

1 Introduction

At present, hybrid methods are often used in qualitative data analysis. To successfully combine methods, 1 must have a clear understanding of the decision rules, or those data areas where these methods are most effective. Many currently used approaches do not give a clear idea of how the decision was made. This creates a number of problems ranging from trust in the quality of the result to the possibility of correcting the method used. These methods include very popular neural networks [1–3]. Based on the very structure of the neural network, for example, deep learning networks, it is not possible to trace the chain of reasoning leading to the result, but you can try to determine the data area on which the trained neural network will be especially effective and where it will be possible to make mistakes. Accordingly, develop correction algorithms and get more fundamental ideas about the data being processed.

2 Materials and Methods

Let's assume that in the area under study, the relationship between features and objects can be represented as a connection by applicants; for data processing, we will use a sigma-pi neural network and logical data analysis.

A. A. Radionov and V. R. Gasiyarov (Eds.): RusAutoCon 2022, LNEE 986, pp. 294–303, 2023.
https://doi.org/10.1007/978-3-031-22311-2_29

The properties of each object is an n- dimensional vector, where n is the number of characteristics used to describe all objects in the area under consideration, j-th coordinate of this vector is equal to the value j-th sign $j = 1, \ldots, n$. Missing data is acceptable.

Formal formulation of the pattern recognition problem [4, 5].

Let $X = \{x_1, x_2, \ldots, x_n\}, x_i \in \{0, 1, \ldots, k_i - 1\}$, where is a set of features. We will assume that each feature is encoded by an arbitrary discrete value $k_i \in [2, ..., N], N \in Z$, $Y = \{y_1, y_2, \ldots, y_m\}$ - set of objects, each object y_i characterized by a corresponding set of features

$$x_1(y_i), \ldots, x_n(y_i) : y_i = f(x_1(y_i), \ldots, x_n(y_i))$$

$$\begin{pmatrix} x_1(y_1) & x_2(y_1) & \ldots & x_n(y_1) \\ x_1(y_2) & x_2(y_2) & \ldots & x_n(y_2) \\ \ldots & \ldots & \ldots & \ldots \\ x_1(y_m) & x_2(y_m) & \ldots & x_n(y_m) \end{pmatrix} \to \begin{pmatrix} y_1 \\ y_2 \\ \ldots \\ y_m \end{pmatrix}$$

Need to find a function $Y = f(X)$.

Very often, when solving such problems, 1 has to deal with data, the main source of which is experience, not theory. Therefore, choosing a formal apparatus for identifying patterns is a difficult task. Let's assume that formalisms that allow any kind of connections on a given set, except for those that contradict the declared experience, are the most complete for working with data. We call the function built on such reasoning a complete data classifier.

As a result of such reasoning, it is possible to construct a Boolean function, from variables, the value of which will be equal to 0, on those sets where there are signs of the object, but the object itself is not, and unity in all other cases $n + m$.

If the training sample consists of elements, then for description using a Boolean function, when k

$$F(x_1(y_i), ..., x_2(y_i), P^\sigma(y_1), ..., P^\sigma(y_n))$$

$$P^\sigma(y_i) = \begin{cases} \overline{P(y_i)} \; if \; \sigma = 0 \\ P(y_i) \; if \; \sigma = 1 \end{cases}$$

This function takes the values "0" on the sets when there is data characterizing the object, and the object itself is negated, i.e. AND "1" on all other a sets

$$x_1(y_i), \ldots, x_2(y_i), \; P^\sigma(y_1), \ldots, P^\sigma(y_n)$$

Example: The following ratios are given in Table 1:

Table 1. Objects and their attributes.

x_1	x_2	y
0	1	a
1	1	b

Let us build a Table 2 that defines a function for variables: $x_1, x_2, P(a), P(b)$

$$f_1(x_1, x_2, P(a), P(b)) == \overline{x_2}\vee P(a)\&\overline{x_1}\vee P(b)\&x_1$$

This is the representation of the function after converting its conjunctive normal form. As can be seen from the example, a variable that has the same value for two objects, being non-informative, stands as a free variable, and objects are identified by a variable that differs [6, 7].

Representing a function in the form of a formula is easier to build a conjunctive normal form, since there will be significantly fewer zeros in the column of function values. As a result, we will get

$$f(X) = \mathop{\&}_{j=1}^{m}\left(\bigvee_{i=1}^{n} \overline{x_i}\vee P(y_j)\right)$$

Given that

$$\bigvee_{i=1}^{n} \overline{x}(y_j)\vee P(y_j),\ j \in [1, ..., m]$$

Can be imagined as

$$\mathop{\&}_{j=1}^{m} x_j(y_i) \rightarrow P(y_i),\ i = 1, ..., l;\ x_j(y_i) \in \{0, 1, ..., k-1\},$$

Table 2. Truth table for all admissible solutions.

x_1	x_2	*P(a)*	*P(b)*	*f(X.Y)*
0	0	0	0	1
0	0	0	1	1
0	0	1	0	1
0	0	1	1	1
0	1	0	0	0
0	1	0	1	0
0	1	1	0	1
0	1	1	1	1
1	0	0	0	1
1	0	0	1	1
1	0	1	0	1
1	0	1	1	1
1	1	0	0	0
1	1	0	1	1
1	1	1	0	0
1	1	1	1	1

where is the predicate $P(y_i)$ takes the value true, i.e. $P(y_i) = 1$ if $y = y_i$, (object corresponding to the given characteristics) and $P(y_i) = 0$, if $y \neq y_i$[four]. We get that this function is a conjunction of production rules that express the feature-object relationship on a given data set [8–10].

In general, in order to separate the objects in this example, 1 clause such as, or is enough. But we will lose some of the values, and in the case of incorrect or noisy data, we will not be able to recognize the most probable of the presented objects. For example, set (1,0).

$$f(x_1, x_2, P(a), P(b)) = P(a)\&\overline{x_1}f(x_1, x_2, P(a), P(b)) = P(b)\&x_2.$$

The constructed function is a complete classifier for the given data. Its properties are considered in detail in [11].

It should be noted that the use of this function makes it possible to obtain a complete partition into all possible classes of the studied data.

The function consists of a finite number of clauses, some of the clauses contain predicates reporting the presence of specific objects, some do not contain any object predicates and consist only of variables. $f(X) = f_1 \vee f_2$; where $f_1(X)$- clauses containing only variables and $f_2(X)$- clauses containing predicates corresponding to the presence of the corresponding object.

The function $f_2(X)$ gives a picture of the division of the initial data into classes. As well as ideas about the special properties of individual objects.

The function $f_1(X)$ contains only variables whose combinations were not used as characteristics of the objects declared in these objects. But this makes it possible to modify the function. Allows you to see what properties the declared objects do not have.

The original function can be represented as follows

$$f(X) = Z_k(q_k(x), P(y_k), X);$$

$$Z_k(q_k(x), P(y_k), X_k) = Z_{k-1}\&(\bigvee_{i=1}^{n} \overline{x_k(y_i)} \vee P(y_k)) \vee q_{k-1}(x)\&(\bigvee_{i=1}^{n} \overline{x_k(y_i)} \vee P(y_k));$$

$$q_k(x) = q_{k-1}(x)\&(\bigvee_{i=1}^{n} \overline{x_k(y_i)});$$

$$q_1(x) = \bigvee_{i=1}^{n} \overline{x_1(y_i)}; \; Z_1 = P(y_1),$$

where $W(X)$ is the complete classifier, Z_j is the connection between objects and their characteristics, Q_j is the current state of the system. The state of the system is a combination of variables that do not participate as a property of the objects under study.

If it is represented in SDNF and successfully reduced, it can represent a compact representation of the data.

Of course, for big data, this approach will be somewhat cumbersome. Therefore, it is interesting to see what part of this function is the solution of neural networks on a given set of data [12].

As an example, we will consider ΣΠ - neuron.

ΣΠ - neuron represents an algebraic model of a neuron that reflects the processes of information processing in the axo-dendritic system of a neuron [12–14], which has the best ability to approximate dependencies.

ΣΠ - neuron in its classical version has a multilinear function of the total signal

$$sp(X) = out\left(\theta + \sum w_k \prod_{i=1}^{n} x_i\right),$$

where θ - constant, *out* - output function.

Teach logic-arithmetic ΣΠ-neuron is possible in 1 pass of the training set.

During training, objects are encoded by natural numbers. For the given example, if $a = 1,\ b = 2$, then $sp = x_2 + x_1 x_2$.

Then, in the case of a request (1,0), the decision of the neuron will give the decision "0", the logical function "b".

If you look at how the neural network solutions will look in the function value column of Table 2, then these will be units of measurement, only in those sets where there is data specified in the table and there is an object corresponding to the matrix. This is shown in Table 3.

Table 3. Truth table for neural network solutions.

x1	*x2*	*P*(*a*)	*P*(*b*)	*f*(*X.Y*)
...	...	...	...	0
0	1	1	0	1
0	1	1	1	1
...	...	...	...	0
1	1	0	1	1
1	1	1	1	1

You can see that the function has 14 possible solutions, the neural network has 4. If our solutions are closed with respect to the data presented, then the possibilities of logical analysis are wider.

For a neural network, a logical corrector can be built, 1 of the ways to build it is described in [14]. After applying it to $p = x_2 + x_1 x_2$, we get the following function

$$f_1(x_1, x_2, P(a), P(b)) = P(a)\&P(b)\&x_2 \vee P(a)\&\overline{x_1} \vee P(b)\&x_1.$$

This function will add the number of solutions to the original table, but will not completely restore it. The request (1,0) will be answered correctly. Because functions

$$f_1(x_1, x_2, P(a), P(b)) = P(a)\&P(b)\&x_2 \vee P(a)\&\overline{x_1} \vee P(b)\&x_1.$$

and different functions

$$f_1(x_1, x_2, P(a), P(b)) = \overline{x_2} \vee P(a)\&\overline{x_1} \vee P(b)\&x_1,$$

but their truth tables intersect on sets where objects are defined. Table 4 shows how many solutions the corrector added.

Table 4. The table shows how many solutions the corrector added.

x1	*x2*	*P(a)*	*P(b)*	f(X.Y)	ΣΠ- n	cor
0	0	0	0	1	0	0
0	0	0	1	1	0	0
0	0	1	0	1	0	1
0	0	1	1	1	0	1
0	1	0	0	0	0	0
0	1	0	1	0	0	0
0	1	1	0	1	1	1
0	1	1	1	1	1	1
1	0	0	0	1	0	0
1	0	0	1	1	0	0
1	0	1	0	1	0	0
1	0	1	1	1	0	1
1	1	0	0	0	0	0
1	1	0	1	1	1	1
1	1	1	0	0	0	0
1	1	1	1	1	1	1

As can be seen from the table, the object part of a function can be restored; only the part where variables act as a disjunctive cannot be restored.

Let's look at a more illustrative example 2, given in Table 5:

Table 5. Objects and features.

x1	x2	x3	x4	y
0	0	1	0	a(2)
0	1	0	1	b(4)
0	1	1	1	c(8)
1	1	1	0	d(16)

$$sp = 2x_3 + 4x_2x_4 + 2x_4x_2x_3 + 14x_1x_2x_3.$$

For any data from the table, the neuron will give the correct solution, and if the data is slightly changed, for example, (1111), (0110), then the results will be doubtful. In the first case we get 14, there is no such object in our data, in the second we get object number 2, although it could well be object number 8. The neural network generally offers the only solution, and in this case, logical methods are more correct, because they offer all possible solutions.

After applying the above method for we get the logical function $sp = 2x_3 + 4x_2x_4 + 2x_4x_2x_3 + 14x_1x_2x_3$.

And the method is to build a tree based on the structure of a sigma pi neuron, Fig. 1, and to find existing connections. The solid line - the property is present. The dashed line - the object has no property.

$$f(x_1, x_2) = x_1x_2 \vee x_1\overline{x_3} \vee \overline{x_4x_3} \vee x_1x_4 \vee \overline{x_2x_3} \vee b\overline{x_3} \vee cx_3x_4 \vee dx_1 \vee bcx_4 \vee acdx_3 \vee a\overline{x_2}.$$

After applying the method of constructing a logic function based on the structure of a sigma pi neuron

$$f(x_1, x_2) = dx_1 \vee a\overline{x_2} \vee b\overline{x_3} \vee cx_3x_4 \vee bcx_4 \vee acdx_3.$$

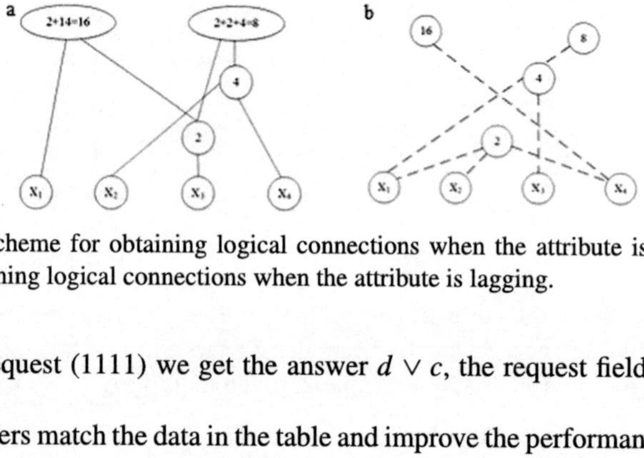

Fig. 1. (a) the scheme for obtaining logical connections when the attribute is present; (b) the scheme for obtaining logical connections when the attribute is lagging.

After the request (1111) we get the answer $d \vee c$, the request field (0110) we get acd.

These answers match the data in the table and improve the performance of the neural network.

However, the function, although it expands the number of solutions, is not a complete data classifier. To obtain a complete classifier, the following algorithm can be proposed:$f_1(x_1, x_2)$.

Let $\{w_1, w_2, \ldots, w_k\}$ is the set of neuron weights. If among the set of recognizable objects there is a numbered element Then we write all the variables corresponding to this weight through disjunction with negation, and those variables that are not as factors of this weight we write without negation, and through disjunction we write the object itself $w_i \in Yw_i$.

For example, this weight corresponds to the object a then we write the following rule:

$$p = 2x_3 + 4x_2x_4 + 2x_4x_2x_3 + 14x_1x_2x_3,$$
$$w = 2,$$
$$f(x_1, x_2) = (x_1 \vee x_2 \vee \overline{x_3} \vee x_4 \vee a).$$

If the next weight value does not correspond to the object in the data, then you need to add reducing weights that contain variables included as factors in this term. If the next weight value does not correspond to the object in the data, then you need to add.

Thus, it is possible get all the rules in the source data, and put a conjunction sign between them to get a function - a complete classifier. For our example, this would be:

$$f(X, Y) = (x_1 \vee x_2 \vee \overline{x_3} \vee x_4 \vee a)\&(x_1 \vee \overline{x_2} \vee x_3 \vee \overline{x_4} \vee b)\&(x_1 \vee \overline{x_2} \vee \overline{x_3} \vee \overline{x_4} \vee c)\&(\overline{x_1} \vee \overline{x_2} \vee \overline{x_3} \vee x_4 \vee d).$$

3 Results and Discussion

Regardless of which approach is applicable to correct the operation of the neural network, we will get a function that has disjunctions that define specific objects, or classes of objects. For certain requests, such a function can have several solutions. 1 of the approaches that will help you choose the most appropriate solution is logical neural networks [15–18].

The inputs of such neural networks will be variables $X = \{x_1, x_2, \ldots, x_n\}$. The output will be a lot of objects. The layer that does the calculationsthere will be disjuncts in which the operations of conjunction and disjunction are replaced by neurons with conjunctures and disjuncts.

Clauses that contain information about objects are uniquely associated with neurons that supply an output signal to those objects that are included in the corresponding subclasses.

As a result of the conjuncture query, it sends a signal of unit weight to each of the objects included in this clause (subclass) [19–21].

The weights of the input signals are set in such a way that, given the values of the input signals, the resulting signal sets would be generated at the output, corresponding to the values of the logical operations of conjunction and disjunction. For example, the input weights of a conjunctor with n inputs are set to $1/n$.

It is clear that at the output of the conjunctor, a logical unit will be generated only when all inputs of such a neuron are equal to one. The weights of all inputs of the disjunctor are set close enough to one, so that to generate a logical one at the output of such a neuron, it would be sufficient to have at least one unit at its inputs.

For any query, the output will be a set of objects that satisfy this query, the weights of which exactly match the result of the frequency analysis procedure used to find the objects that most fully respond to this query.

4 Conclusions

A logical approach to the analysis of the results of neural networks, providing a unique opportunity to analyze and correct the resulting solutions. As has been shown, neural network solutions are only part of the possibilities that are present in the data. To identify all possible, logically consistent solutions, it is necessary to supplement the trained neural network with correctors, which will allow to establish logical connections between recognizable objects. This significantly improves the quality of the resulting solutions.

In the method under consideration, it is proposed to identify logical connections between the objects of the training sample, this can be d1 even without a training sample by analyzing the network structure. Unlike the Attention method, the proposed approach reveals logical rather than statistical patterns in the data. Knowledge of logical patterns, in contrast to statistical 1s, will allow formalization and more accurate understanding of the nature of the analyzed data.

References

1. Zhuravlev, Y.: On an algebraic approach to solving problems of recognition or classification. Probl. Cybern. **33**, 568 (1978)
2. Shibzukhov, Z.M.: Correct algorithms for aggregation of operations. Pattern Recognit. Image Anal. **24**(3), 377–382 (2014)
3. Naimi, A.I., Balzer, L.B.: Multilevel generalization: an introduction to super learning. Europ. J. Epidem. **33**, 459–464 (2018)
4. Lyutikova, L.A.: Construction of a logical-algebraic corrector to increase the adaptive properties of the ΣΠ-neuron. J. Math. Sci. **253**(4), 539–546 (2021). https://doi.org/10.1007/s10958-021-05251-3
5. Dmitrichenko, D.P., Zhilov, R.A.: The method of construction of logical neural networks on the basis of variable-valued logical functions. J. Math. Sci. **253**(4), 500–505 (2021). https://doi.org/10.1007/s10958-021-05246-0
6. Haoxiang, W., Smys, S.: Big data analysis and perturbation using a data mining algorithm. J. Soft Comput. Parad. (JSCP) **3**(1), 19–28 (2021)
7. Vijesh Joe, C., Raj, J.S.: User recommendation system dependent on location-based orientation context. J. Trends Comput. Sci. Smart Technol. (TCSST) **3**(1), 14–23 (2021)
8. Grabish, M., Marichal, J., Mesiar, K., Pap, E.: Aggregation functions (Encyclopedia of Mathematics and its Applications). Cambridge University Press, Cambridge (2009)
9. Calvo, T., Belyakov, G.: Aggregating functions based on penalties. Fuzzy Sets Syst. 161(10), 1420–1436 (2010)
10. Mesiar, R., Komornikova, M., Kolesarova, A., Calvo, T.: Fuzzy aggregation functions: a revision. In: Sets and Their Extensions: Representation, Aggregation and Models. Springer-Verlag, Berlin (2008)
11. Yang, F., Yang, Z., Cohen, W.: Differentiable learning of logical rules for reasoning in the knowledge base. In: Advances in the Field of Neural Information Processing Systems, pp. 2320–2329 (2017)
12. Flach, P.: Machine Learning: The Art and Science of Algorithms That Give Meaning to Data. Cambridge University Press, Cambridge (2012)
13. Akhlakur, R., Sumaira, T.: Ensemble classifiers and their applications: a review. Int. J. Comput. Trends Technol. **10**, 31–35 (2014)
14. Dyukova, E.V., Zhuravlev, Y., Prokofiev, P.A.: 2015 Methods for improving the efficiency of logical correctors. Mach. Learn. Data Anal. **1**(11), 1555–1583
15. Lyutikova, L.A., Shmatova, E.V.: Building logical operations to identify patterns in data. E3S Web of Conferences 224 (2020)
16. Lyutikova, L.A., Shmatova, E.V.: Analysis and synthesis of pattern recognition algorithms using variable logic. Inf. Technol. **22**(4), 292–297 (2016)
17. Burges, C.J.: A tutorial on support vector machines for pattern recognition. Data Min. Knowl. Disc. **2**(2), 121–167 (1998)
18. Aladiev, V.: Computational Algebra System. mopl. Software Library

19. Prepare, S.A., Cook, S.A., Rekhov, R.A.: Relative efficiency of systems of proof of statements. J. Symb. Log. **44**(1) (1979)
20. Shibzoukhov, Z.M.: On a constructive method for synthesizing families of majority-correct algorithms. In: Proceedings of the VII International Conference on Pattern Recognition and Image Analysis 1:113–115 (2004)
21. Dimitrichenko, D.P.: A method for diagnosing a robotic complex using logical neural networks apparatus. In: 2021 International Russian Automation Conference (RusAutoCon), pp. 907–911 (2021)

Processing of Analyzed Textual Information to a Meaningful Structure

R. E. Semenov(✉)

MIREA – Russian Technological University, 78, Vernadskogo Ave., Moscow 119454, Russia
Crosonline@gmail.com

Abstract. The results of studies of the morphological analysis of the text are demonstrated. Application of the technology of automatic processing of Russian-language texts to determine the parts of speech presented in the digital image. Definition of significant and non-essential information in the text. Identification and construction of a concept for implementation into an intelligent automated text recognition and analysis system. The correspondence of each word of the text to a certain part of speech according to a predetermined structure. The automatic selection of the necessary algorithms for bringing the text to the final form is carried out to get a real-time response. The designed system uses a certain structure, according to which, at the same time, text analysis is carried out to identify the main idea of a group of proposals, and presents the information received in an ordered form. The system can be upgraded and receive constant support in the form of updates to the relevance databases and calculation of criteria for weight characteristics. It can be implemented in the systems of Internet resources, used as an independent project supporting autonomous operation, or for implementation in intelligent systems.

Keywords: Morphological analysis · Computational linguistics · Automatic text processing · Automated recognition · Analysis of Russian-language text

1 Introduction

Automation of various fields of activity allows us to identify new areas for research and scientific development. Artificial intelligence algorithms are used in many industries. This makes it easier and faster to perform the necessary actions. The needs for the development of artificial intelligence pose the task of the most comfortable interaction and communication of a person with a computer. Developments in the field of analysis of oral human speech and text analysis allow you to process words and perform certain actions with them. Due to the popularization of work on digital media, there is a lot of publicly available information, but along with a useful and significant part of any text, insignificant information becomes. When reading and processing the text, there are situations when the meaning is lost against the background of irrelevant information. This phenomenon is called information noise, which makes it difficult to read and assimilate information, but, on the other hand, it may happen that the text will not have sufficient information to understand.

A. A. Radionov and V. R. Gasiyarov (Eds.): RusAutoCon 2022, LNEE 986, pp. 304–313, 2023.
https://doi.org/10.1007/978-3-031-22311-2_30

2 Conceptual Model

To structure and formalize information, it is necessary to isolate a concept from it as a kind of logical structure [1]. At the same time, the conceptual structure should allow assessing the completeness and adequacy of information, forming a knowledge base at the program level. To build such a structure, a morphological text analyzer should be developed, which identifies various parts of speech and correlates them with various elements of the structure [2].

The conceptual structure of situational analysis presented in Fig. 1 represents the interrelationships of the elements.

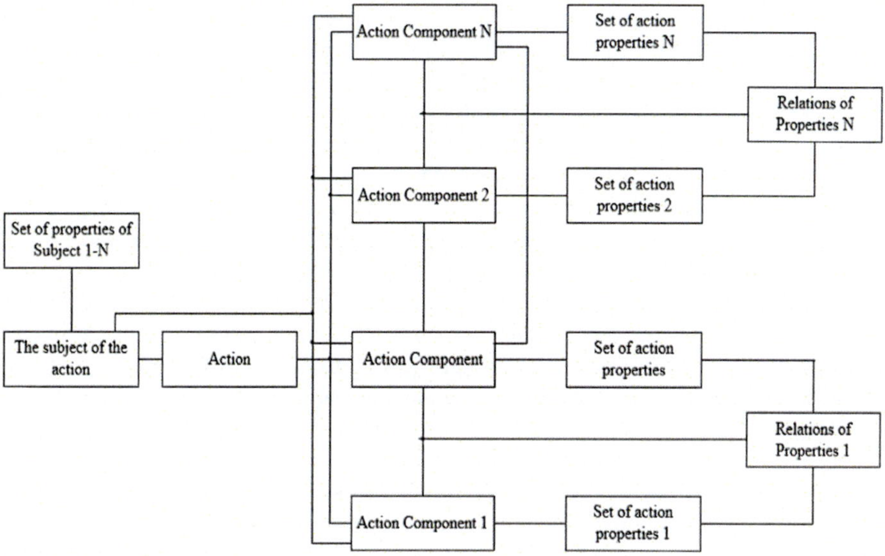

Fig. 1. Structure for interaction of analysis elements.

The figure below shows two types of semantic binary bidirectional relations. The first are called interactions, which exist between the subject, the action itself, the components and the object of the action, called links [3]. Others are dependencies of relationships that are called constructs and link components and exist between an object and components [1].

Each element of the structure is mapped to a specific class of words. The word class refers to parts of speech united by a common grammatical meaning, syntactic role in a sentence and a set of morphological features [4]. The analysis of each sentence in Russian is a laborious process due to its characteristic features, therefore simplified algorithms for processing and analyzing Russian-language texts will be inaccurate. The Russian language is characterized by such features as: arbitrary word order in the presentation, declensions, types of verbs, and many more nuances for the literate writing of the text. Until now, the rules of the Russian language are being corrected and are subject to changes [5]. Therefore, morphological text processing plays an important role for use in intelligent systems.

3 Morphological Construction

Let's analyze the morphological features of the words of the Russian language. The word classes are divided into interjective and non-interjective. The former is not used in conceptual construction because they have emotional content. Another class of words is unmeaning, which is divided into: service and independent parts of speech. The official ones include prepositions, particles and conjunctions. Independent parts of speech are divided into: nominal and verbal. The nominal parts of speech include: nouns, adjectives, numerals and pronouns. Verb parts of speech are called verbs, participles, adverbs and adverbs [6]. When comparing the conceptual structure, independent parts of speech will be used to compile the most specific concept shown in Fig. 2.

The representation of parts of speech as elements of the conceptual structure is not unambiguous, therefore, to ensure a more accurate definition of a part of speech, a combination of dictionary and wordless morphological analyzers will be used later. Morphological analysis determines the normal form of a word, called a lemma, from which the analyzed word form was formed, including its accompanying parameters [7].

A pair of characteristics <parameter name, parameter value> is called a Morphological parameter. For the parameter name, gender, number, tense, declension and other signs of words accepted in the language used are applicable. The value of the parameter is a specific value that can accept this attribute. So, for example, gender can be masculine, feminine, or neuter. At the same time, a situation is possible in which several parameters can be compared to one word form [8]. When developing an analyzer, the concept of homonymy plays a significant role. Homonyms are called units of the language that are different in meaning, but identical in spelling [5]. In terms of computational linguistics, homonymy can be defined as a situation in which several meanings for one characteristic can be applied to one word form. To unambiguously recognize the characteristics of a part of a speech, it is necessary to compile an additional dictionary of homonyms, which will contain the most frequently used word forms that have the property of homonymy [9].

Morphological analysis is carried out on the basis of two methods, wordless and wordless. Dictionary morphology implies the presence of a dictionary that contains a set of morphological parameters for each word form, which makes it possible to obtain its normal form [4]. To implement dictionary morphology, a dedicated memory area is needed to store the dictionary. In the case when a word that is not in the dictionary occurs during the analysis, the analysis is not carried out. To process homonyms, a separate homonymy dictionary should be compiled, which will contain the most frequently used word forms with these properties. It will be possible to increase the unambiguity of the analyzed parts of speech when constructing a conceptual structure and generating a logical conclusion [1]. It is necessary to use a semantic-syntactic analyzer, which will simplify conflict resolution. The analysis of an ambiguous word and context is carried out, it can also be applied to the conceptual structure of speech parts.

When analyzing, it should be borne in mind that words can be grouped by paradigms. These are rules that allow you to get all the forms of a word in a lexeme for a given

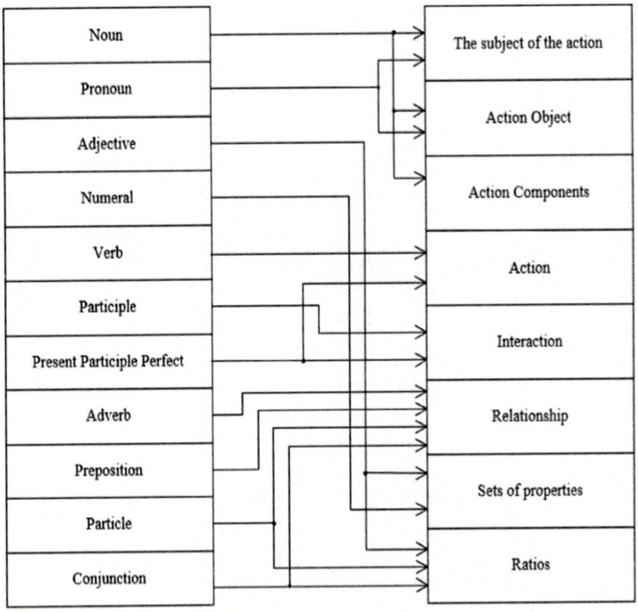

Fig. 2. Comparison of morphology and structure.

stem or to get all the forms of a word in a set of word forms of one word for a given immutable part of a word. Wordless morphologies store only paradigms of words. Also, a set of prefixes and suffixes with morphological information tied to them can be stored in wordless morphology [9]. The greatest advantage of wordless morphologies is that they can predict the morphological characteristics of almost any word if its paradigm of change falls under one of these. Wordless morphological analyzers allow you to reduce the consumption of RAM. Unlike dictionary morphologies, the speed of analysis is higher, and the volume of stored databases is significantly reduced.

The main disadvantage is the high percentage of errors. For example, in the word "swimming", which is a noun, there the "ing" at the end of the word, which can be taken as the ending of the verb in the continuous form [10]. To avoid inconsistent results, special algorithms have been developed to identify endings and postfixes in a certain sequence, in this regard, the percentage of errors has been reduced.

A system based on the analysis of words by the immutable part of the word is a prerequisite for a more accurate analysis [11]. Such a system works on the basis of stemming rules that allow you to give a morphological search for the basis of a word. During this analysis, the unchangeable part of the word is considered and used as a normal form, while morphological information is not taken into account. For example, analyzing the word "crown", the stem will be the string "crown". For further consideration of the part of speech and the relationship of words in a sentence and bringing it to the concept, this initial form will be used. At the next stages, independent morphological information will be assigned to this form.

4 Designing a Model-Parametric Structure

The main subject of scientific interest is the automation of the process of studying the parameters of a given text and designing meaningful response structures. In the subject area of research of complex objects, which is the projected system, the categories of such abstractions are defined: parameters and models [12]. At the same time, a parameter is understood as a formal image of a property modeled in a computational environment, and a model is any relationship between parameters.

Note that any model formally is: $M_j : R(P_{1j}, P_{2j}, ..., P_{ij})$, where: j is index (identifier) of the model; P_{ij} is the i-th parameter of the j-th model; R is the relationship between the parameters. This relation can be represented in any form.

In Fig. 3. The parameter category is represented in two systems "creator - user" and "computational - formal – external environment".

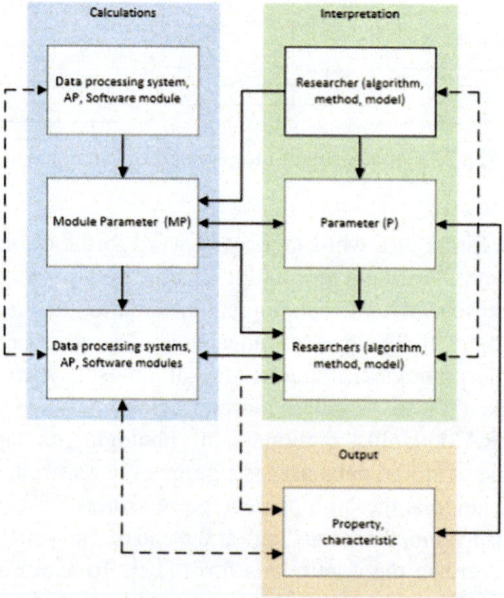

Fig. 3. Structure of the logical model of the system.

The parameter is an image of the simulated property. The result of mapping a parameter to a computing environment is called a parameter model. Both the parameter and its model always have a creator. Note, however, that these may be different specialists (researchers, designers, designers of new complex objects) [13]. To construct a parameter (determine its characteristics), the "creator" uses his knowledge of the laws of the simulated subject area. Implicitly, algorithms, methods, models, and even software complexes can be considered parameter constructors. In fact, these "intelligent agents" only make demands on the properties of the parameters they use, always determined by a researcher with "natural intelligence". In this case, the creator of the algorithm, method, model, etc. After determining the parameter, he can determine its model in the computing environment.

Users of parameter values are divided into three classes: researchers, methods (algorithms, models) and software tools. The latter use parameters in a computing environment. Note that any computational module is related to the synthesis-analysis of models based on any methods and/or algorithms [14]. Intelligent agents work with parameter models. Researchers (designers, designers), algorithms and models use parameters because they work in an environment of formalization (synthesis of models) and interpretation (their analysis). The parameter P is the formalization of the studied properties/characteristics of the studied/projected complex object. Model M is any relation between the parameters (properties, characteristics) of a complex object that is of interest from the point of view of research. Space is defined as a set of homogeneous objects (objects, phenomena, states, variables, etc.) between which there are spatially similar relations [15]. Often words homogeneous and spatially similar are omitted and space is defined as a tuple.

$(M, A_1, A_2, \ldots, A_n)$, where: M is some set; $A_1, A_2, \ldots, A_n$ are relations between its elements.

Sometimes space is spoken of as a set M, between the elements of which some relations are implied.

Since all operations and processes are carried out at the level of models and parameters, it is legitimate to call such a space model-parametric [16]. We will denote and call it further $< M, P >$ -space.

The model-parametric space will be the set of all models, parameters, and relationships between them that characterize the properties of the projected system.

The most suitable apparatus for describing and studying the structure of $< M, P >$ -space is graph theory. The elements (atomic components) $< M, P >$ are: models (set M), parameters (set P) and relations between them, i.e. $M = \{M_j\}, j \in J, P = \{P_i\}$, $i \in I$, where the sets of indices I and J are determined by the ones considered in each in a particular case, objects, explicitly and implicitly defining them as integral structures. The objects include various aggregates, nodes, functional subsystems, components of the designed complex product, its whole and the system describing its behavior and functioning in the external environment [17]. Thus, in the most general case, the $< M, P >$ -space is defined on the set of the direct Cartesian product of the sets M and P, i.e. $< M, P > \subseteq M \times R$. Arcs-relations between models and parameters, depending on the context of consideration, are attributed a different meaning. But, by default, it is assumed that the arrow pointing from the parameter to the model means that this parameter is independent in this model, and the arrow pointing from the model to the parameter corresponds to the dependence of the parameter on the model.

Thus, the $< M, P >$ -space is represented in the form of an oriented graph, the vertices of which correspond to models and parameters, and the arcs correspond to the relations between them [15].

These arguments motivate the expediency of excluding non-directional models from consideration. Indeed, any model is designed to simulate any properties, characteristics of the designed product, i.e. it should always have input and output parameters. The orientation of the digraph and the specificity of the $< M, P >$ -space lead to the presence of the following properties of its topology [18]:

- The parameter may have no outgoing arcs. In this case, it is not used in other models and, as a rule, integral parameters have this property
- If it has several outgoing arcs, then its values are used in the corresponding number of models
- The parameter may not have incoming arcs. In this case, in all models it is used as an independent one, and its values are determined only by the user or measured by some sensor
- If it has several incoming arcs, then its values can be obtained on the basis of different models, and in this case it is necessary to specify the context of their use
- Any model must have at least one outgoing arc corresponding to the calculated parameter, otherwise there is no consequence in the causal relation (the text of the model)
- If the model has several outgoing arcs, then we can talk about an integrated model, since several parameters are calculated with its help
- Any model must have at least one incoming arc corresponding to the input parameter, otherwise there is no cause in the causal relation (the text of the model).
- If the model has several incoming arcs, the values of the output parameters are calculated based on the values of the corresponding number of independent (input) parameters (arguments)

5 Defining Parameter Values

Next, we can talk about the subspace of parameters associated with models, or about the subspace of models associated with parameters. And then we get weighted digraphs $< M, P >$. It is possible to distinguish other subspaces in $< M, P >$ -space by various signs [19].

To do this, it seems advisable to introduce into consideration some structures of a larger order than individual elements (models, parameters, their relationships). An example of such structures is the method of determining various properties of a complex product being designed. Figure 4 shows a graphical image of a fragment of such a technique for determining the values of the (integral) parameter P1 based on the values of P5, P9, P11, P12, P13. The next hierarchical level is the appearance (generalized model–parametric spaces of various subsystems, aggregates).

The neighborhood $< M, P >$ of the first-order space with respect to the model M_j will be called the set of parameters P_i directly related to the model M_j. Denote this neighborhood. All elements of the neighborhood of the model, in view of statement, are parameters, and are separated by M_j by a unit length.Thus, when constructing $< M, P >$ -neighborhoods at the logical level, some model or some parameter is put in the center of attention [20]. Everything else $< M, P >$ -space is considered relative to this element, i.e. its ordering (sorting) is carried out in relation to the parameter under consideration or the model under study.

For the system being designed, such a model will make sense in the morphological ordering of the elements of the Russian language. To determine the constructions of

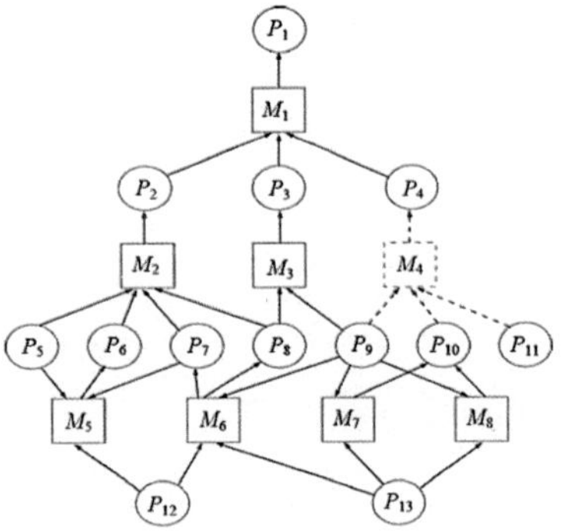

Fig. 4. Structure of the logical model of the system.

text analysis and modeling of responses. According to the surroundings of the designed models, it will be possible to determine the importance values of certain features. Based on the parameters, a sample will be generated for training the neural network of text analysis.

6 Application of the Obtained Models

To test the simulated structures, I used an edited existing language model. There are two strategies for applying pre-trained language representations to subsequent tasks: a function-based one that contains separate pre-trained representations for specific tasks. The second is point tuning – it is often found in generative transformers, which assume a more precise adjustment of the weights of the entire neural network for the required task. These two approaches have the same objective function during pre-training. There are two stages in the BERT (Bidirectional Encoder Representations from Transformers) structure: pre-training and fine-tuning. During pre-training, the model is trained on unlabeled data using various pre-training tasks. For fine tuning, the BERT model is first initialized with pre-trained parameters, and all parameters are fine-tuned using labeled data from subsequent tasks. Each subsequent task has separate customized models, even if they are initialized with the same pre-trained parameters. Fine-tuning based on the self-awareness mechanism in Transformer allows BERT to simulate many subsequent tasks - regardless of whether they include single text or text pairs - by replacing the corresponding inputs and outputs. The transformer is the first transduction model that relies entirely on the direction itself to compute representations of its input and output data without using recurrent neural networks, aligned sequences, or convolutions.

The efficiency of the neural network is shown in Table 1, where 2 classes (Positive – interesting, Negative – uninteresting) of documents are presented. Based on the requested information, the analyzer is able to determine the importance of the term entered in the documents under consideration based on the predefined structure of the conceptual representation of each document under consideration.

The result of the work carried out is successful and can be used in further developments. According to the information provided, 6 percent of the total number of documents

Table 1. The result of the neural network working with models.

Class	Quantity	Percent
Positive	4340	93.94%
Negative	280	6.06%
Total	4620	100%

does not correspond to the requested information after processing algorithms. Manual verification shows that the quality of text selection according to the conceptual structure is high, and the missing percentage is false positive positives. The same number of false negatives, which is from 1 to 3 percent of the total number of analyzed documents. The result of the work is a software capable of processing documents on the structures of the conceptual representation and identifying the most significant information. After structuring the documents, the request for the relevant information is classified according to the content of the documents and documents that do not correspond to the request are discarded. The presented solution will significantly accelerate the existing algorithms of document processing and is aimed at the subsequent creation of an autonomous system for analyzing the relevance of the requested information with the possibility of obtaining an instant response.

7 Conclusion

Therefore, template models of sentence structures will be used for the described system. The presented models will be the basis for future developments. The obtained results of research on the interaction of text analysis methods and structure allow us to start creating a text processing system and bringing information to the concept. This design of the system algorithms allows you to get a response from the system in real time [21]. The designed system automatically selects the necessary algorithms for bringing the text to the final form. The decision support system, based on certain methods, conducts an appropriate analysis of terms and, based on the results obtained, assigns an index to the document to identify the main idea of the text and the relevance of the search query.

The results of the study represent improved methods and designs that, as a result of technological improvements, have become available for use in the system. The organization of a multi-level text analyzer allows you to uniquely determine the characteristics of words for morphological analysis of the Russian language, in further research it is possible to modernize support for any language. The system can be implemented in large systems of Internet resources, in intelligent systems, or used as an independent project that supports autonomous operation by implementing dictionaries of words and rules with the possibility of updating databases.

References

1. Anferov, M.A.: Genetic clustering algorithm. Russ. Technol. J. **6**, 134–150 (2019)
2. Feldman, S.E.: The answer machine. Synthesis Lectures on Information Concepts, Retrieval, and Services. Morgan & Claypool Publishers **4**, 1–137 (2012)
3. Andreychikov, A.V., Andreychikova, O.N.: Intelligent digital technologies of conceptual design of engineering solutions: textbook. In: INFRA-M, p. 511 (2019)
4. Hoffman, M.D., Blei, D.M., Bach, F.R.: Online learning for latent Dirichlet allocation, pp 856–864. NIPS, Curran Associates, Inc. (2010)
5. Nejdl, W., Krestel, R., Fankhauser, P.: Latent Dirichlet allocation for tag Recommendation. In: Proceedings of the third ACM conference on Recommender systems, pp. 61–68. ACM (2009)
6. Khandelwal, U., He, H., Qi, P., Jurafsky, D.: Sharp nearby, fuzzy far away: how neural language models use context. In: Proceedings of the 56th Annual Meeting of the Association for Computational Linguistics **1**, 284–294 (2018)
7. Sorokin, A.B., Lobanov, D.A.: Conceptual design of intelligent systems. Inf. Technol. **1**(24), 3–10 (2018)
8. Bolotova, L.S.: Artificial intelligence systems: models and technologies based on knowledge. Finance and Statistics, p. 663 (2012)
9. Sorokin, A.B., Smol'yaninova, V.A.: Conceptual design of expert systems. Inf. Technol. **9**(23), 634–641 (2017)
10. Koval, S.A.: Linguistic problems of computer morphology, p. 151. IS.-Peterburg, SPb (2005)
11. Beale, M.H., Hagan, M.T., Demuth, H.B.: Neural Network Toolbox. User's Guide. Math Works, Natick, p. 406 (2015)
12. Tsitulsky, A.M., Rogov, I.S., Ivannikov, A.V.: Intellectual analysis of the text, student 6:476–483 (2020)
13. Boyarskij, K.K.: Introduction to Computational Linguistics: tutorial, p. 72. NIU ITMO, SPb (2013)
14. Grebeneva, Yu.N.: Dictionary of homonyms and homoforms of the Russian language, p. 351. Ajris-press (2011)
15. Ticher, S., Meyer, M., Vodak, R., Vetter, E.: Methods of text and discourse analysis, p. 356. Publishing house "Humanitarian Center" (2017)
16. Sosnina, E.P.: Introduction to Applied Linguistics: study guide. UlGTU, Ulyanovsk, p. 110 (2012)
17. Krongauz, M.A.: Semantics: textbook, p. 352. Izdatel'skij centr "Akademiya" (2005)
18. Chen, D.-Y., Zhao, H., Zhang, X.: Semantic mapping methods between expert view and ontology view. J. Softw. **31**(9), 2855–2882 (2020)
19. Chesalin, A.N., Grodzenskiy, S.Y., Van, T., Nilov, M.Yu., Agafonov, A.N.: Technology for risk assessment at product lifecycle stages using fuzzy logic. Russ. Technol. J. **8**(6), 167–183 (2020)
20. Bol'shakova, E.I.: Automatic processing of natural language texts and computational linguistics, p. 272. MIEM (2011)
21. Amaeva, L.A.: Comparative analysis of data mining methods. Innov. Sci. **2**(1), 27–29 (2018)

The Building a Patent Landscape for Technological Forecasting Tasks

D. Korobkin(✉), M. Saveliev, G. Vereschak, and S. Fomenkov

Volgograd State Technical University, 28, Lenin Avenue, Volgograd 400005, Russia
dkorobkin80@mail.ru

Abstract. Methods and technologies for solving the problem of patent landscape visualization based on cluster analysis of the patent array are considered and used. Algorithms for downloading patent archives, parsing patent documents, clustering patents and visualizing the patent landscape have been developed. A software for clustering patent documents based on the Latent Dirichlet allocation model and visualization of the patent landscape on clustering data using the gensim, PySpark, and sklearn libraries has been implemented. The implemented software has been tested on patents issued by the US Patent and Trademark Office. The accuracy of classification of patents by category has been achieved - 84%. Possible ways to improve the system: (a) add the ability to save the LDA model trained on certain categories of patents to predict the category of a new patent document (b) improve the quality of preprocessing of the patent description text; (c) develop a software on the customer's resources (computing cluster), in order to fully test the Apache Spark framework; (d) saving a large amount of data on the customer's resources (computing cluster) using HDFS technology (Hadoop Distributed File System).

Keywords: Patents · Information extraction · Clustering

1 Introduction

The process of generating new technical solutions, the search for analogues for the developed technology and the analysis of technological trends can be simplified by analyzing the existing patent base [1, 2]. The patent document fixes the key features of the invention that make it unique, and also describes the processes necessary for its reproduction. According to statistics, the number of patented technologies is only increasing every year, so it is becoming increasingly difficult to find among the many patent documents those that have a technology with high innovative potential and prospects.

Speaking about the task of determining the prospects and innovative potential of the technology enshrined in the patent, it can be argued that there is no standard for solving and automating such a task.

A fairly common way to analyze the technology presented in the patent is the SAO method. In [3], the authors verify the proximity of the technology to the technological trend by extracting matrix vectors consisting of SAO structures. And in [4], term-document matrices are created based on SAO structures, where patents are documents

A. A. Radionov and V. R. Gasiyarov (Eds.): RusAutoCon 2022, LNEE 986, pp. 314–324, 2023.
https://doi.org/10.1007/978-3-031-22311-2_31

and SAO structures are terms. In this case, the innovation potential is represented by the relative size of the thematic cluster. Similarly, the SAO model for extracting information from English-language patents is used in [5–7]. The paper [8] need to study the structure of a patent document to improve the quality of data extraction.

In [9], the authors proposed an "Innovative radar". This is a method of determining the innovation potential in the form of a questionnaire on a patent. By filling out the questionnaire, the authors receive an assessment of the potential of the patent on a 100-point scale. In [10], the authors analyzed user reviews of the patent to determine the innovation potential. From user reviews, you can extract information about the technology specified in the patent and the color of the user's reaction. The authors [11] have created a methodology for determining technologies with high potential - Technology Opportunity Discovery. In this work, clustering of patents by keywords is used. According to [11], the authors are able to single out a technology with a high potential from a multitude.

The aim of the work is to develop a software for clustering USPTO patent documents based on the Latent Dirichlet allocation (LDA) model and visualization of the patent landscape to improve the efficiency of identifying new trends and ideas in the field of synthesis of technical solutions.

2 Materials and Methods

To perform the task of clustering patent documents and visualization of the patent landscape, it is necessary to perform a number of preparatory actions. Such actions include downloading and analyzing patent arrays.

2.1 Algorithm for Downloading Patent Archives

The input data of the algorithm is a link to the web page with the patent archives of the United States Patent and Trademark Office (USPTO) for the selected year [12] and the absolute path to the directory where the patent archives need to be downloaded and unzipped. The web page is a list of links to patent archives that are published weekly. A patent archive is a ZIP file containing a patent array in an XML file. The web page searches for certain HTML tags containing links to download patent archives. Each found link is added to the list of links for further processing. For links in the generated list, operations are performed to download the archive, unzip it and save it to the specified directory. The output data of the algorithm are unzipped weekly patent arrays in XML files.

2.2 Patent Array Parsing Algorithm

The path to the directory containing weekly patent arrays obtained during the operation of the algorithm for downloading patent archives is received at the input of the algorithm for parsing patent documents. A patent array is an XML file containing descriptions of patents in XML format, where before each patent there is a declaration that it is XML. Such a declaration is allowed only at the beginning of the XML file, so the initial XML file of the patent array is not syntactically correct and there is a need to divide the patent array into single patent documents.

The specified directory is analyzed for the presence of subfolders with patent arrays. The found patent arrays are divided into separate patent documents in the form of XML files. Parsing is performed for each patent document.

The patent is checked for the presence of a classification-ipcr tag containing a description of the patent class according to the International Patent Classification (IPC). If a class description is found, further processing of the document is carried out, parsing of such patent fields as: section, class, subclass, main-group, description is performed. To match the patent to the specified category, the section and class fields are checked, and if the check is successful, the data of the subclass, main-group and description tags are extracted. The set of nested tags of the p element description is a description of the patent. The text nodes of the nested p tags are combined to form a single text with a description of the patent. The extracted information is recorded in a csv file, which is a collection of filtered patents. This analysis is carried out with each patent document.

2.3 Algorithm for Generating a Data Set of Patent Documents

This algorithm is designed to generate a data set of patents filtered by the necessary categories for subsequent clustering, and their number by slice of each category. The algorithm reads voluminous input data, the size of which can exceed several gigabytes, and then converts them to parquet format.

The input data is a collection of patents in csv format, obtained as a result of the parsing algorithm of patent documents. The output data is a parquet format file with the following columns: the group of this patent ("main-group" in the IPC classification), the subclass of this patent ("subclass" in the IPC classification), the text of the patent description, the number of words in the patent description.

For this algorithm to work, it is necessary to have a column with the text of the patent description. A number of actions are required before converting to parquet: input data rows are checked for zero values, user-defined functions are performed over the columns, patents are sorted in descending order of the number of words in the patent description text. The resulting file can be filtered by belonging to the classification category and/or the number of patents in the selected category, after which a new parquet file will be created. The algorithm also provides for combining several parquet output files to obtain a composite data set.

2.4 Algorithm for Constructing the Patent Landscape

To build a patent landscape, it is necessary to normalize texts describing patents, cluster patent datasets using the LDA method, and visualize the results of the LDA model.

Normalization of description texts includes tokenization of the text, lemmatization of tokens, removal of stop words and vectorization of tokens that make up the text of the patent description. The result of normalization is a dictionary of tokens and a vectorized representation of a set of patents. Normalization must be performed so that the LDA model can cluster data correctly.

The results of the machine learning model are visualized by composing a patent landscape from a set of graphs. The input data for the LDA model is a dataset with a column of vectorized documents and a dictionary with tokens. The patent landscape includes a table of the distribution of patent categories by clusters, a graphical representation of clusters on a coordinate plane, a graph of the identified keywords in each cluster and their weight, a graph of the distribution of the number of tokens in each document, a graph of the distribution of the number of tokens in each document by cluster slice, a graphical representation of clusters on a coordinate plane using the data visualization method high dimension t-SNE. The t-SNE visualization method, a stochastic embedding of neighbors with a t-distribution, is a nonlinear dimensionality reduction technique well suited for embedding high-dimensional data for visualization into a low-dimensional space.

To build a complete patent landscape, the LDA model of the gensim library is used [13]. The LDA model provided by the libraries PySpark [14] and sklearn [15] is used for clustering graphs on the coordinate plane, keywords of each cluster, the weight of keywords of each cluster, as well as a summary table with distribution by topic.

3 Results

The software is implemented in the Python and Scala programming languages and includes 4 blocks:

1) Patent archive download module.
2) Patent array parsing module.
3) The module of formation of a data set of parent documents.
4) The module of patent clustering and visualization of the patent landscape.

The BeautifulSoup library is used for parsing patent documents [16]. The gensim, PySpark, and sklearn libraries are used for clustering. The clustering model visualizations are based on the libraries pyLDAvis [17] and matplotlib [18].

The developed software was tested on patent arrays for 2019–2021. Figure 1 shows the distribution of the number of words in the document.

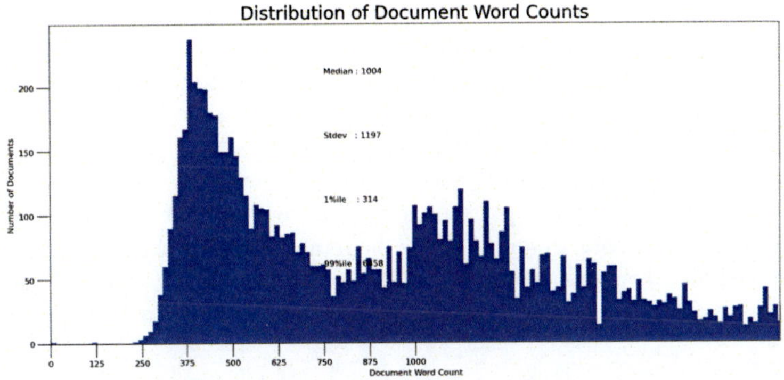

Fig. 1. Distribution of the number of words in the document.

Figures 2 and 3 show histograms with the number of words and their weight across a slice of topics.

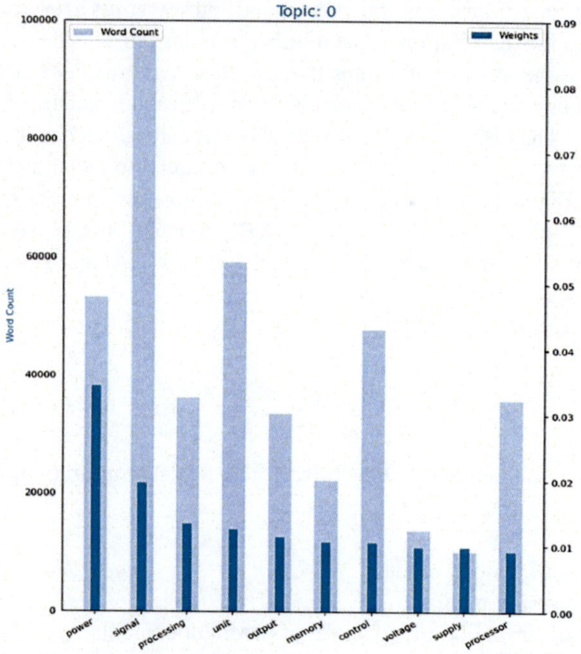

Fig. 2. Number of words and their weight according to the slice of the first topic.

Figure 4 shows a stochastic embedding of neighbors with a t-distribution using the example of 2 categories.

Figure 5 shows a stochastic embedding of neighbors with a t-distribution using the example of 6 categories. The machine learning algorithm for visualization shown in Figs. 4 and 5 is a nonlinear dimensionality reduction technique well suited for embedding high-dimensional data for visualization into a low-dimensional space. In particular, the method models each high-dimensional object with a two- or three-dimensional point in such a way that similar objects are modeled by closely spaced points, and dissimilar points are modeled with a high probability by points far apart from each other. The t-SNE algorithm is often able to detect well-separated clusters.

Figure 6 and Fig. 7 show fragments of an interactive map of a set of topics. This functionality is provided by the pyLDAvis plugin. Figure 6 shows the most common words included in the topic. Figure 7 shows the topics arranged in the form of circles.

Figure 8 shows the number of words according to the slice of topics of the sklearn library.

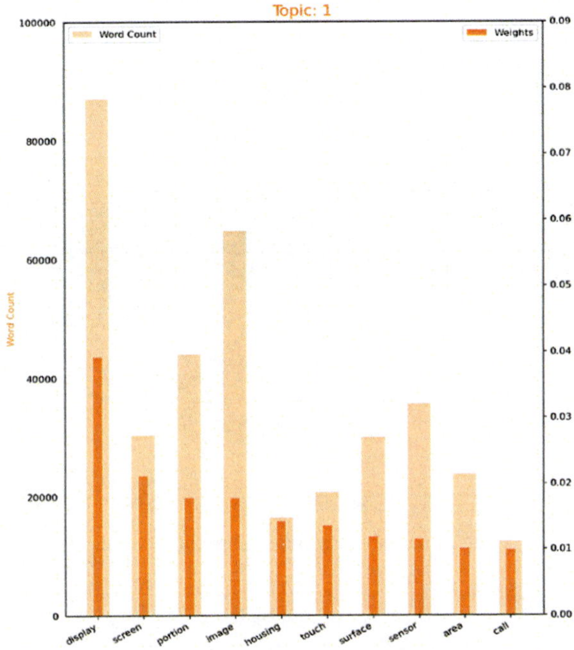

Fig. 3. Number of words and their weight according to the slice of the second topic.

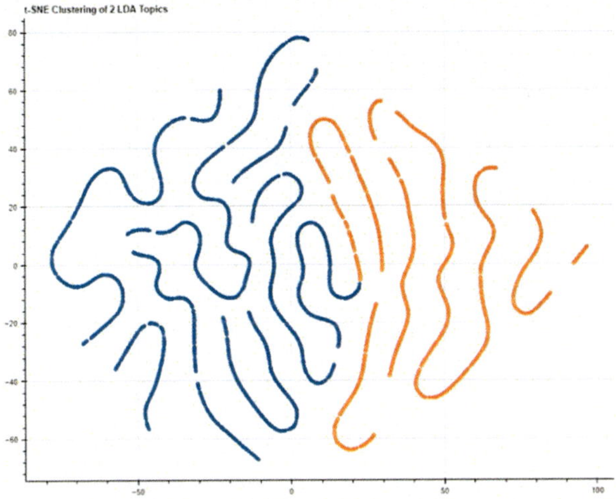

Fig. 4. Stochastic embedding of neighbors with a t-distribution on the example of 2 categories.

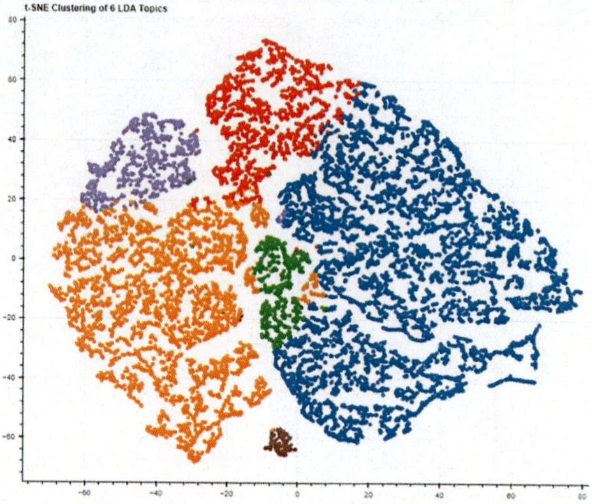

Fig. 5. Stochastic embedding of neighbors with a t-distribution on the example of 6 categories.

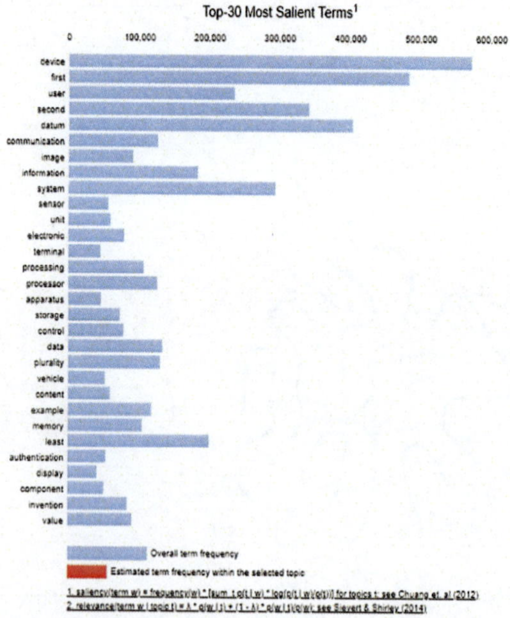

Fig. 6. Most common words included in the topic.

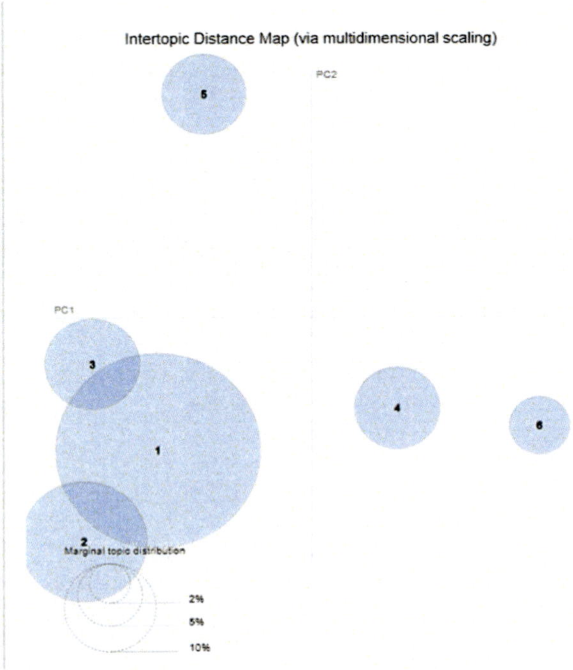

Fig. 7. Topics arranged in the form of circles.

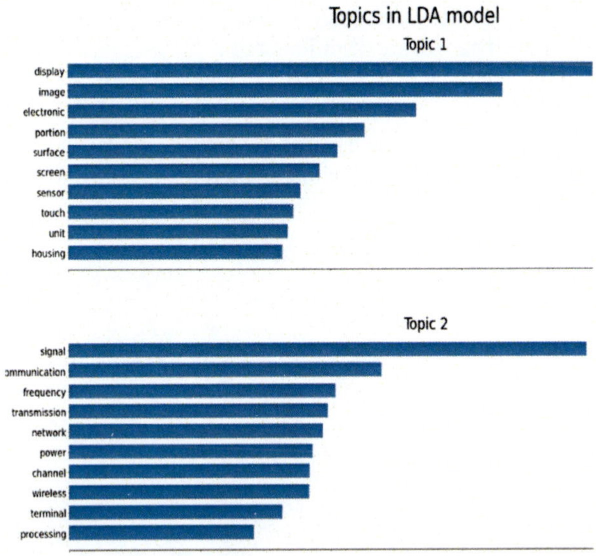

Fig. 8. Number of words acording to the slice of topics of the sklearn library.

Let's look at six different tests for the visualization module. The purpose of this testing is to determine whether the patent landscape is correctly visualized with different input data, as well as how accurate the clustering of patents is Test 1 (Table 1). This test is designed to find out how the volume of the dataset affects the operation of the software module and the accuracy of clustering. Two very similar subcategories of patents were selected.

Table 1. Test 1.

Type of clustering	Clustering accuracy		
	Expected result	Result for a dataset of 2000 patents	Result for a dataset of 22000 patents
Clustering by Gensim	70%	50%	53%
Clustering by PySpark	70%	50%	53%
Clustering by sklearn	70%	67%	77%

The conclusions at this stage are as follows:

1) PySpark and gensim are not doing their job well enough, in the future it is necessary to find out the reason;
2) the functionality of sklearn has proven itself as the best of the three clustering algorithms, in the future it is necessary to find out the reason;
3) it is possible that the data sets are insufficient in volume for this clustering operation, which affected the accuracy of the result.

Test 2 (Table 2). This test is designed to find out how the model works on two different categories of patents that have a common theme (and therefore similar keywords).

Table 2. Test 2.

Type of clustering	Clustering accuracy		
	Expected result	Valid result, 10000 patents	Valid result, 100000 patents
Clustering by Gensim	70%	63%	69%
Clustering by PySpark	70%	62%	
Clustering by sklearn	70%	68%	74%

The conclusions at this stage are as follows:

1) the gensim and PySpark libraries began to show better results. Accuracy has increased by 10–15% on a data set of 10,000 patents, the categories of which are

somewhat different than on a twice as large data set, but whose categories have greater semantic coherence with each other (from test 1);

2) sklearn is more accurate compared to other libraries;
3) on a volume of 100k patents on a local computer, PySpark is not able to perform a clustering operation. More resources are needed for the checks.

In Table 3 you can see the result of the clustering model on two different categories of patents that do not have a common theme. A data set of 10,000 patents was used.).

Table 3. Test 3.

Type of clustering	Clustering accuracy
Clustering by Gensim	80%
Clustering by PySpark	84%
Clustering by sklearn	80%

4 Conclusion

The study develops and describes algorithms for downloading USPTO patent archives, parsing patent documents, clustering and visualization of the patent landscape. The algorithms are implemented as a software. The software was tested on USPTO patent documents for 2019–2021. The use of this developed software will facilitate the process of making strategic decisions in the field of patenting a certain area at the level of the inventor or organization and increase the effectiveness of identifying new trends and ideas in the field of synthesis of new technical solutions. The software can be used to research patents, search for relationships between patents and build visual representations of groups of patents by category.

Possible ways to improve the system: (a) add the ability to save the LADA model trained on certain categories of patents to predict the category of a new patent document (b) improve the quality of preprocessing of the patent description text; (c) develop a software on the customer's resources (computing cluster), in order to fully test the Apache Spark framework; (d) saving a large amount of data on the customer's resources (computing cluster) using HDFS technology (Hadoop Distributed File System).

Acknowledgments. The study was supported by the grant of the Russian Science Foundation No. 22-21-20125, https://rscf.ru/en/project/22-21-20125/, and the Administration of the Volgograd Region.

References

1. Korobkin, D., Fomenkov, S., Fomenkova, M., Vayngolts, I., Kravets, A.G.: The software for computation the criteria-based assessments of the morphological features of technical systems. In: Cyber-Physical Systems, pp. 161–172. Springer, Cham (2021)

2. Korobkin, D., Fomenkov, S., Vereschak, G., Kolesnikov, S., Tolokin, D., Kravets, A.G.: The formation of morphological matrix based on an ontology "patent representation of technical systems" for the search of innovative technical solutions. In: Cyber-Physical Systems, pp. 149–160. Springer, Cham (2021)
3. Yang, C., Huang, C., Su, J.: An improved SAO network-based method for technology trend analysis: A case study of graphene. J. Informet. **12**(1), 271–286 (2018)
4. Kim, S., Park, I., Yoon, B.: SAO2Vec: Development of an algorithm for embedding the subject–action–object (SAO) structure using Doc2Vec. PLoS ONE **15**(2), e0227930 (2020)
5. Guo, J., Wang, X., Li, Q., Zhu, D.: Subject–action–object-based morphology analysis for determining the direction of technological change. Technol. Forecast. Soc. Chang. **105**, 27–40 (2016)
6. Wang, X., Qiu, P., Zhu, D., Mitkova, L., Lei, M., Porter, A.L.: Identification of technology development trends based on subject–action–object analysis: The case of dye-sensitized solar cells. Technol. Forecast. Soc. Chang. **98**, 24–46 (2015)
7. Yang, C., Zhu, D., Wang, X.: SAO semantic information identification for text mining. Int. J. Comput. Intell. Syst. **10**(1), 593 (2017)
8. Souili, A., Cavallucci, D., Rousselot, F., Zanni, C.: Starting from patents to find inputs to the problem graph model of IDM-TRIZ. Procedia Eng. **131**, 150–161 (2015)
9. De Prato, G., Nepelski, D., Piroli, G.: Innovation radar: Identifying innovations and innovators with high potential in ICT FP7, CIP & H2020 projects. JRC Scientific and Policy Reports–EUR 27314:11–15 (2015)
10. Roh, T., Jeong, Y., Jang, H., Yoon, B.: Technology opportunity discovery by structuring user needs based on natural language processing and machine learning. PLoS ONE **14**(10), e0223404 (2019)
11. Feng, L., Niu, Y., Liu, Z., Wang, J., Zhang, K.: Discovering technology opportunity by keyword-based patent analysis: a hybrid approach of morphology analysis and USIT. Sustainability **12**(1), 136 (2019)
12. Bulk Data Storage System. United States Patent and Trademark Office. https://bulkdata.uspto.gov. Accessed 10 Aug 2022
13. gensim – Topic Modelling in Python. https://github.com/RaRe-Technologies/gensim/. Accessed 10 Aug 2022
14. PySpark Documentation. https://spark.apache.org/docs/latest/api/python/. Accessed 10 Aug 2022
15. scikit-learn: Machine Learning in Python. https://scikit-learn.org. Accessed 10 Aug 2022
16. BeautifulSoup. https://www.crummy.com/software/BeautifulSoup/. Accessed 10 Aug 2022
17. pyLDAvis. https://github.com/bmabey/pyLDAvis. Accessed 10 Aug 2022
18. Matplotlib: Visualization with Python. https://matplotlib.org. Accessed 10 Aug 2022

Developing E-Learning Courses in a Gaming Environment with an Integrated Assistant Bot for Secondary School Students

A. N. Varnavsky(✉), K. A. Karmanova, R. A. Kurochkin, and D. A. Popov

HSE University, 20, Myasnitskaya St, Moscow 101000, Russia
avarnavsky@hse.ru

Abstract. Currently, information technologies are mainly used in higher education. At the same time, there are problems in the process of teaching secondary school students. Therefore, it is necessary to offer technologies that would increase the effectiveness of the learning process. Given the popularity of the Minecraft game and the development of the use of the chatbot, it is proposed to integrate these technologies. The paper investigates the use of various information technologies by secondary school students. As a result, conclusions are drawn about the need for the development of information technologies. The development of electronic training courses with an integrated assistant bot for secondary school students of courses on the basics of digital circuitry and programming in Python is described. The behavior of the assistant bot during training and completing course tasks is described. It is assumed that such an approach will increase the involvement of students in the learning process and homework.

Keywords: Gamification · Assistant bot · Minecraft · Integration

1 Introduction

Currently, there is a development of information technologies that can improve the efficiency of the educational process. Such technologies can be used at all levels of education. They were most widely used in higher education.

Such technologies include feedback management and gamification.

Feedback allows you to manage the learning process throughout this entire course. This allows you to focus learning on the joint action of the teacher and the student.

Gamification of the learning process allows you to increase user engagement in the educational environment. Gamification can be presented in different forms: accumulation of points/coins, passing a quest during which the user learns information, etc.

Integration of these technologies will improve the effectiveness of training. Therefore, it can be suggested to study digital disciplines (for example, digital circuitry and programming in Python) to use the most popular game of the 21st century Minecraft with an integrated assistant bot that will give up-to-date information about the correctness of

A. A. Radionov and V. R. Gasiyarov (Eds.): RusAutoCon 2022, LNEE 986, pp. 325–334, 2023.
https://doi.org/10.1007/978-3-031-22311-2_32

executed commands in Python. During the training, the user will be able to receive feedback from the assistant bot, which will give up-to-date information about the correctness of the executed commands in Python.

The purpose of the work is to substantiate the need to develop e-learning courses with an integrated assistant bot for secondary school students and to develop courses on the basics of digital circuitry and programming in Python with an integrated assistant bot.

2 A Study on the Use of Assistant Bots by High School Students

To conduct a study of the use of assistant bots by secondary school students, a questionnaire was developed. The questionnaire contained the following 10 questions.

1. How often do you miss the preparation for classes and do not complete homework in individual subjects?
2. Are you fully or partially preparing for classes or doing homework?
3. How often do you have questions that you can't find an answer to while preparing homework or studying the material? What is the nature of such questions?
4. What technologies do you often use to study the material and do homework?
5. How often do you use the Internet to do homework and prepare for classes in various subjects?
6. What social networks and messengers do you use?
7. Which chatbots do you use for studying?
8. Do you use voice assistants?
9. What social networks and messengers do you use and for what do you use them?
10. What information technologies of assistance in studying, and preparation for classes are effective?

62 students of Russian secondary schools in grades 9–11 took part in the survey. The results of the survey were as follows.

Figure 1 presents the results of the answer to the first question of the questionnaire about shipping preparation for classes and homework. At the same time, only 12.9% of respondents complete the task completely, 71% - about three-quarters, the rest – half or less.

In the process of preparing homework, 45.2% of respondents periodically have questions. Moreover, these issues are both organizational and in terms of material.

16.1% of respondents prepare homework without using computer equipment, 32.2% use it sometimes, and the rest always. At the same time, it can be noted that less than 10% do not use a smartphone for preparation, the rest either periodically or always use it.

The use of the Internet for homework preparation prevails in physics and chemistry, the Internet is rarely used in the process of preparing for the Russian language and biology.

It can be noted that almost all students use some kind of social network or messengers, and a small part of students use social networks for educational purposes. However,

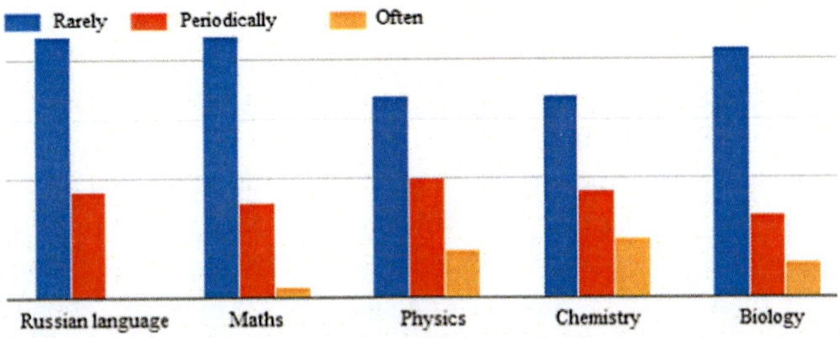

Fig. 1. The results of the answer to the question about the frequency of skipping preparation for classes and not doing homework in various subjects.

only 16.1% of respondents use chatbots. There are other statistics about voice assistants: 77.4% of respondents use them.

Figure 2 presents the results of the answer to the question of which information technologies are the most effective.

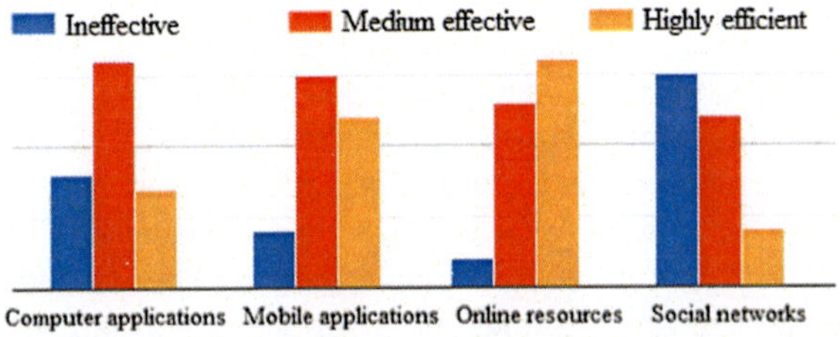

Fig. 2. The results of the answer to the question about the effectiveness of various information technologies to help prepare homework.

The following conclusions can be drawn from the results of the survey.

There is a problem with preparing homework for secondary school students. Moreover, the problem is mainly observed in technical compulsory disciplines. We can say that there will also be problems with such optional disciplines. Students periodically have questions in the preparation process.

Almost all students use social networks and messengers, but not all students use voice assistants. Chatbots are used by a very small number of students.

Thus, the use of gamification technologies, along with feedback, would solve the problem. The synergistic effect of gamification and feedback would be quite effective. Therefore, it is possible to follow the path of developing e-learning courses in a gaming environment with an integrated assistant bot for secondary school students.

3 Integration of Gamification Technologies and Assistant Bots

Modern education systems are primarily aimed at creating a comfortable, interesting, and engaging educational environment for students, a feedback model that allows improving interaction between teachers and students [1].

Many people still have the opinion in their minds that the practice of computer games is an addiction and a pointless waste of time. Despite this, Russian researchers have proved that the negative impact of computer games is exaggerated in society, and the positive impact has not yet been proven. Foreign researchers insist on thinking about the integration of gaming computer mechanics into the educational process [2].

Already, some foreign schools use game systems: students accumulated points for completing a task, then the points are converted into badges or achievements. The behavior of students also affects their “game rating”: the number of points increases for exemplary behavior, and decreases for undesirable behavior. This affects the final ranking at the end of the year. However, only the presence of game elements does not significantly affect the involvement of students. In this case, the context of application plays an important role – children should be able to experiment with rules, opportunities, and emotions, taking part in the educational process using gamification [3].

Gamification (gamification) is often considered a way to motivate users in the process of both electronic [4] and classical [5] learning. It does not carry direct educational goals but is only a tool with which the educational process acquires a gaming atmosphere. Gamification facilitates the learning process [6].

Assistant bots are used in various spheres of human activity, their use is especially effective in such areas as sales, marketing, technical support, and customer service [7]. The use of assistant bots in the field of education is now actively developing. The authors of the article [8] are convinced that the use of bots in education can provide significant assistance in the learning process. At the moment, there are many different chatbots, one way or another related to education. For example, there are bots with which students can learn various references or organizational information, such as the schedule or the list of the group [9, 10]. However, we are more interested in developments that are directly involved in the learning process. The articles [11, 12] consider educational chatbots that act as educational consultants and reference systems, but some of them can also conduct testing and output statistics on solving tasks. The use of chatbots allows not only to reduce labor costs for the execution of the feedback process, but also to speed up this process, because the bot will be able to instantly inform the user about successes and errors when performing work, and it also provides ample opportunities for statistical analysis of the success of the course. Among other things, assistant bots provide opportunities for the organization and effective use of information resources through the use of information technology.

It is also worth noting that the issue of the need to automate the verification of students' programs in various programming courses is raised in [13, 14] because such technology allows you to save the teacher's time for more fruitful interaction with students instead of manually checking solutions to problems. In [13, 14], the creation of automatic verification systems is described, but the approach used in these works is not suitable for our case due to the specifics of the practical tasks of our programming course.

By combining gamification technologies and assistant bots, you can get a product that will be competitive in terms of student engagement and activity in comparison with the usual online and offline courses.

4 Training Courses - Analogs

CODDY Programming School for children has developed a course on programming in Minecraft [15]. The course aim to learn the basics of the Python programming language using the Minecraft game world. The program consists of four modules, each of which has 4 training days. The course content captures all the main topics, as well as some more complex topics for in-depth study.

The course is aimed at children aged 7–10 years. The authors' website states that during the course, students will be able to get acquainted with the Python programming language, learn syntax, learn how to link program code with game objects, create their game worlds and build complex structures using algorithms. The duration of the course is from 4 months.

There is also a similar course from the educational portal Geek Brains IT Education [16]. The course program consists of 14 lessons on the main programming topics. The duration of the course is 3 months.

The online school Kodabra has several courses on Minecraft at once. The courses have different levels of difficulty and age recommendations.

The course "Minecraft: Computer Literacy and Mechanisms" consists of 20 classes, which include the development of basic computer skills and digital literacy [17]. Students are introduced to the basic concepts and computer skills. Also, this course is aimed at developing creativity in children, within its framework, students will be able to create their world in the game Minecraft with various simple mechanisms and buildings. In the end, the participants' works are defended in front of other students and a group of teachers. Classes are held 1–2 times a week, the course is aimed at children aged 7–8 years.

The course "Minecraft: Development of mods and 3D models" consists of 20 classes in which students master the MCreator modification for Minecraft, which allows you to create your own simplest modifications [18]. Also in this course, students will get acquainted with the BlockBench program, which allows you to create your models of game items and other entities. In the end, the work is also defended in front of other students and teaching staff. The course is aimed at children aged 9–10 years.

The course "Minecraft: Creating Worlds and mini-games" consists of 20 classes in which children learn the basics of Python and C#, pass various algorithms, logical expressions, and data types, working with variables, and random numbers [19]. Just like in the previous course, students study the MCreator modification, as well as the World Painter modification, which allows you to edit the world map inside the game using various brushes, which can also be changed, and adjust various parameters inside them. The course is aimed at children aged 7–9 years.

The course «Minecraft: JavaScript Programming» consists of 20 classes, including the study of block programming, random numbers, logical expressions, loops, variables, and functions [20]. During the course, students create their projects on a local Minecraft server. The guys are also learning how to set up their server, connect to the server and work with the server. The course is aimed at children aged 10–14 years.

Comparing various analogs, we can conclude that the program of programming courses is often very similar, sometimes completely identical.

Courses from the online school Kodabra teach children mainly some initial computer control skills or basic knowledge of the game. These courses do not carry a large information load, they are more aimed at extracting commercial benefits for the conducting organization.

From the point of view of using a bot assistant, the analog of development is [21]. This paper discusses the organization of a programming course in Python using a digital assistant. The assistant implemented by the authors generates personalized tasks for students, checks solutions, and also keeps progress statistics. However, it is worth noting that despite the wide functionality of the solution under consideration, the practical part of the course is classical, unlike the solution we propose, where all theoretical information is fixed in practice inside the Minecraft game.

There are no analogs of courses on the basics of digital circuitry. At the moment, there are many different video content on the Internet related to the creation of schemes on binary code in Minecraft, but there is no full-fledged course with a training program.

The closest analog of the course on digital circuit engineering is an online course on the openedu.ru portal from the St. Petersburg State Electrotechnical University "LETI" named after V.I. Ulyanov - "Digital circuit Engineering" [22]. The course takes place in the format of video lectures with test thematic tasks and the final examination test. There are no practical tasks for constructing various schemes.

5 Proposed Courses with Assistant Bot

The developed courses on the basics of digital circuitry and programming in Python can be conducted both in a distance format and in a full-time format in computer classes. To use them, you need to install the following programs on your computer: Minecraft, Python, Java Developer Kit, Minecraft Python API, and Minecraft Spigot server.

When studying a topic of the course on the basics of digital circuitry, the student first turns to the methodological instruction, which describes the theoretical information on the topic, studies the material, and then performs laboratory work, which also examines in detail the device and principles of the game mechanism (for example, Figs. 3 and 4). In the control task, the course participants are invited to create a prototype, a simplified version of a machine engine, using previously acquired knowledge and created mechanisms.

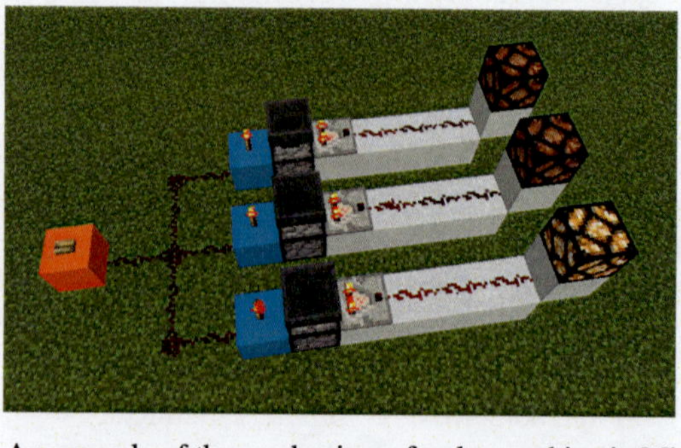

Fig. 3. An example of the mechanism of a slot machine in Minecraft.

Methodological guidelines with laboratory work were also developed for the Python programming course to consolidate the material covered.

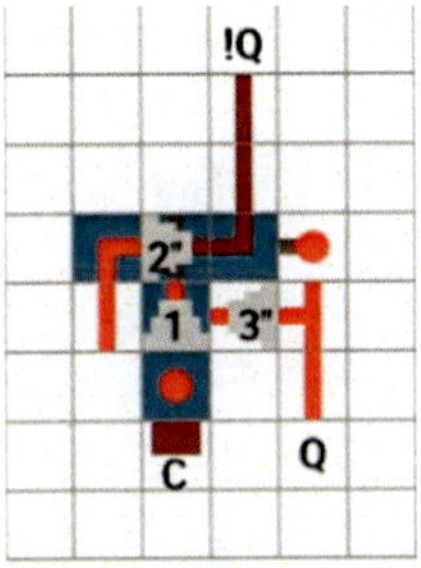

Fig. 4. The scheme of the T-trigger in Minecraft.

The bot assistant solves the problem of automatic verification of the work done by students.

By running a pre-created script, the student will be able to see the status of the completed task in the game chat: "The task was completed successfully" or "The task was completed incorrectly, check the correctness of your decision".

Thus, the student will be able to get feedback without asking the teacher for it. This immerses students in an environment of self-study and self-reflection. At the same time, students do not lose the opportunity to seek help from a teacher who can help sort out the difficulties and misunderstandings that arise (Fig. 5).

A significant limitation in the development of the assistant bot was that students perform tasks on local game servers that cannot be accessed from the outside, so a decision was made to compile a general checking script that the student should run independently after completing the task. For the convenience of users, the checking program contains functions for checking all tasks of the course, so no matter what task the student completed, the checking program is the same for all tasks. After launching, the program outputs a message to the game chat, asking the student for the topic number and the task number, and then launches the corresponding verification function.

The structure of the checking program is as follows: in its main part, which begins its work immediately after the user starts, there is communication with the student, during which the program learns the number of the completed task and the number of the topic to which it relates. After that, the program analyzes the received numbers using conditional operators and, based on the fulfillment of certain conditions, for example, "the topic number is 1 and the task number is 1", calls the verification function that is responsible for this task. Based on the results of the check, this function generates feedback about the correctness of the task, and displays it in a chat with the player, after which the work of the checking program is completed.

It is worth noting that the Minecraft Python API does not include the functionality for processing player messages in the chat, there is only a method for sending a text to the chat using the program. To solve this problem, system files are used that store information about the operation of the server, that is, logs, from where it seems possible to process player messages in the chat.

Fig. 5. An example of a student's communication with an assistant in a game chat.

The verification program contains an individual verification algorithm for each task. Below is a generalized scheme of the task verification algorithm, in which the student needed to build an object from blocks of wood in certain coordinates.

This algorithm checks the material of the blocks in the coordinates where the student needed to build a parallelepiped of wood. The program separately analyzes the unique identifier of each necessary block for compliance with a given value, in our case, wood.

More complex algorithms were also written, for example, for tasks where students needed to use conditional constructions. In one of these tasks, it was necessary to build a house with a "secret" door, which appears only when the player is directly in front of it. While checking this task, the program at the beginning remembers the current location of the player and moves him away from the house to check whether the student has built the house correctly. An additional check of the blocks inside the house is performed - they must have the identifiers of the air blocks because the house is a hollow structure. Then the program moves the game character to the coordinates in front of the "secret" door and analyzes the ids of the blocks in the place where the passage for the player should have appeared. Upon completion of the check, the program returns the character to its original location and displays the results of the task check.

Thus, it is assumed that the bot assistant is an additional channel for receiving feedback, which increases the involvement of children in the learning process and also contributes to the development of the ability to independently analyze the proposed solution and find mistakes in it, without resorting to the help of a teacher.

6 Conclusion

The integration of gamification into the educational process will soon become a trend, as analysts predict. New approaches to learning will help to increase the involvement and interest of students in learning the material.

After analyzing the current analogs of Minecraft courses, we can conclude that there are no analogs of the course on the basics of digital circuitry implemented in the game, which makes this work unique. In conjunction with the Python programming course, it is possible in the future to create a course that will include various data elements of the two courses. For example, a student will need to complete a circuit engineering task (build some kind of logic circuit) using programming skills and interacting with the game using commands.

Assistant bots can also be integrated into the educational process, especially in such courses, where support and feedback for students are constantly required. Such bots will become indispensable assistants in conditions when the teacher cannot immediately tell the student whether he has completed the task correctly or not, which will push the children to independently study problems and analyze their work.

Acknowledgments. The publication was prepared within the framework of the Academic Fund Program at the HSE University in 2022 (grant № 22–00-052).

References

1. Varenina, L.P.: Geymificatcia v obrazovanii (Gamification in education). Hist. Socio-Educ. Thought **6**(6-2), 314–317 (2014). (In Russian)
2. Idrisova, A.: Chto takoe geymificatcia? (What is gamification?). https://skolki-project.com/blog/chto-takoe-gejmifikatsija. Accessed 4 Jun 2022 (2020)
3. Lee, J., Hammer, J.: Gamification in education: what, how, why bother? Acad. Exch. Q. **15**(2), 1–5 (2011)
4. Karpenko, O.M., Lukyanova, A.V., Abramova, A.V., et al.: Geymificatcia v electronnom obuchenii (Gamification in e-learning). Territory of new opportunities. Distance and virtual learning 4:28–43 (2015). (In Russian)
5. Bakhmetyeva, I.A., Yaylaeva, R.N.: Geymificatcia v obrazovanii (Gamification in education). Colloquium-journal. – Holoprystan Reg. Employ. Cent. **21**(45), 32–34 (2019). (In Russian)
6. Orlova, O.V., Titova, V.N.: Geymificatcia kak sposob organizatcii obucheniya (Gamification as a way of organizing learning). Bull. Tomsk. State Pedagog. Univ. **9**(162), 60–64 (2015). (In Russian)
7. Tsyvarev, I.V.: Issledovanie i razrabotka chat-botov tselevogo naznacheniya (Research and development of targeted chatbots). Youth school-seminar on control problems in technical systems named after A.A. Vavilov **1**, 86–87 (2021). (In Russian)
8. Molnár, G., Szüts, Z.: The role of chatbots in formal education. In: IEEE 16th International Symposium on Intelligent Systems and Informatics (SISY), IEEE, Subotica, 13–15 September 2018
9. Korovkin, V.A., Boyko, I.D.: Chat-bot tsifrovoy assistant studenta (Chat-bot-digital student assistant). In: Proceedings of the XVIII International scientific and practical conference of students, graduate students and young scientists, Tomsk Polytechnic University, Tomsk, 22–26 March 2021 (2021). (In Russian)
10. Kamynin, K.V., Pavlyukov, D.A., Garmatiy, N.I. et al.: Chat-bot "Vkontakte" ili informatsionniy pomoschnik dlya vuza (Chat-bot "Vkontakte" or information assistant for the university). In: Proceedings of the XX International scientific and technical conference of graduate students and students. Automation of technological objects and processes, Donetsk National Technical University, Donetsk 26–28 May 2020 (2020). (In Russian)

11. Kireeva, N.A., Rodionov, A.S., Farkhutdinov, R.I., et al.: Razrabotka chat-bota po istorii dlya primeneiya v tekhnicheskom vuze (Development of a chat-bot on history for use in a technical university). Electrotech. Inf. Complexes Syst. **14**(3), 73–79 (2018). (In Russian)
12. Gavronskaya, Y.Y., Korsakova, E.A.: Khimik-bot: virtualniy pomoschnik v podgotovke k ege (Chemist-bot: a virtual assistant in preparing for the exam). Chemistry at school **6**, 51–54 (2021). (In Russian)
13. Minkin, I.V.: Sistema avtomaticheskoy proverki laboratornykh rabot po programmirovaniyu (System for automatic testing of laboratory work on programming). In: Proceedings of the VI All-Russian Scientific and Technical Conference of Students, Postgraduates and Young Scientists, Siberian Federal University, Krasnoyarsk, 20–25 April 2010 (2011). (In Russian)
14. Serobabov, D.S.: Avtomaticheskaya proverka laboratornikh rabot po programmirovaniyu (Automatic verification of programming laboratory works). In: Proceedings of the VI International Scientific and Practical Conference. Consolidation of intellectual resources as a foundation for the development of modern science, Petrozavodsk, 28 October 2021 (2021). (In Russian)
15. Minecraft programming for kids. https://coddyschool.com/courses/minecraft-programming. Accessed 4 Jun 2022
16. Minecraft: Programming with Python. https://gb.ru/courses/geek-school/minecraft. Accessed 4 Jun 2022
17. Minecraft: Computer literacy and mechanisms. https://codabra.org/minecraft/komputernaya-gramotnost-i-mehanizmy-lvl0-7-8-let. Accessed 4 Jun 2022
18. Minecraft: Development of mods and 3D models. https://codabra.org/minecraft/razrabotka-modov-i-3d-modeley-lvl2-9-10-let. Accessed 4 Jun 2022
19. Minecraft: Create worlds and mini-games. https://codabra.org/testminecraft1. Accessed 4 Jun 2022
20. Minecraft: JavaScript Programming. https://codabra.org/minecraft/programmirovanie-na-javascript-lvl2-10-14-let. Accessed 4 Jun 2022
21. Andrianova, E.G., Demidova, L.A., Sovetov, P.N.: "Tsifrovoy assistent prepodavatelya" v massovom professionalnom obuchenii ("Digital teaching assistant" in mass professional training for the digital economy). Russ. Technol. J. **10**(3), 7–23 (2022). (In Russian)
22. Digital circuit engineering. https://openedu.ru/course/eltech/DSD/. Accessed 4 Jun 2022

On the Interpretability of the Deep Learning Models Including in the Process of Neural Network Modeling of Dangerous Hydrometeorological Events to Provide a Stable Operation in Industry

L. Haritonova(✉)

Volgograd State Technical University, 28, Pr. Named After V.I. Lenina, Volgograd 400074, Russia
haritonova410@yandex.ru

Abstract. This study deals with a number of aspects for the interpretability of nonlinear models which even after the complete training by neuron networks restrict the researcher's ability to obtain information about processes within these models and treated as black-boxes. Dangerous hydrometeorological events bring great damage to industry and economy of a country and the world in general. Some results of the research are given implemented for the developed neural models including the multi-layer neural networks. Dangerous hydrometeorological events are considered for all federal regions in the Russian Federation (from the year 2010 to 2018). Herewith, as an input, 16 parameters have been taken. The regression analysis was performed for the predicted results and the target outputs, the author obtained regression lines equations and the correlation coefficients. As the training function, the method of Levenberg-Marquardt was used, as the adaption learning function, the researcher used the training function of the gradient descent with momentum, as the transfer function, the hyperbolic tangent, as the Performance function – the Mean Squared Error. Some practical recommendations are given to designers and specialists using artificial neural networks in industry and other branches of economy. Conclusions are specified and some future perspectives are explained as a form of conclusion.

Keywords: Interpretability · Deep Learning · Neural modeling · Artificial neural networks · Regression analysis · Dangerous hydrometeorological events · Stable operation in industry

1 Introduction

The problem of studying model interpretability is very urgent today and it is safe to say that this issue will be further developed in the near future.

A. A. Radionov and V. R. Gasiyarov (Eds.): RusAutoCon 2022, LNEE 986, pp. 335–345, 2023.
https://doi.org/10.1007/978-3-031-22311-2_33

2 The Topicality, Scientific Importance, Review of Literature (Publications)

In a considerable number of works, the prospects of the ANN application have been indicated [1–8].

By way of example, in [1] the multi-layer neural networks are treated in order to forecast air leakage in the automobile heat exchangers. The results of modeling and simulation of artificial neural networks are given in the paper. The comparison of various network parameters is made for increasing the efficiency of forecast and the issued errors.

The neural networks are a strong technique in working with non-linear relations. Along with this, even the neural network with the needed training does not provide a researcher with information about the inner structure of the data and the revealed relations. In its essence, the neural network is a set of the weight matrix. There is no explanation of these weights. This requires the clarification of the interpretability problem for the neural networks.

The ANN can forecast the output variables. However, the processes and mechanisms in the network geometry are often ignored, and the ANN are considered as black-boxes [9].

Study [10] deals with the issue of application of the formerly known method of forecast interpretation using the influence functions.

The work [11] is devoted to the "black-box" models interpretation problem. Let us consider a question what can be understood under the notion of models interpretation. As such, we will understand "degree with which a person can understand the reasons of various solutions". Herewith, it can be possible to find answers to questions of how a model with training will present us with the forecasts; how various parts of the model influence the forecasts; how a model will make various decisions for a group of objects. Herewith, models are treated the interpretation of which do not present problems due to the model geometry – these are the "naturally interpreted models" [11]. These are regressions, for example, (linear and logistic ones), decision trees, et al.

However, the explanation of the internal design of some model types, for instance, Deep Learning networks presents an unsolved challenge. Herewith, the following methods are known to create explanatory mechanisms (model-agnostic methods):

- model class reliance – study of model certainty for the model-independent version of indicator importance algorithm (which provides with the assessment of the dependence for the mean error of the model under study upon dispersion values of each particular indicator);
- local interpretable model-agnostic explanations (the LIME method). Behind this method, there is the training of the local interpretable models by forecasts of that model which is under study aiming at explanation of particular forecasts;
- partial dependency graphs presenting a limitary influence of particular indicators upon the resulting value, etc.

As is noted in [12] the models based on artificial neural networks show excellent results, however, it turned out problematic to state their ability to reason on abstract notions. In this connection, the study presents an approach to measure abstract reasoning in the training machines and a number of ideas is given on the nature of the generalization itself. ANN can be applied to an increasing number of real world problems of considerable complexity. They are used for solving problems that are too complex for conventional technologies or for those types of problems which do not have an algorithmic solution.

The climate presents a data problem. The application of an artificial intelligence does not only provide with understanding the patterns and laws of climate change but also to forecast it withstanding the negative effects of the change.

Scientists from the World Meteorological Organization addressed the UNO and governments of different countries with a request to announce an extraordinary meteorological situation due to the global warming. The rise of temperatures by 2° before the end of the century will result in a failure in the ecosystem and lead to suffering of the most part of the planet's population that can become unfit for human life.

The study [13] presents informations on prototyping dairy IoT edge devices which based on the artificial intelligence (AI) models allow to find solutions to prevent impacts of climate change.

The research [6], based on the data given in [14], the neural modeling has been performed for the evaluation of dangerous hydrometeorological phenomena with the case of maintaining reliable operation of transportation systems. As the input indicators, the following values were used – x_1 – year (1998 through 2015) and x_2 – months (January through December).

3 Problem Setting

Study of hazardous phenomena (events) in all federal regions of the state is of undoubted interest as well as taking various kinds of these dangerous hydrometeorological phenomena as the input parameters.

Apart from that, while designing the geometry of new neural network structures for various science and technology branches, researchers often point out that the number of the hidden internal layers in the networks and the number of neural processing elements were taken experimentally out of the intuitive notions, (apart from the neuron number taken as their value calculated by the known empiric formulas, for example, given in [15, 16]). The following question has not been studied – how the number of hidden layers and the number of neurons in each layer influence the value of correlation coefficients and the Mean Square Error. Apart from this, there are no clear answers to the question: what number of training and iterations will be sufficient, since their higher values require extra time. It is supposed to look for answers to the given questions for specialists in various industry fields by the example of neural network modeling for hazardous hydrometeorological events for all federal regions in the Russian Federation.

4 Theoretical, Results

Let us take as the input parameters:

Some examples of how your references should be listed are given at the end of this template in the 'References' section, which will allow you to assemble your reference list according to the correct format and font size.

x_1 – year (from 2010 to 2018)
x_2 – federal districts (Central; Volga District; Northwestern; Southern; North-Caucasian; Ural; Far Eastern; Sibirian);
x_3 – strong wind;
x_4 – heavy rainfall;
x_5 – snowstorm and snowfall;
x_6 – mixed precipitation;
x_7 – whirlwinds;
x_8 – severe frost;
x_9 – anomalous cold weather;
x_{10} – intense heat;
x_{11} – anomalous hot weather;
x_{12} – hail showers;
x_{13} – ice-forming phenomena;
x_{14} – windchill;
x_{15} – fog;
x_{16} – complex of weather events.

The information from The Report on the climate features on the territory of the Russian Federation was taken as the output data (that is an official source from the Federal Service for Hydrometeorology and the Environmental Monitoring (Roshydromet)) affirms that in 2018, hazardous hydro-meteorological events the were observed by 133 events more than in 2017 totally on the territory of Russia.

According to available observations, there is a significant rise in the total number of such hazardous events yearly [17].

A one-way network (feed – forward back prop NN) was applied as a network type.

The studied neural networks consisted of:

- the input layer (out of 16 neurons) – one to each input variable;
- from one to eight hidden layers out of several neurons (from 4 to 32);
- one input layer out of one neuron representing the output variable.

As a target, the total aggregate of hazardous weather events in all the federal districts of the Russian Federation during the years under study.

An example of one of the studied geometries (16-8-8-8-1-1) containing three hidden layers out of 8 neurons in each is shown in Fig. 1.

Since the study [8] showed that the best results can be obtained when using the following algorithms in the neural networks aided research of hazardous weather events: as the training function, the technique of Levenberg – Marquardt, as the adaption learning function – the learning function of the gradient descent with momentum, as the transfer function – the hyperbolic tangent, as the Performance function – Mean Squared Error, they were used in the proposed study.

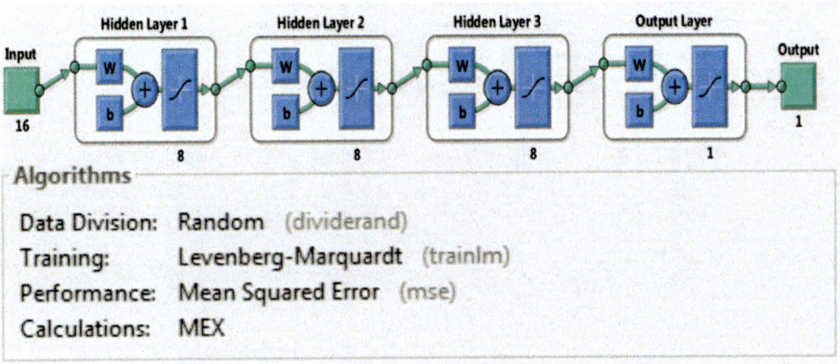

Fig. 1. An example of one of the studied geometries (16-8-8-8-1-1).

A great deal of experiments were carried out, some of the results are presented in Figs. 2, 3, 4 and 5.

The rise in the number of Hidden Layers enhances the complexity degree. As it is shown in [18], when the number of hidden layers equals 6, the complexity level reaches an almost infeasible solution; one of the reasons for that – the problem of the gradient disappearing.

This fact is proved by the results presented in Figs. 6 and 7It follows that with the number of Hidden Layers in excess of 6, the value of the total correlation coefficient drops sharply with the corresponding sharp increase of the Mean Squared Error.

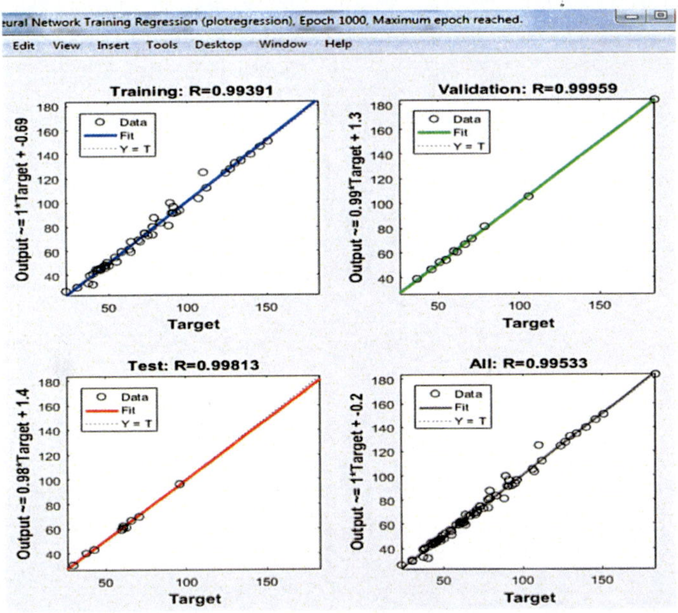

Fig. 2. Correlation coefficients and the regression lines equations in training, testing, validation and the total and Regression equations for the outputs for the network Structure 16-8-8-8-1-1.

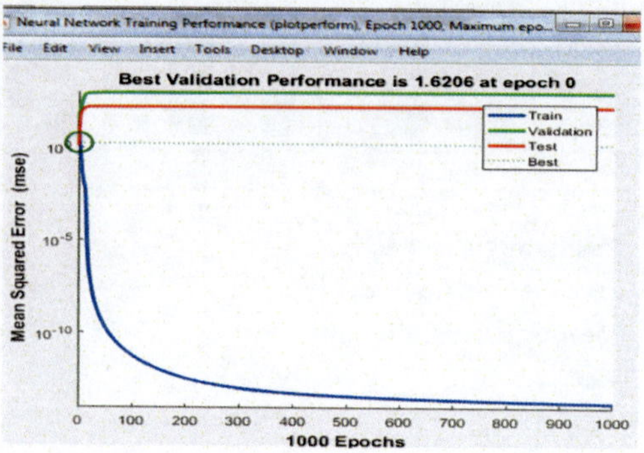

Fig. 3. Value of the variable Mean Square Error (MSE) for the case shown in Fig. 1.

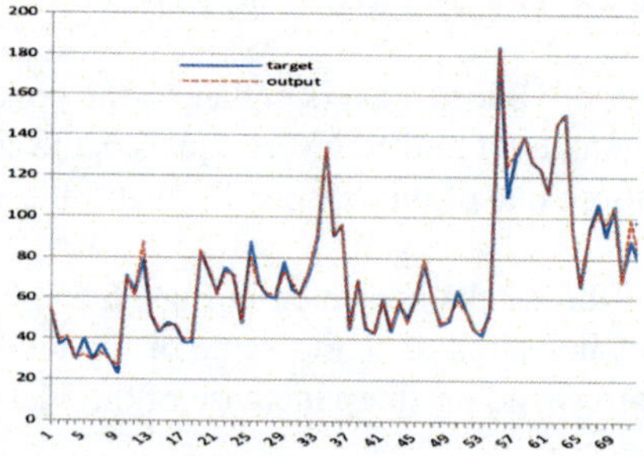

Fig. 4. Graph of comparing the target data with the output values calculated using neural network 16-8-8-8-1-1.

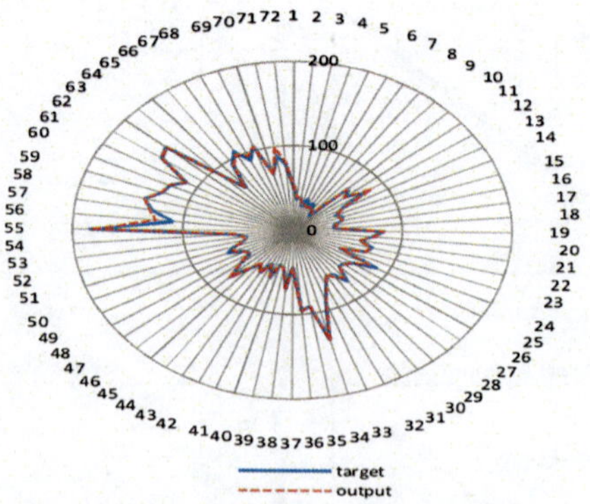

Fig. 5. Circle graph comparing the target data with the output values calculated using the neural network 16-8-8-8-1-1.

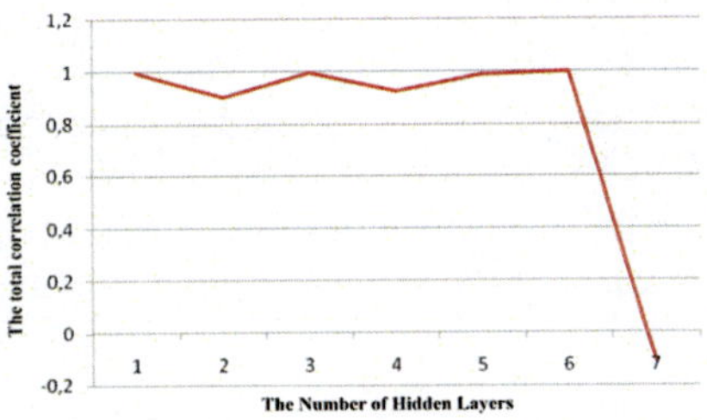

Fig. 6. Values of the total correlation coefficient depending on the number of Hidden Layers with the number of trainings equaling 15.

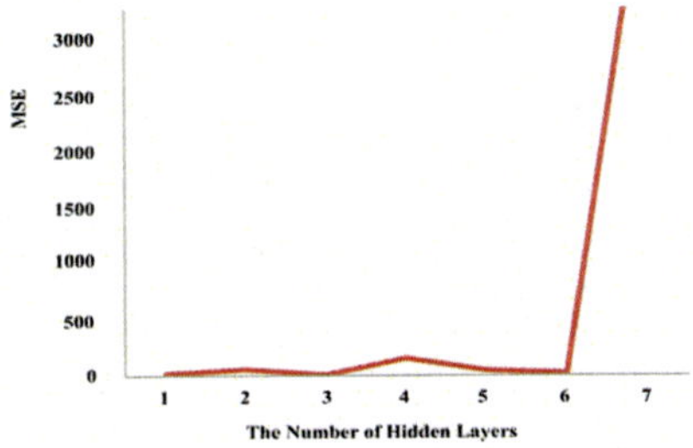

Fig. 7. Mean Square Error values depending on the number of Hidden Layers with the number of trainings equal to 15.

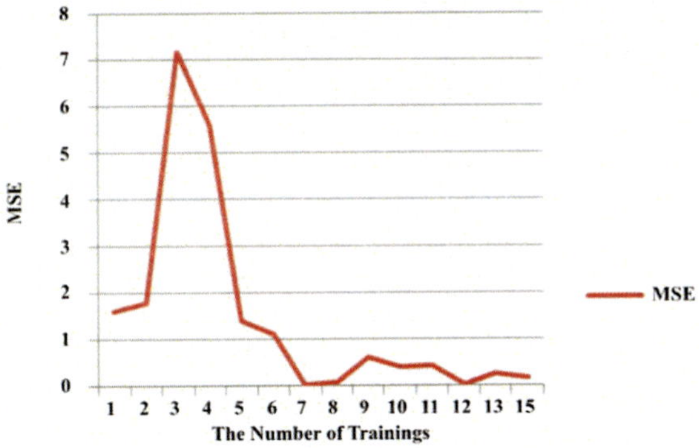

Fig. 8. Dependency of the MSE value upon the number of trainings at the following neural network geometry: 1 hidden layer with 8 neurons.

The values of the MSE at the various numbers of trainings with one Hidden Layer drops both in case with 8 neurons and with 6 neurons. But if in the first case (Fig. 8) after the brief rise the Mean Square Error drops quite considerably and remains at the quite low value after 7 trainings, in the second case (Fig. 9) with the general tendency to decreasing, this drop is not so warranted. However, in general, 13…15 trainings can be recommended.

Figures 10, 11 and 12 presents a comparative study of the success of application for various numbers of neurons at different geometry of the neural network.

It follows from the analysis of graphs that the tendency to the change (rise or drop) of the MSE preserves its form at various numbers of iterations. However, the rise of trainings number from 5 to 15 allows to decrease the Mean Square Error absolutely. Herewith, the rise of the number of iterations up to 1000 at the simultaneous (rise of trainings number boosts the rise of the total correlation coefficient (Fig. 12).

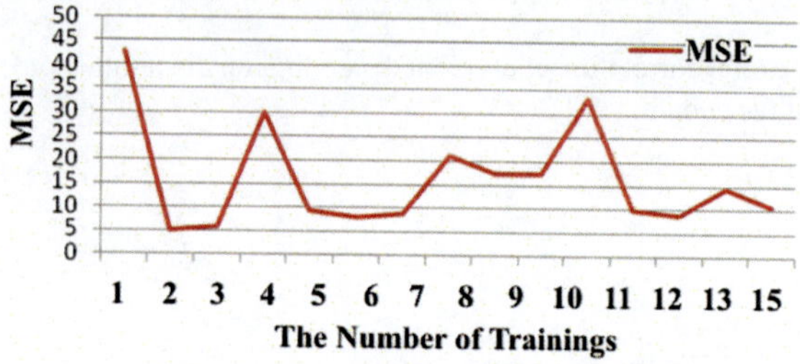

Fig. 9. Dependency of the Mean Square Error value upon the number of trainings with the following geometry of neural network: 1 hidden layer with 6 neurons.

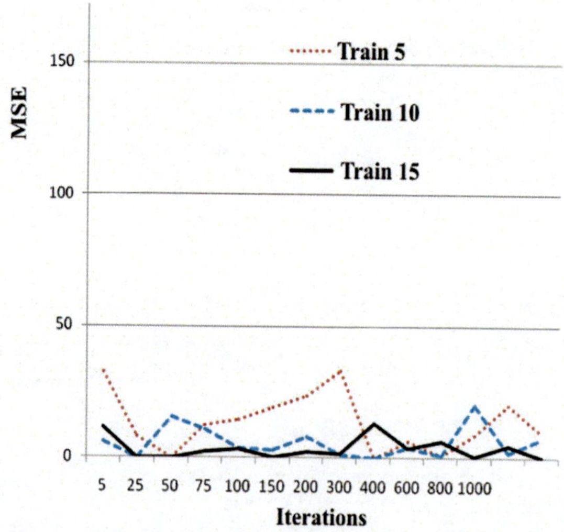

Fig. 10. Dependency of the Mean Square Error value upon the number of iterations at the following geometry of neural network: 1 hidden layer with 8 neurons with various number of trainings (5; 10; 15 correspondingly).

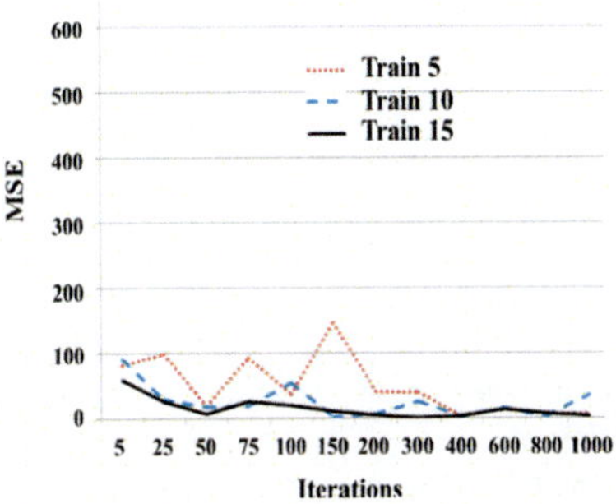

Fig. 11. Dependency of the Mean Square Error value upon the number of iterations at the following geometry of neural network: 2 hidden layers in 4 neurons in each with various number of trainings (5; 10; 15 correspondingly).

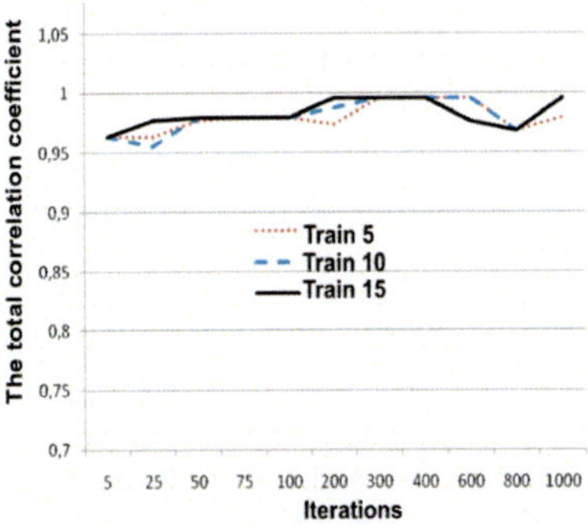

Fig. 12. Dependency of the total correlation coefficient value upon the number of iterations at the following geometry of the neural network: 2 hidden layers with 4 neurons in each with different number of trainings (5; 10; 15 correspondingly).

5 Conclusions, Practical Recommendations, Future Prospects

Thus, to achieve higher values of the total correlation coefficients at smaller values of the Mean Square Error, it is reasonable to perform about 1000 iterations at a fairly large number of trainings in the neural network (about 15). It does not make sense to increase the number of the Hidden Layers in most cases more than 6.

The given article presents only part of the obtained results because of the space restriction. Further work continuation along with the presented data will enable to decrease heavy costs (due to the forecast improvements) and rise in the stability of industrial enterprises for various kinds of production.

Research embracing many neurons [19, 20] in different layers of artificial neural networks is very important for simulating human brain operation. Hence, the development prospects are so important. Herewith, excellent results are to be expected when bringing artificial neurons into more widespread use realized on an actual quantum processor [21] and the so-called quantum neural networks [22].

References

1. Peta, K., Zurek, J.: Prediction of air leakage in heat exchangers for automotive applications using artificial neural networks. In: 9th IEEE Annual Ubiquitous Computing, Electronics and Mobile Communication Conference, EMCON 2018; New York City, United States, November 2018, pp. 721–725, 8796666 (2008)
2. Ribeiro, J.X.F., Liao, R., Aliyu, A.M., Liu, Z.: Prediction of pressure gradient in two and three-phase flows in vertical pipes using an artificial neural network model. Int. J. Eng. Techno. Innov. **9**(3), 155–170 (2019)
3. Haritonova, L.: Neural modeling of the energy efficiency factor for recuperators (heat exchangers) using impinging jets for the metallurgical productions and machine building. In: Radionov, A.A., Kravchenko, O.A., Guzeev, V.I., Rozhdestvenskiy, Y.V. (eds.) ICIE 2018. LNME, pp. 1269–1278. Springer, Cham (2019). https://doi.org/10.1007/978-3-319-95630-5_133
4. Haritonova, L.: Application of the neural modelling method for the Solar radiation analysis in a number of cities in the Russian Federation for the solution of environmental problems. In: E3S Web of Conferences 126:000772019 (2019). https://doi.org/10.1051/e3sconf/201912600077
5. Haritonova, L.P.: On the issue of neural modeling of some dynamic parameters of earthquake. Int. J. Eng. Technol. **7**(23), 440–442 (2018). https://doi.org/10.14419/ijet.v7i2.23.15328
6. Haritonova, L.P.: On the of neural modeling of some dynamic parameters of earthquakes and fire safety in high-rise Construction. In: E3S Web of Conferences, vol. 33, HRC2017, EDP Sciences, Les Ulis, France (2010). :https://doi.org/10.1051/e3sconf/20183302072
7. Haritonova, L.P.: On neural modeling of heat exchange in heat exchangers (recuperators) with the systems of plane-parallel impingement jets for machine building and metallurgical productions. procedia engineering, In: International Conference on Industrial Engineering, vol. 206, pp. 1002–1008 (2017). https://doi.org/10.1016/J.PROENG.2017.10.585
8. Haritonova, L.: the neural modeling for the assessment of hazardous hydro meteorological phenomena including the case of the providing a sustainable work (reliability) on transport. In: Murgul, V., Popovic, Z. (eds.) EMMFT 2017. AISC, vol. 692, pp. 700–709. Springer, Cham (2018). https://doi.org/10.1007/978-3-319-70987-1_74
9. Gevrey, M., Dimopoulos, I., Lek, S.: Review and comparison of methods to study the contribution of variables in artificial neural network models. Ecol. Model. **160**, 249–264 (2003)
10. Koh, P.W., Liang, P.: Understanding black-box predictions via influence functions. In: Proceedings of the 34th International Conference on Machine Learning, PMLR, vol. 70, pp. 1885–1894 (2017)
11. Molnar, C.: Interpretable machine learning (2018). https://christophm.github.io/interpretable-ml-book/. Accessed 6 Sept 2018
12. Islam, M., Chen, G., Jin, S.: An overview of neural network. Am. J. Neural Netw. Appl. **5**(1), 7–11 (2019)
13. Kedari, S., Vuppalapati, J.S., Ilapakurti, A., Kedar,i S., Vuppalapati, R., Vuppalapati, C.: The role of supervised climate data models and dairy IoT edge devices in democratizing artificial intelligence to small scale dairy farmers worldwide. Advances in intelligent systems and computing. In: Yang, X.S., Sherratt, S., Dey, N., Joshi, A. (eds), Fourth International Congress on Information and Communication Technology, vol. 1 027, Springer, Singapore (2020). https://doi.org/10.1007/978-981-32-9343-4_31
14. The Report on the climatic features in the Russian Federation for the year 2015, Roshydromet, Moscow, 2016

15. Zhang, G., Patuwo, B.E., Hu, M.Y.: Forecasting with artificial neural networks: the state of the art. Int. J. Forecast. **14**, 35–62 (1998)
16. NeuroShell 2 User's Manual, Ward Systems Group, Inc., Frederick (1996)
17. A Report of climate features on the territory of the Russian Federation in: Roshydromet. Moscow **2019**, 45–49 (2018)
18. Jan, B., et al.: Deep learning in big data analytics: a comparative study. Comput. Electr. Eng **75**, 275–287 (2019)
19. Koroteev, M.V.: Review of some contemporary trends in machine learning technology. E-Manag. **1**(1), 26–35 (2018)
20. Murphy, K.P.: Machine Learning: "A Probabilistic Perspective." Massachusetts, The MIT Press, Cambridge (2012)
21. Tacchino, F., Macchiavello, C., Gerace, D.: An artificial neuron implemented on an actual quantum processor. NPJ Quant. INF **5**, 26 (2019)
22. Zhao, J., Zhang, Y.-H., Shao, C.-P., Yu-Chun, W., Guo, G.-C., Guo, G.-P.: Building quantum neural networks based on a swap test. Phys. Rev. A **100**, 012334 (2019)

Flexible Manufacturing Systems

Demand Side Management in the Smart Electricity Industry

I. Kuptsov[1](✉), Yu. Zatsarinnaya[1], and R. Gainullin[2]

[1] Kazan State Power Engineering University, 51, Krasnoselskaya Street, Kazan 420066, Russia
ilya.cuptzov@yandex.ru

[2] Kazan National Research Technological University, 72, Karl Marx Str, Kazan 420015, Russia

Abstract. Currently, one of the priority tasks in the energy strategy of the Russian Federation until 2035 is to improve the management of the processes of production, transmission, distribution, and consumption of all types of energy resources, which contributes to an increase in the efficiency of the functioning of the fuel and energy complex. In addition to improving energy efficiency in the field of production, transformation, and consumption of energy, it is worth paying attention to the area of planning, organization, and control of the process of circulation of energy resources. During the intensive introduction of digitalization and increasing the importance of information and communication resources, one of the most promising and effective tools for increasing economic efficiency used in world practice is managing the demand for electricity from end consumers. The presented tool is actively used in Western countries, countries of Southeast Asia and has successfully proven itself over the past 30 years of its use. Demand management is most often used by reducing electricity consumption using energy storage devices, etc., adjusting the intensity of the operation of pumping system engines and settings for air conditioning and/or refrigeration systems, changing the production cycle or partially turning off lighting, etc.

Keywords: Demand response · Aggregator · Cost of services · System operator · Electricity consumption

1 Introduction

Now there are about 100 thermal power plants, coal-fired and gas-fired plants in Russia. About 20 of them are in Moscow. These power plants have been loaded at 48–50% for the last few years. The market and consumers pay for energy consumption as for 100%. This problem is systemic in many countries of the world, not only in Russia. On the scale of our country, the consumer pays 500 billion rubles annually [1].

The problem is the uneven consumption of electricity. At night, people sleep, businesses do not work, and electricity is consumed in minimal amounts. Then there are two peaks: morning and evening, which also correlate with seasonality. At its minimum, the RF power system consumes 80 GW during a warm summer night, and at its maximum, 150 GW during a cold winter evening. The situation is complicated by the fact that

A. A. Radionov and V. R. Gasiyarov (Eds.): RusAutoCon 2022, LNEE 986, pp. 349–359, 2023.
https://doi.org/10.1007/978-3-031-22311-2_34

nuclear power operates in the base part of the load schedule, it cannot be switched on or off quickly. At the same time, when the consumption grows it is necessary to quickly turn on the power plants to meet the demand for electricity. Technologically, this can only be done by thermal power plants: coal or gas. Consequently, when demand increases, these plants are switched on to the grid to ensure the peak shown in green in Fig. 1.

To provide this peak for a few hours a day, and sometimes only in winter, certain stations are activated. These stations are used to provide peak load for several weeks a year. At the same time, the cost of their maintenance, staffing, and presence in the market are paid by the end users. This is the main problem, the essence of which is the demand for a limited number of power plants a few days a year [2].

Let's assume that enterprises which are loaded at peak load on the power system will limit their consumption to a specified level if they have the technical capability to do so. This condition will be met by the enterprise with a certain motivation. By not consuming N number of kW, the enterprise will receive money for it, as if it did not generate energy. What is the benefit of this?

Now when a consumer or group of consumers is switched off, at that moment the peak, the most expensive power plant will not need to switch on, because the consumption peak will be cut off. And those who are turned off will produce "virtual electricity," because they will have their own "virtual power plant," producing electricity exactly as much as they were cut off at the peak [3, 4].

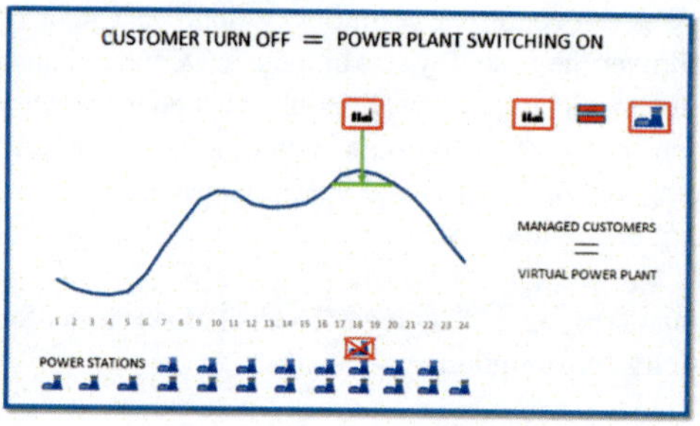

Fig. 1. Power plant shutdown.

2 Principle of Operation of the Demand Response Management Mechanism

There is a system operator, which manages the load in the network, and an aggregator company. The aggregator concludes contracts with different consumers who are willing to conduct a demand management event by a certain amount of kW. The system operator gives a command to reduce consumption and a certain time, and the consumer stops his technological equipment.

Working procedure (Fig. 2):

- The system operator gives the command to reduce

- The aggregator transmits commands to consumers
- Consumers execute commands
- The system operator settles accounts with the aggregator
- Based on the execution of commands the aggregator settles with consumers

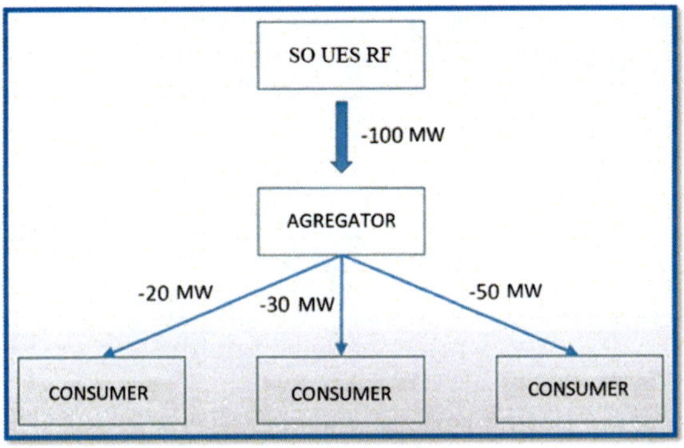

Fig. 2. The role of aggregators.

Due to objective reasons, not every consumer can unload at a certain moment. According to available statistics, to provide a demand management event for 100 MW, it is necessary to conclude a contract for 250–300 MW. Constantly and steadily disconnected about 30% of customers from the total number of applications filed.

Then the system operator pays money to the aggregator, the aggregator gives part of the money to the consumer, part of the money keeps itself in the form of profit [5, 7].

3 Participants and Their Roles in the Demand Management System

Let us distinguish the main participants of the demand management market.

System operator is a specialized organization, which solely performs centralized operational and dispatch control in the Unified Energy System of Russia.

Demand-side management aggregators - organizations, which purchase services from retail consumers, combine their ability to change consumption and convert it into goods and services in the power, capacity, and system services markets.

Consumers - the final link in the retail electricity market chain that participates in-group load change management [6, 8] (Figs. 3).

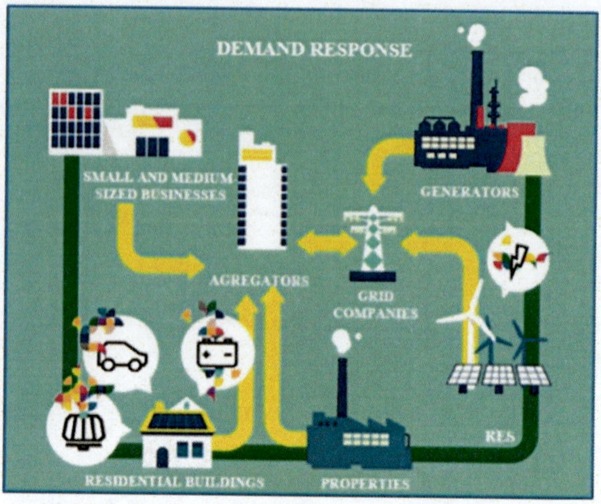

Fig. 3. Participants of the demand management system.

4 The Role of Load Aggregators

Load aggregators - organizations that purchase services from retail consumers, consolidate their ability to change consumption and convert it into goods and services in the electricity, capacity and system services markets, and broadcast a portion of the effect obtained in the wholesale market to consumers.

A load aggregator enters into contracts with retail market consumers for the provision of services to change the load of their equipment. When appropriate economic conditions arise, the aggregator receives a signal from the wholesale market infrastructure organizations to reduce the load in a certain volume, distributes this volume among the participating consumers and gives them a signal to change the load. The consumer's load changes by a given amount over a given time interval. The aggregator receives payment in the electricity and capacity market (or system services) for reducing electricity consumption or increasing generation. The consumer receives payment for consumption change services from the aggregator in accordance with the service agreement concluded between them [6, 8].

Load Aggregators' Tasks:

- to conclude service contracts with retail market consumers to change the load of their equipment by a specified amount, the number of times specified in the contract by a notice issued in advance;
- act as an agent in the wholesale market - the consumer does not need to understand complex rules, establish interaction with the wholesale market infrastructure, go through complicated admission procedures;
- to search for consumers, who have the technology to change the consumption without affecting the main production.

5 Calculation Model for Determining the Cost of Consumer Services

The demand management mechanism is designed to solve the problem of demand for a limited number of power plants a few days a year.

The practical solution to this problem is proposed in the form of a calculation model. The basic formulas and calculation data were taken from the website of SO UES RF.

The presented calculation model will allow the enterprise itself to determine the possibility and expediency of introducing demand management technology. Also, this model will allow calculating the profit, which in the long run can be obtained by the enterprise when using the technology of DSMC [9, 10].

Considering the target audience (consumers of the retail market) such properties of calculations become fundamentally important as:

- simplicity;
- reproducibility of calculation results.

In addition, studies conducted abroad show that even a significant increase in the complexity of the control methods, as a rule, leads only to an insignificant increase in the accuracy of the methods.

Considering the above remarks, let us proceed to the formulation of the calculation model.

Stage 1. Data Entry. Defining the Price Zone
At this stage, data necessary for further calculations are collected. These include:

1) PR - price of rendering services per month.
   Currently there are two price zones in the Russian Federation, which can be conditionally divided into the European - price zone 1, and the Ural - price zone Electricity prices for each price zone for demand management events are set by the System Operator. As of 01.06.2022, the electricity price is:

   - 1 PZ – 585279 rubles;
   - 2 PZ – 166911 rubles.

2) P_P-declared volume of reduction of consumption of the controlled object.
   This value represents the volume by which it is planned to decrease every hour "under ideal conditions".
3) pt-volume of control object consumption reduction in hour t.

This value characterizes the actual "non-ideal" decrease in consumption of the object for each hour during the day in which the demand management event occurs [6, 8].

An example is presented in Table 1.

Stage 2. Calculation of the Planned Volume of Services for the Month
At this stage, the planned volume of services for the month is calculated.

Table 1. The amount of reduction in consumption of the control object during the day.

Decrease by the hour, 1 day	
1 h	3110
2 h	3050
3 h	3200
4 h	3170

The following formula is used to determine the planned volume of services per month:

$$V_{PLAN} = P_P \cdot k_{DURAT} \quad (1)$$

where V_{PLAN} - planned volume of provision of demand management services, MW; P_p - declared volume of reduction of consumption of the control object, by which the enterprise plans to unload during each hour in each period of a day "under ideal conditions", MW. Taken from stage 1; k_{DURAT} - reduction coefficient, which depends on the number of hours per day for which the object is ready to unload.

Two variants of unloading are assumed (Table 2):

Table 2. Unloading options.

Period duration per day, hour	Decrease coefficient
2	0,5
4	1

Stage 3. Calculation of the Actual Provision of Services for the Month

The actual volume of services rendered is determined for each billing period for each management facility by the formula:

$$V_{FACT} = V_{PLAN} \cdot (1.25 \cdot k_{READ} \cdot k_{FACT} - 0.175 \cdot k_{READ} - 0.075)$$

where V_{FACT} - volume of demand management services rendered, MW; V_{PLAN} - planned volume of demand management services, MW; k_{READ} - coefficient of readiness of the control object to reduce consumption in the calculation period; k_{FACT} - coefficient of confirmation of the volume of consumption decrease by the control object.

The calculation of Vfact can be divided into four steps.

Let us consider each element of this formula separately.

1) The planned volume of rendered services, V_{PLAN}, is calculated in the second step and substituted in the above formula.

2) Readiness factor

$$k_{READ} = N_{READ}/N_{WD}$$

where k_{READ} - coefficient of readiness of the control object to reduce consumption in the calculation period; N_{READ} - number of working days in the calculation period, when the readiness of the control object to reduce consumption is confirmed; N_{WD} - number of working days in the calculation period.

In fact, the readiness coefficient can be characterized as a "risk fee". The more working days the company will be ready to conduct a demand management event, the more profit it will be able to make. In accordance with the latest changes in the contract for the provision of demand management services, the number of days in the calculation period when the readiness of the control object to reduce consumption is confirmed must be at least 7.

3) Decrease confirmation coefficient

$$k_{FACT} = 1/N_{EVENT} * \sum P_T/P_P$$

where N_{EVENT} is the number of demand management events in the calculation period;

In accordance with the contract on provision of demand management services the number of events does not exceed 5.

P_T - final volume of decrease in consumption of the object of management for the time of the demand management event. Calculated in the following paragraph.

P_P - Declared volume of decrease in consumption of the controlled object. Taken from stage 1.

4) Total reduction in consumption

$$P_T = 1/T_{DURAT} \cdot \sum \min(p_t, P_P)$$

where P_T - final volume of reduction of the control object consumption, MW; T_{DURAT} - duration of the period of reduced consumption of the controlled object, h; p_t - volume of reduction of the control object consumption per hour, MW; t - serial number of the hour of the demand control event, in which the reduction of the control object consumption was to be carried out; P_P - declared volume of decrease in consumption of the control object, MW.

An example of calculations is presented in Table 3.

In accordance with the terms of the contract for the provision of demand management services, the obligations to reduce the consumption of the object of management are fulfilled if during each hour of the period of reduced consumption of the claimed activity $P_t \geq P_P$*0.75.

An example is presented in Table 4.

5) Then all obtained values are substituted into the formula for calculating V_{FACT}.

Table 3. Calculation of the final volume of consumption reduction.

Decrease by the hour, 1 day	
1 h	3110
2 h	3050
3 h	3200
4 h	3170
Pт1	3.1325

Table 4. Verification of reduction liabilities.

Verification of reduction liabilities
Executed

After calculating V_{FACT}, we go on to calculate the cost of services rendered.

Stage 4. Determining the Cost of Services Rendered
We will use the formula to determine the cost of consumer services:

$$S = PR \cdot V_{FACT}$$

PR - Price of services per month, rubles/MW; V_{FACT} - The amount of demand management services rendered, MW.

Stage 5. Determining the Net Profit Received by the Consumer
In providing its services, the aggregator is also interested in making a profit. Having analyzed the existing aggregators at the market of demand management services, it was found that the cost of aggregator services is no more than 10% of the enterprise's profit from the use of demand management technology for a month [6, 8].

The final formula, including the deduction of the cost of aggregator services:

$$SP = S - (0,05 \cdot S)$$

where *SP* - net profit of the consumer, rubles; *S* - the cost of services rendered by the consumer, rubles. Taken from stage 4.

A view of the program developed based on MS Excel is shown in Fig. 4.

Having analyzed the above data, we can conclude that such a program can be distributed by SO UES to consumers to independently determine profits from the use of electricity demand management technology. Such a step will accelerate the process of implementing this technology at other enterprises, thereby fulfilling the priority objectives of the energy strategy of the Russian Federation until 2035.

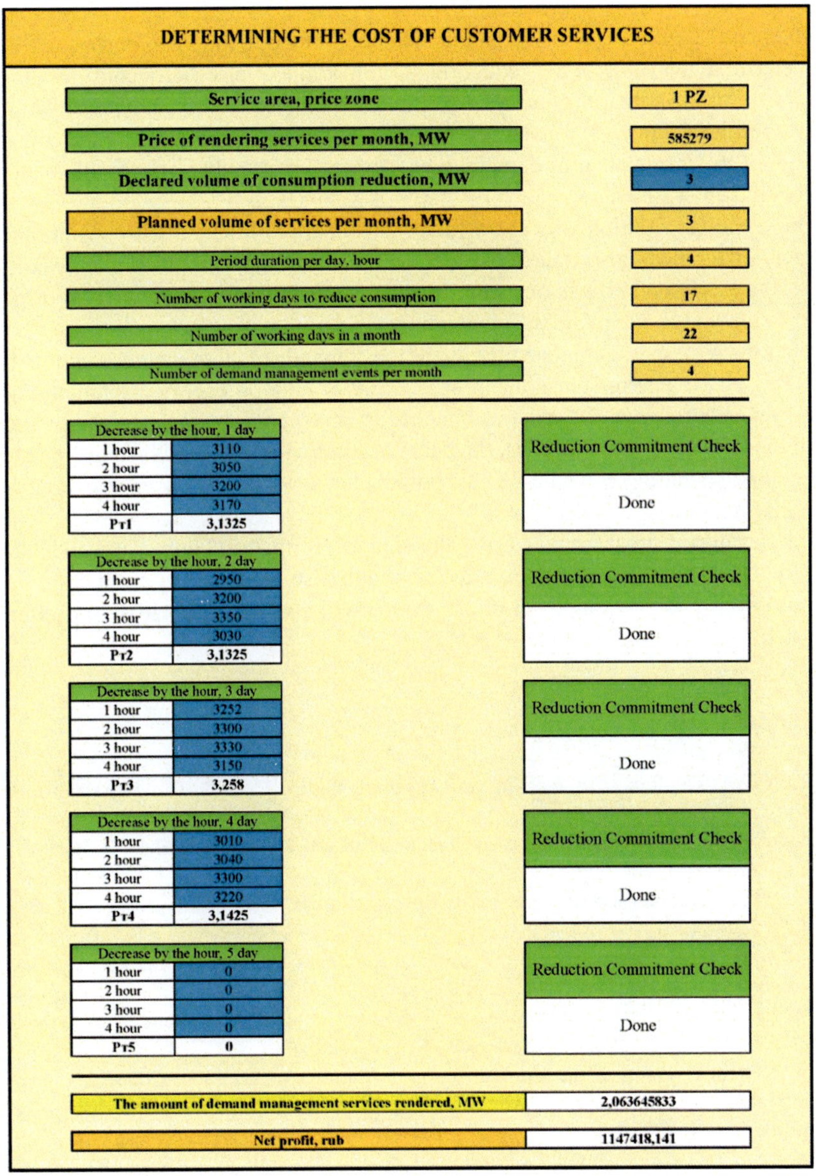

Fig. 4. Program for calculating the cost of services rendered.

6 Conclusion

The aim of the work was to develop theoretical and methodological foundations and to develop a mathematical apparatus based on Microsoft Excel to manage the demand for energy resources of industrial enterprises.

The main attention is focused on the calculation model, the main purpose of which is to determine the value of the reduction in the volume of electricity consumption and the value of the resulting profits.

The result of the work was the developed mathematical apparatus in the form of a program of the mechanism of demand management for its subsequent implementation in industrial enterprises.

The technical solution will allow retail consumers to reduce their consumption of electricity, which will also reduce the load on the power system in times of peak loads. One of the associated effects of using the technical solution is the reduction of power transmission losses due to an even load on the grid.

After the widespread implementation of the demand response methodology, System Operator will be able to start collecting an array of data on energy consumption over several years, during which the demand management technology will become an integral part of the RF power system. Thus, the data obtained will allow the System Operator to start gradually shutting down unclaimed power plants, which have been in operation for several days a year.

The developed program will bring the solution to this problem closer, which will not only have a positive economic effect, but will also reduce CO_2 production due to the shutdown of unclaimed power plants, which will also have a positive impact on the ecosystem.

References

1. FLEXIWATTS: Report of the Working Group 4: GO15. Reliable and Sustainable Power Grids, p. 8. Moscow (2013)
2. Sidorovskaya, N.: Demand side management on global electricity markets. J. Energorynok **7**, 1–7 (2015)
3. Posypanko, N.: Demand response in the Russian market: barriers and prospects. J. Energorynok **7**, 1–4 (2017)
4. Zhikharev, A.B., Posypanko, N.: Demand response in the Russian market: barriers and prospects. Research "VYGON Consulting, pp. 31–34 (2018)
5. Decree of the Government of the Russian Federation No. 699 "On Amendments to the Rules of the Wholesale Electricity and Capacity Market" (2010)
6. JSC System Operator of the Unified Energy System (SO UES JSC): Concept, Concept of functioning of distributed energy resource aggregators as part of the Unified Energy System of Russia, p. 46 (2018)
7. Decree of the Government of the Russian Federation of March 20, 2019 no. 287: On amendments to some acts of the Government of the Russian Federation on the functioning of aggregators of electricity demand management in the Unified Energy System of Russia, as well as improving the mechanism of price-dependent reduction of electricity consumption and system reliability services. Collection of Laws of the Russian Federation. 01.05.2019. № 13. Art. 1403 (2019)
8. Sidorovich, V.V., Kuleshov, M.A., Chausov, I.S.: Expert analytical report. Demand side management in Russian electric power industry: opening opportunities, pp. 7–50 (2019)

9. Dzyuba, A.P.: Management of demand for energy resources of industrial properties: theory and methodology. Dissertation, Chelyabinsk, p. 542 (2021)
10. Decree of the Government of the Russian Federation of 08.02.2021 no. 132: On introducing amendments to some acts of the Government of the Russian Federation on the functioning of aggregators to manage the demand for electricity in the Unified Energy System of Russia, as well as improving the mechanism of price-dependent reduction in electricity consumption and repealing certain provisions of certain acts of the Government of the Russian Federation (2021)

Industrial Automation Systems Cybersecurity

Lattice-Based Ordered Multisignature for Industrial IoT

E. Aleksandrova, A. Fedichev, and A. Yarmak(✉)

Peter the Great St. Petersburg Polytechnic University, 29, Polytechnicheskaya Street, St. Petersburg 195251, Russia
yarmak.av@ibks.spbstu.ru

Abstract. In this paper lattice-based ordered multisignature was proposed. The proposed signature uses the lattice's basis delegation mechanism and provides not only data authenticity, but also the signature generation chain verification, which can be applied in various systems with a hierarchical structure, including the industrial Internet of Things. A security analysis of this scheme was carried out, in which 5 attack scenarios were considered, taking into account the attacker's various capabilities. Methods for improving the scheme security are proposed. Ways for improving the scheme performance are also suggested. To conduct performance testing, two scheme versions were launched: with and without optimization. It was found that when using the optimized variant, the time of parameters generation increases, but signature generation and verification time and the total time for all algorithms decreases. This research can be used as a starting point for further study of the lattice's basis delegation use to constructing ordered multisignature schemes.

Keywords: Lattice · Group-oriented authentication · Ordered signature · Multisignature · Basis delegation · Industrial IoT

1 Introduction

The Industrial Internet of Things (IIoT) is an example of a complex system, the architecture of which includes a large number of interconnected various elements that involve working with data. The access rights of users and nodes in such systems are often organized in a hierarchy, which is due, for example, to the use of access control models.

As with any other device that supports networking, ensuring information security in the IIoT is essential [1–4]. For IIoT-devices, this is especially critical, since in the event of an information confidentiality violation or device authentication, the entire production work at the enterprise can be disrupted, which can lead to huge losses for the company. Violation of the transmitted information confidentiality leads to possible data leaks from one company to a competitor company, and authentication violation allows an intruder to integrate into a network organized between devices and transmit their own control commands to other IIoT-devices, which in turn leads to a logic violation of the entire production system.

A. A. Radionov and V. R. Gasiyarov (Eds.): RusAutoCon 2022, LNEE 986, pp. 363–373, 2023.
https://doi.org/10.1007/978-3-031-22311-2_35

Digital signature can be used to ensure data authenticity and trust. However, in the case of industrial IoT, which can be an object of critical infrastructure, a problem may arise in which it is necessary to provide not only control over the data authenticity, but also verification of the path along which this information has passed. For example, when initializing devices, the head node must be sure that each of the controlled devices is ready for operation (i.e., has signed a message), and at the same time, each node located in the hierarchy below took part in the signature chain generation from the nodes. In electronic document management systems, when agreeing on an order, it may be necessary to sign it by several participants - from the initiating employee to the general manager - in a strictly defined order. Ordered digital multisignature schemes can be used to solve such problems.

2 Related Work

The concept of an ordered multisignature was first proposed in 1994 in the paper [5]. Later, schemes were developed based on RSA and ElGamal cryptosystems [6], used discrete logarithm and factorization problems [7], as well as on GDH (Gap Diffie–Hellman) [8] and CDH (Computational Diffie–Hellman problem) [9] problems. In 2007, a scheme was presented [10], based on the use of identity approach [11].

Optimal schemes of ordered digital signatures must satisfy requirements [12]:

- Fixed size of the signature. The size of the signature should not depend on the number of members.
- Impossibility of falsification. Adding a partial signature to the chain without knowing the private key corresponding to the public key should be a computationally hard problem.
- Scheme scalability. New members of the group should be able to participate in the process of collective signature generation.
- The verification process for a collective signature should be identical to the verification process for an individual signature.

To organize an ordered digital signature on lattices, the basis delegation algorithm proposed in 2009 in the [13] can be used. It is lattice-based cryptosystems that are the most popular candidates for the standardization of post-quantum cryptosystems. At the moment, the existing algorithms for solving hard problems of lattice theory have more than polynomial complexity both on a classical and quantum computer.

Let $a_1, a_2, \ldots, a_n$ are linearly independent vectors in $\mathbb{R}^m$. A lattice Λ is a set of linear combinations of vectors: $a_1, a_2, \ldots, a_n$ with coefficients from $\mathbb{Z}$:

$$\Lambda = \{z_1 a_1 + \ldots + z_n a_n | z_i \in \mathbb{Z}, i = 1, \ldots, n\}.$$

The vectors $a_1, \ldots, a_n$ are called the basis of the lattice, and the number n is called the dimension (or rank) of the lattice. For $m = n$ the lattice is called full-rank. Below are some properties of lattices:

- The lattice is a discrete set, that is, any bounded set of $\mathbb{R}^m$ contains a finite number of lattice points.

- The lattice basis is not unique.
- For a lattice of dimension greater than 1, there are infinitely many different bases.
- If $a_1, \ldots, a_n$ and $b_1, \ldots, b_n$ are lattice's basis, then there is a matrix U such that $[a_1, \ldots, a_n] = U[b_1, \ldots, b_n]$.
- The transformation matrix between lattice bases is unimodular, that is, its determinant is equal to ± 1.

Since the lattice is a discrete set, it contains a nonzero vector of minimum length. Such vector is called the shortest lattice vector, and its norm is denoted $||\Lambda||$. The q- ary lattice Λ_q is defined as follows [14]:

$$\Lambda_q(A) = \left\{ v \in \mathbb{Z}_q^m | Av = 0 \bmod q \right\}.$$

Hierarchical identity-based encryption (HIBE) scheme is described in [15] and [16]. Such scheme involves users who organize themselves into a tree. Each entity (tree node, user) is given a private key from its parent, which it can then delegate to its child entities, and the public key is the identity of this node. With this approach, each entity can decrypt messages intended for itself and its child entities, but it cannot do so for any other parent nodes. In this case, each node, knowing its private key, will not be able to restore the parent node's private key. The first hierarchical encryption schemes were based on bilinear maps [16–18]. Subsequently, schemes based on hardness assumptions from lattice theory were proposed [13, 19]. There are also schemes in which a sequence of entities is used instead of a tree structure [20].

As noted in [21], a hierarchical identity encryption scheme can be converted into a signature scheme. Thus, in article [13], a signature scheme was proposed based on a hierarchical encryption scheme, which consisted in using the secret master key as the message M signature key, and the delegated secret key for the identifier M as the signature itself (as the identifier *id* take the binary representation of the message M). Signature verification then requires encrypting a random message for an identifier and trying to decrypt it with a generated signature (that is, a delegated secret key for the same identifier M). This signature scheme is not an ordered digital signature scheme and therefore requires modification to organize hierarchical authentication.

3 Ordered Multisignature Based on Basis Delegation

The basis delegation algorithm was first described in the [13] in the context of HIBE scheme. The basis delegation is defined as follows.

Let A be a matrix from $\mathbb{Z}_q^{n \times m}$, and let T_A be a "short" basis of $\Lambda_q(A) = \left\{ v \in \mathbb{Z}_q^m | Av = 0 \right\}$. Using a deterministic algorithm, it is necessary to generate a matrix B from A and a random basis T_B for $\Lambda_q(B)$, so from A, B and T_B it is difficult to recover any short bases of $\Lambda_q(A)$. In 2010, an alternative method of basis delegation was proposed without increasing the dimension of the resulting matrix [20].

The proposed signature is based on the HIBE hierarchical encryption scheme [13]. Let l be the number of signature participants (in this case, $l = d$, where d is the depth from the HIBE scheme), and n be the security parameter. Then the parameters are chosen based on the following relationships:

$$m = 6n^{1+\delta} \quad q = m^{\frac{3}{2}l+2}\omega(\log^{2l+1} n).$$

where δ is such that $n^{\delta} > \lceil \log q \rceil$. Here m is rounded up to the nearest integer and q is rounded up to the nearest prime number.

Let each member be matched with an identifier $id_i = \{0, 1\}^*$. Let's denote $id_{h|l} = (h, id_1, id_2, ..., id_l)$. An ordered signature scheme is proposed, which includes the following procedures:

- Setup is the procedure for generating the public parameters, the signing master key, and the user id's and sequence of these id's that will be used to verify the signature.
- Signature generation is the procedure for generating a signature by the i-th member, which takes as input the message M and the current value of the partial signature σ_{i-1} from the previous participant. At the output, the procedure returns a new value of the partial signature σ_{i+1}, or an error if the input data is incorrect or if an attacker's signature is embedded in the procedure.
- Signature verification is a procedure that takes as input a message M and a signature value σ. The sequence of identifiers of the signature authors in the given order together with the hash value of the message is used as the signature verification key.

3.1 Setup

1. Run the *HibeSetup* algorithm to generate public parameters $PP = \{A, u_0\}$ and a master key $MK = \{T_A\}$. The matrix T_A will be the basis of the lattice $\Lambda(A)$. In this case, the master key is issued only to the first participant. Thus, the first member has the highest privileges in the chain.
2. For each user, generate an $id_i = \{0, 1\}^*$. Form a sequence $\{id_1, id_2, ..., id_l\}$, that will be used to generate and verify the signature. In this case, the order of user identifiers in this sequence should correspond to the order in which the signature is generated.

3.2 Signature Generation

The following algorithm is used to generate the signature of message M. The first member of the group does the following:

1. Calculate the $h \leftarrow H(m) \in \mathbb{Z}_q^{m\times m}$, where $H(m)$ is a hash function from the HIBE scheme.
2. Calculate its delegated basis based on the master key MK and $id_{h|1}$, using the *HibeExtract* algorithm. Result is S_1.
3. Consider the partial signature equal $\sigma_{i-1} = \{S_1\}$ and pass it to the next participant in the chain.

Participant with number i performs the following actions:

4. Get a partial signature $\sigma_{i-1} = \{S_{i-1}\}$ from the previous participant.
5. Perform delegated basis check S_{i-1}:

    a) Select a random message M'.
    b) Encrypt the given message for the $id_{h|i-1}$, using the *HibeEncrypt* algorithm.
    c) Decrypt the received ciphertext using the *HibeDecrypt* algorithm and delegated basis S_{i-1}.
    d) If the decrypted message matches the original one, it continues to execute the steps of the algorithm, otherwise it returns an error and stops the work.

6. Calculate the $h \leftarrow H(m) \in \mathbb{Z}_q^{m \times m}$, where $H(m)$ is a hash function from the HIBE scheme.
7. Calculate its delegated basis S_i, using the HibeDerive algorithm, inputting parent $id_{h|i-1}$ and child $id_{h|i}$, as well as basis S_{i-1}. Result is S_i.
8. Consider its partial signature equal to $\sigma_i = \{S_i\}$.

The last participant takes the signature as the signature of the message $\sigma_l = \{S_l\}$.

3.3 Signature Verification

To verify the signature of a message M, the verifier performs the following steps.

1. Calculate the $h \leftarrow H(m) \in \mathbb{Z}_q^{m \times m}$, where $H(m)$ is a hash function from the HIBE scheme.
2. Select a random message M'.
3. Encrypt the given message for the identifier $id_{h|l}$, using the *HibeEncrypt* algorithm.
4. Decrypt the received ciphertext using the *HibeDecrypt* algorithm and signature S_l.
5. If the decrypted message matches the original one, considers the message signature to be correct and returns "1", otherwise it returns "0".

4 Analysis of Proposed Scheme

Let's consider some possible attacks on the ordered signature scheme.

1. *Injection into the signature chain of an attacker who is not a member of the group.* To continue the chain, the attacker need have his own identifier, which is in a predetermined list of valid identifiers.
2. *The attacker generates his own identifier and uses it for signature generation.* Then for the participant following the attacker (or for the verifier, if the attacker is embedded at the end of the chain), the matrix F_{id}, calculated at steps 5 b) and 5 c) of the *Signature generation* algorithm (it is denoted as F_{id_1}), will not coincide with the matrix F_{id}, calculated by attacker at step 7 of the *Signature generation* (it is denoted as F_{id_2}) (or at steps 3 and 4 of the *Signature verification*, if the attacker is embedded

at the end of the chain, in this case, in the calculations presented below, replace *l* since in the first case it will be equal to $F_{id_l} = A \cdot H(h) \cdot H(id_1) \cdot \ldots \cdot H(id_{i-1})$, and in the second $H(id_k) \neq F_{id_l}$Ï.

3. Changing the order of signing a message by one participant. There are three possibilities for this attack.

   a) *if the attacker is inside the chain.* Let the number of the attacker be k, but it is embedded in the chain in the place under the number i. In this case, the matrix $F_{id} = A \cdot H(h) \cdot H(id_1) \cdot \ldots \cdot H(id_{i-1}) \cdot H(id_k)$, calculated by the attacker when delegating the basis will not match the $F_{id} = A \cdot H(h) \cdot H(id_1) \cdot \ldots \cdot H(id_{i-1}) \cdot H(id_i)$, calculated by the participant following the attacker at the stage of partial signature verification. Then this participant will not be able to correctly decrypt the random message M' and will detect the embedding of the attacker.
   b) *if the attacker is at the beginning of the chain.* In this case, the attacker needs to know the secret master key of the signature, without which he will not be able to initiate the signature generation process. If the attacker generates a random master key $T_A^{'}$, then it is obvious that the participant following him will not be able to confirm the correctness of the partial signature generated by the attacker, since the matrix $T_A^{'}$ will not be a trapdoor for the lattice $\Lambda_q(A)$.
   c) *if the attacker is at the end of the chain.* In this case, the situation is similar to the previous one, except that instead of the next participant, the verifier will find the mismatch (*i* replace with *l*), and then the signature will also not be considered correct.

4. Changing the order of signing a message by a group of participants. With such an attack, three scenarios are also possible.

   a) *if a group of attackers is inside the chain.* This situation is similar to the situation from case 3): an order violation will be detected by the participant following this group of attackers: let the group of attackers have numbers $j, j+1, \ldots, j+k$ and want to fit into places with numbers $i, i+1, \ldots, i+k$. Then the matrix computed by the last attacker will be equal to

$$F_{id} = A \cdot H(h) \cdot H(id_1) \cdot \ldots \cdot H\left(id_j\right) \cdot H\left(id_{j+1}\right) \cdot \ldots \cdot H\left(id_{j+k}\right).$$

The matrix calculated when verifying a partial signature by the participant following the attackers (that is, having the number $i + k + 1$), will be equal to

$$F_{id} = A \cdot H(h) \cdot H(id_1) \cdot \ldots \cdot H(id_i) \cdot H(id_{1+1}) \cdot \ldots \cdot H(id_{i+k}).$$

Therefore, it will detect a mismatch and the signature generation process will be stopped.

   b) *if the group of attackers is at the beginning of the chain.* Then it is necessary that one of the attackers knows the secret master key of the signature. If this condition is met, then the participant following the attackers will also detect a mismatch, similar to the situation from the previous paragraph.

c) *if the group of attackers is at the end of the chain.* This situation is similar to the previous paragraph, except that instead of the participant with the number $i + k + l$ the verifier will find the mismatch:

$$A \cdot H(h) \cdot H(id_1) \cdot \ldots \cdot H\big(id_j\big) \cdot H\big(id_{j+1}\big) \cdot \ldots \cdot H\big(id_{j+k}\big) \neq$$
$$\neq H(h) \cdot H(id_1) \cdot \ldots \cdot H(id_{l-k}) \cdot H(id_{l-k+1}) \cdot \ldots \cdot H(id_l).$$

5. *Changing the order of signing by all participants.* If all participants collude and decide to change the order of the signature, the verifier will detect this, since the operation of matrix multiplication is non-commutative. That is, the result of matrix multiplication when calculated F_{id} at the signature generation stage will not coincide with the result of matrix multiplication when calculated F_{id} at the signature verification stage, therefore, the verifier will not be able to correctly decrypt a random message M'.

It should be noted that since only a sequence of identifiers is needed to generate a signature by the i-th participant (where $i = 2, \ldots, l$) violation of its integrity will be critical. Since the right to sign a participant is determined by the presence of his identifier in the list of identifiers $id_{|l}$, this list should be protected. In addition, when changing the order of identifiers in the list the security of the signature will be violated: if the offender can replace the list $id_{|l}$ with his own $id'_{|l} \neq id_{|l}$ for all participants in the signature and for the verifier and embed in the right place when generating the signature in accordance with $id'_{|l}$, then the signature will be considered correct. Therefore, it is necessary to provide integrity control of this list.

The following approaches can be used to improve scheme security.

- Optionally authenticate the sequence of identifiers $id_{|l}$ at the stage of keys and parameters generation any digital signature scheme that is resistant to attacks using quantum computing. Let there be a key issuing authority that generates keys and parameters, and also sets the sequence of participant identifiers $id_{|l}$, which will correspond to the order of signature generation. Also, this issuing center will choose a signature scheme that is resistant to quantum computer attacks and generate public and private keys of this scheme. Using the private key, the issuing center will generate a signature for $id_{|l}$, and the public key will be declared as a public parameter of the scheme. Then, in the process of forming an ordered signature, before checking partial signatures by participants, they will also first verify the signature for $id_{|l}$, and if a substitution is detected, the process of forming an ordered signature will be stopped. The verifier will do the same. In this way, it will be guaranteed that each signer and verifier has a genuine list of identifiers $id_{|l}$.
- When generating and verifying a signature, use the corresponding ciphertext instead of a list of identifiers in plaintext. Let the issuer, at the stage of keys and parameters generation, choose an encryption scheme that will be used to encrypt the identifiers' list in the required sequence. The encrypted list will be the signature's public key, and the private key will be the key with which this list can be decrypted. Thus, in order

for each member to determine the next participant, it needs to decrypt the encrypted sequence of identifiers using its own key.

Also, when forming a message M signature the first participant can upload the calculated value $H(M) \in \mathbb{Z}_q^{m\times m}$ to the same server, and the following participants can use this value to form their partial signature. At the same time, the requirements for computing resources of the following participants will decrease sharply. However, for this approach, it is necessary to provide additional protection for the value $H(M)$ uploaded to the server, since if the first participant is the offender, he will be able to replace the value $H(M)$ with the value $H\left(M'\right)$ for some other message M'.

It can be seen from Tables 1, 2, 3 and 4 that in the variant with optimization, the process of keys and parameters generation takes more time than in the variant without optimization (tested device specifications: Quad-Core Intel Core i7, 1.2 GHz, 16 Gb RAM). This is due to the fact that in the optimized version of the proposed scheme, the values $H\left(id_{|i}\right)$ for $i = 1, \ldots, l$ are calculated during scheme initialization, so the running time of the *Setup* algorithm has increased. At the same time, the time for generating and verifying a signature in the optimized version has greatly decreased.

Table 1. Parameter generation time.

Group size	Generation time of parameters and signature keys, s	
	without optimization	without optimization
1	0.2077	0.7523
5	0.0128	3.3006
10	0.0142	6.2111
15	0.0157	9.3806
20	0.0158	12.4418

Table 2. Signature generation time.

Group size	Signature generation time, s	
	without optimization	without optimization
1	3.9951	3.3026
5	186.7782	28.5388
10	644.3985	62.5724
15	1369.0535	100.4092
20	2352.8346	142.4214

Table 3. Signature verification time.

Group size	Signature verification time, s	
	without optimization	without optimization
1	25.1344	3.7277
5	66.1368	4.2143
10	116.0183	4.8237
15	165.4240	5.5710
20	215.0205	6.2874

Table 4. The total running time.

Group size	The total running time of the algorithms, s	
	without optimization	without optimization
1	29.3371	7.7825
5	252.9279	36.0537
10	760.4311	73.6073
15	1534.4933	115.3608
20	2567.8709	161.1506

Based on Fig. 1, it can be conclude that, despite the increase in the parameters' generation time in the optimized version compared to the version without optimization, the total running time of all algorithms has significantly decreased.

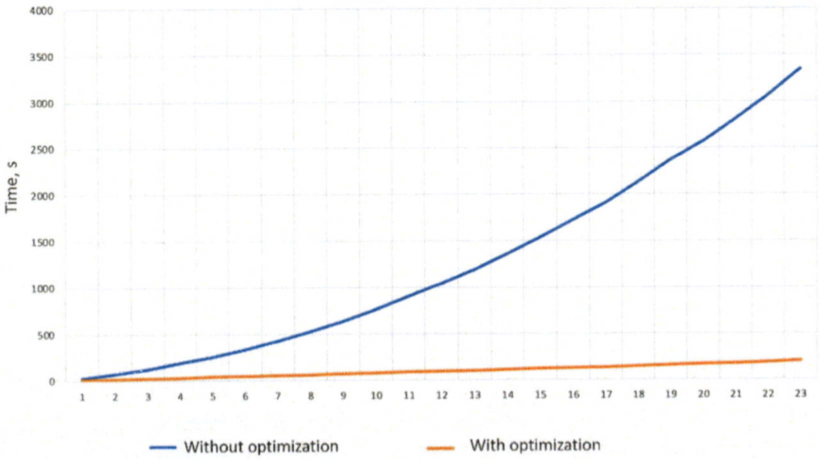

Fig. 1. Dependence of the total running time for the Setup, Signature generation and Signature verification algorithms on the group size.

It was also revealed that the operation that takes the longest time is the *SamplePre* algorithm. This is because this algorithm requires the transformation of the lattice's basis into an orthonormal system of vectors (the Gram–Schmidt orthogonalization process). To speed up this operation, the optimized *TrapGen* and *SamplePre* algorithms [22] can be used, thereby increasing the performance of the entire scheme.

5 Conclusion

Thus, in this paper, an ordered signature scheme based on the lattice basis delegation was proposed, as well as possible attacks on this scheme. This scheme satisfies requirements of fixed size of the signature, impossibility of falsification and scheme scalability, since the size of the signature is fixed and depends only on the parameters of the scheme, and not on the size of the group of participants, the addition of unauthorized participants to the group is detected at the verification stage, and the forgery signature is a difficult task. In addition, to add a new member, it is enough to update the sequence of identifiers $id_{|l}$.

In this case, requirement 4 (see Sect. 2) ceases to be satisfied, since the signature verification is equivalent to the process of verifying the participant's partial signature. It was also found that the security of the proposed scheme could be improved by adding encryption or signing of a sequence of identifiers $id_{|l}$ to it. In addition, this scheme can be further optimized by adding an offline stage - calculating values $H\left(id_{|i}\right), i = 1, \ldots, l$ during the initialization of the scheme.

Acknowledgments. The reported study was funded by RFBR, project number 20-37-90106.

References

1. Pavlenko, E., Poltavtseva, M.: Mathematical methods for implementing homeostatic control in digital production systems. In: Schaumburg, H., Korablev, V., Ungvari, L. (eds.) TT 2020. LNNS, vol. 157, pp. 1–9. Springer, Cham (2021). https://doi.org/10.1007/978-3-030-64430-7_1
2. Zegzhda, P.D., et al.: Approach to the evaluation of the efficiency of information security in control systems. Autom. Control Comput. Sci. **54**, 864–870 (2020). https://doi.org/10.3103/S0146411620080362
3. Krundyshev, V., Kalinin, M.: Generative adversarial network for detecting cyber threats in industrial systems. Smart. Innov. Syst. Technol. **220**, 1–13 (2021). https://doi.org/10.1007/978-981-33-6632-9_1
4. Lavrova, D.S.: Maintaining cyber sustainability in industrial systems based on the concept of molecular-genetic control systems. Autom. Control Comput Sci **53**, 1026–1028 (2019). https://doi.org/10.3103/S0146411619080145
5. Doi, H., et al.: Multisignature scheme with specified order. Paper presented at the Symposium on Cryptography and Information Security, Japan, pp. 27–29 January 1994
6. Burmester, M., et al.: A structured ElGamal-Type multisignature scheme. In: Imai, H., Zheng, Y. (eds.) PKC 2000. LNCS, vol. 1751, pp. 466–483. Springer, Heidelberg (2000). https://doi.org/10.1007/978-3-540-46588-1_31

7. Mitomi, S., Miyaji, A.: A multisignature scheme with message flexibility, order flexibility and order verifiability. In: Dawson, E.P., Clark, A., Boyd, C. (eds.) ACISP 2000. LNCS, vol. 1841, pp. 298–312. Springer, Heidelberg (2000). https://doi.org/10.1007/10718964_25
8. Lin, C.Y., Wu, T.C., Zhang, F.: A structured multisignature scheme from the gap Diffie-Hellman group. In: Cryptology ePrint archive (2013). https://eprint.iacr.org/2003/090.pdf. Accessed 2 Jan 2021
9. Li, X., Chen, K., Zhang, L., Li, S.: Proxy structured multisignature scheme from bilinear pairings. In: Cao, J., Yang, L.T., Guo, M., Lau, F. (eds.) ISPA 2004. LNCS, vol. 3358, pp. 705–714. Springer, Heidelberg (2004). https://doi.org/10.1007/978-3-540-30566-8_82
10. Boldyreva, A., Gentry, C., O'Neill, A., Yum, D.H.: Ordered multisignatures and identity-based sequential aggregate signatures, with applications to secure routing. In: Proceedings if ACM Conference on Computer and Communication Security, pp. 276–285 (2007). https://doi.org/10.1145/1315245.1315280
11. Gangishetti, R., Gorantla, M.C., Das, M.L., Saxena, A.: Identity based multisignatures. Informatica **17**, 177–186 (2006). https://doi.org/10.15388/INFORMATICA.2006.131
12. Saxena, A., Soh, B.: One-way signature chaining: a new paradigm for group cryptosystems. Int. J. Inf. Comput. Secur. **2**, 268–296 (2008). https://doi.org/10.1504/IJICS.2008.020606
13. Cash, D., Hofheinz, D., Kiltz, E., Peikert, C.: Bonsai trees, or how to delegate a lattice basis. In: Gilbert, H. (ed.) EUROCRYPT 2010. LNCS, vol. 6110, pp. 523–552. Springer, Heidelberg (2010). https://doi.org/10.1007/978-3-642-13190-5_27
14. Ajtai, M.: Generating hard instances of the short basis problem. In: Wiedermann, J., van Emde Boas, P., Nielsen, M. (eds.) ICALP 1999. LNCS, vol. 1644, pp. 1–9. Springer, Heidelberg (1999). https://doi.org/10.1007/3-540-48523-6_1
15. Horwitz, J., Lynn, B.: Toward hierarchical identity-based encryption. In: Knudsen, L.R. (ed.) EUROCRYPT 2002. LNCS, vol. 2332, pp. 466–481. Springer, Heidelberg (2002). https://doi.org/10.1007/3-540-46035-7_31
16. Gentry, C., Silverberg, A.: Hierarchical ID-based cryptography. In: Zheng, Y. (ed.) ASIACRYPT 2002. LNCS, vol. 2501, pp. 548–566. Springer, Heidelberg (2002). https://doi.org/10.1007/3-540-36178-2_34
17. Boyen, X., Waters, B.: Anonymous hierarchical identity-based encryption (without random oracles). In: Dwork, C. (ed.) CRYPTO 2006. LNCS, vol. 4117, pp. 290–307. Springer, Heidelberg (2006). https://doi.org/10.1007/11818175_17
18. Waters, B.: Dual system encryption: realizing fully secure IBE and HIBE under simple assumptions. In: Halevi, S. (ed.) CRYPTO 2009. LNCS, vol. 5677, pp. 619–636. Springer, Heidelberg (2009). https://doi.org/10.1007/978-3-642-03356-8_36
19. Agrawal, S., Boneh, D., Boyen, X.: Efficient lattice (H)IBE in the standard model. In: Gilbert, H. (ed.) EUROCRYPT 2010. LNCS, vol. 6110, pp. 553–572. Springer, Heidelberg (2010). https://doi.org/10.1007/978-3-642-13190-5_28
20. Agrawal S, Boneh D, Boyen X (2010) Lattice basis delegation in fixed dimension and shorter-ciphertext hierarchical IBE. In: Lecture Notes in Computer Science (including subseries Lecture Notes in Artificial Intelligence and Lecture Notes in Bioinformatics), pp 98–115
21. Boneh, D., Franklin, M.: Identity-based encryption from the Weil pairing. SIAM J. Comput. **32**, 586–615 (2003). https://doi.org/10.1137/S0097539701398521
22. Alwen, J., Peikert, C.: Generating shorter bases for hard random lattices. Theory Comput. Syst. **48**, 535–553 (2011). https://doi.org/10.1007/S00224-010-9278-3

Applying a System of Automatic Threat Modeling for APCs at Information Infrastructure Facilities

D. Chernov(✉), A. Sychugov, and A. Sovgir

Tula State University, 92, Lenin Avenue, Tula 300012, Russia
cherncib@gmail.com

Abstract. Modern technological process automation means inextricably linked with using information technologies. The increasing of number of automated process control systems posits sharply the problem of providing information security at all levels of automation means functioning. Due to the presence of dynamically emerging threats to information security, specialists need to perform their modeling, then evaluate and minimize the negative consequences caused by threats. In this paper the author analyzes the obtained results of using the developed automated system for modeling threats to information security of automated process control systems, tested at various information infrastructure facilities. The analysis of the results is carried out on the basis of qualitative and quantitative estimation criteria obtained from information security specialists authorized to carry out the processes of modeling information security threats of the above-mentioned objects. In the framework of the analysis the author presents functional schemes and algorithms that describe general principles of an automated system functioning. In addition, the author fulfills a description of the existing methods and means of automating the processes of modeling threats to information security of automated process control systems in a formalized form and also presents their algorithms and functional schemes.

Keywords: Automated process · Control systems · Information security · Threat · Threat estimation · Threat model · Information infrastructure

1 Introduction

One of the main and the most important directions providing information security in the field of state and public security is the increase of critical information infrastructure safety and functioning sustainability, developing mechanisms for detecting and preventing information threats and eliminating the consequences of their manifestation, increasing safety of citizens and territories from the consequences of emergencies caused by information and technical impact on critical information infrastructure facilities [1]. Automated process control systems (hereinafter APCs) refer to the information infrastructure objects and are meant to implement a set of measures to provide information security [2–4].

A. A. Radionov and V. R. Gasiyarov (Eds.): RusAutoCon 2022, LNEE 986, pp. 374–385, 2023.
https://doi.org/10.1007/978-3-031-22311-2_36

As one of the priority measures providing information security of information infrastructure facilities the document [5] indicates the necessity of detecting possible penetration points for an external intruder into information infrastructure facilities of and industrial economy sector. To fulfill the above-mentioned measure for APCs it is necessary to perform information security threat modeling processes which include detecting potential intruders, actual threats as well as protection measures proposed to minimize the consequences of implementing APCs information security actual threats [6, 7].

Information security threat modeling processes have a high-level actuality in implementing, configuring and functioning SOC (Security Operations Center) [8]. In compliance with the document [9], one of the 11 Strategies of a World-Class Cyber Security Operations Center is Cyber Threat Intelligence (cyber threat intelligence CTI). CTI - refers to the collection, processing, organizing, and interpreting of data into actionable information or products that relate to capabilities, opportunities, actions, and intent of adversaries in the cyber domain to meet a specific requirement determined by and informing decision-makers. In compliance with the given wording, we can conclude that APCS information security threat modeling process is a private case of CTI and can be automated in accordance with the requirements to the process of information infrastructure protection.

Some best practices in use of automated systems CTI include the following:

- Have a good feedback loop: Is the CTI actionable such that it can be integrated into the various SOC tools? Is it yielding useful results to the SOC team?
- Quality control: If the data is inaccurate, feedback should be used to adjust the SOC reliance.
- Track the pedigree: Is the CTI derived or evolving over time? Where did it originate? What type of data was used to derive it? Is it a reliable source? Also, ensure any caveats are tracked (such as proprietary, cannot be forwarded, etc.).
- Define and maintain standards for attribution/association: attribution is difficult, adversary association is more attainable: know ahead of time what is "good enough" for the SOC and drive consistency in language and thresholds used to articulate analytic certainty.
- Define and maintain standard for annotation and action: understand ahead of time when to watch based on activity, and when to act based on hits.

In conjunction with the above-mentioned criteria a system of automatic APCs information security threat modeling must comply with the following requirements:

- Give reliable and relevant information for detecting actual IS actual threats at all levels of APCs;
- Have a common information space for storing coherent knowledge about the processes of providing security to a multi-level APCs;
- Analyze a situation and develop recommendations for making decisions about countering actual threats to APCs information security, to give a grounded explanation of the recommended decision;

- Belong to the class communication Driven—a system based on an expert group decision making (a system for facilitating opinion exchange and calculation the average expert values);
- Comply with the principle of data unification, since there many types of information exchange and transmission systems;
- Support the principle of data processing scalability. An automated threat modeling system must equivalently process both small data amounts and its arrays. By equivalency we mean uniformity in processing principles [10].

Taking into account the above-mentioned criteria and requirements the author has developed an automated system of information security threat modeling (AS «ISTM» APCs).

2 Data Requirements

The current state of the problem of APCs information security threat modeling results from the lack of software complexes for automation of the mentioned processes. The current experience of APCs information security threat modeling is based on the recommendations detailed in the works [10–12] and has no automation means for determent actual threats posed by both external and internal intruders.

The aim of this work is the approbation of the software developed by the author - an automated system of information security threat modeling (hereinafter AS "ISTM") [13]. The approbation of APCs AS «ISTM» is held in the conditions of functioning of information infrastructure facilities of industrial organizations that use APCs and aimed at determent of the possible increase of efficiency of information security threat modeling using the developed software at each of the levels of APCs, namely: the upper one, the middle one and the field level [14]. The effectiveness assessment is done in comparison with the existing automated, normative, and methodological means [7, 11] with respect to the following quantitative characteristics:

- the total number of actual threats to information security at each APCs level;
- the total number of protective measures offered to minimize the consequences of implementing APCs information security actual threats;
- the total number of person-hours spent on obtaining the results of APCs information security threat modeling.

3 The Existing Means of APCs Information Security Threat Modeling

OCTAVE and TRIKE [15, 16] refer to the number of the most commonly used automation means for APCs information security threat modeling.

The automation means OCTAVE performs threat modeling using the following steps, presented in graphic notation IDEF0 in Fig. 1:

- identifying concern areas;

• identifying threat scenarios.

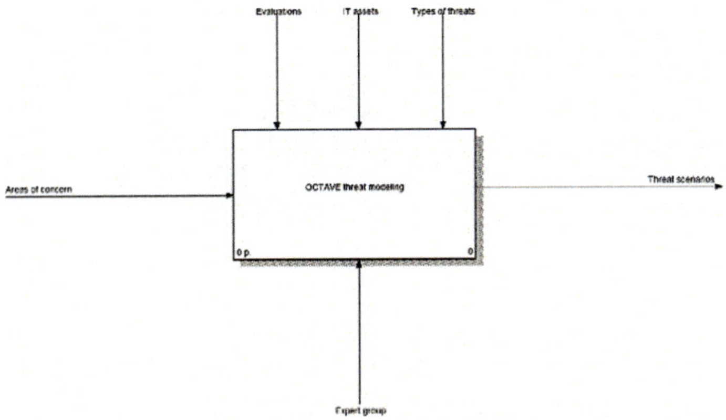

Fig. 1. IDEF0 – OCTAVE threat modeling diagram.

Step 1: The given step is the beginning of risk identification process through brain storm of project group participants. At this stage high-level areas and types of risks actual to the analyzed IT assets are determined.

Step 2: determines several threat types:

• human factor and using technical means;
• human factor with using physical access;
• technical and other problems.

For every threat type and for every IT-asset threat implementation scenarios can be determined, including description of their impact-assets and implementation probability. In doing so the probability is estimated by the following criteria: high; average; low. To simplify this laborious process OCTAVE methodology contains special questioning lists that allow visualizing expert assessments of participants in the process of APCs information security threat modeling [17], which is reflected in the decomposition of this step in Fig. 2.

Thus OCTAVE structure is divided into three definite stages related in output to the decomposition presented above, namely: creating threat profiles based on high-level assets, identifying high-risk vulnerabilities of an IT-infrastructure, developing a strategy and security plans based on a threat impact level. Assuming this, the procedure of APCs information security threat modeling based on OCTAVE can be presented as a tuple (1).

$$A = \langle B, C, D \rangle \quad (1)$$

where A - threat scenarios; B - high-level areas of concern; C - risks to IT assets; D - threat impact level.

TRIKE methodology allows determining APCs vulnerabilities and applying certain measures to eliminate them. It is important to mark that in accordance with the considered

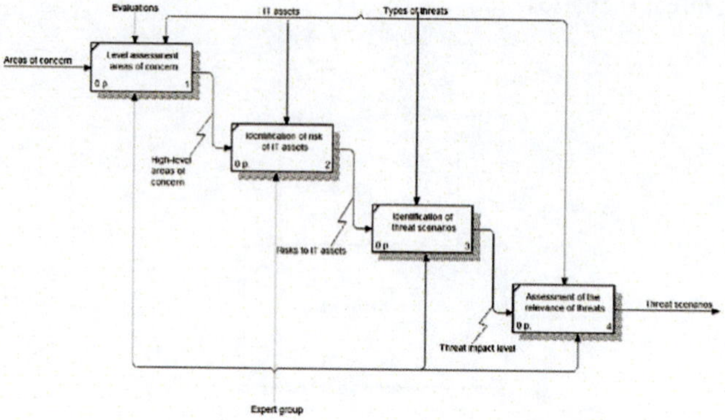

Fig. 2. IDEF0 – Decomposition of OCTAVE.

methodology – a system vulnerability is a complete way through an attack tree from one or more leaves to a threat, where all conditions for each attack are met.

As the key peculiarity of the considered methodology we mark the presence of repeated paths of the attack graph, which allows us to unite them into attack libraries in order to minimize calculations when building graphs and increase the efficiency of using graphs when identifying actual threats to APCs information security threats [18].

An attack tree, first expressed by Bruce Schneier as a "formal, methodical way of describing the security of systems based on varying attacks," is typically a graphical representation of vulnerabilities within an IT system, although it can be expressed in a text format which is not graphical in nature. Whether graphical or text format is represented, there are some similarities between these two representations. Every path through the attack tree represents a unique attack vector through the organization and shows what vulnerabilities an attacker can leverage to obtain access to sensitive material. At every level or leaf of the attack tree, there are different vulnerabilities, some of which must be combined to cause a breach and some which are capable of causing breaches by themselves. These are known respectively as "and" and "or" decompositions, and are represented differently based on whether the graphical, text, or combination format of the attack tree is used. It is up to the creator of the attack tree as to how much information is shown and in what formats the attributes are represented, as there are no formal standards in place. Despite the lack of formal standards, there are some recommendations for attack tree generation, such as creating the trees, measuring the probabilities of each leaf, removing or marking improbable attack paths, generating countermeasures, and applying the most appropriate countermeasures to leaves. Each independent level is called a "leaf" node, and vulnerabilities are moved through during an attack from the bottom to the top, which is called the "root" node. For this attack tree, straight horizontal lines represent "and" decompositions, and curved horizontal lines represent "or" decompositions [19].

The security model presented in Fig. 3 describes in detail the stages of information security threat modeling which include: obtaining a requirement model, obtaining an implementation model, developing a list of actual threats as well as creating an attack library.

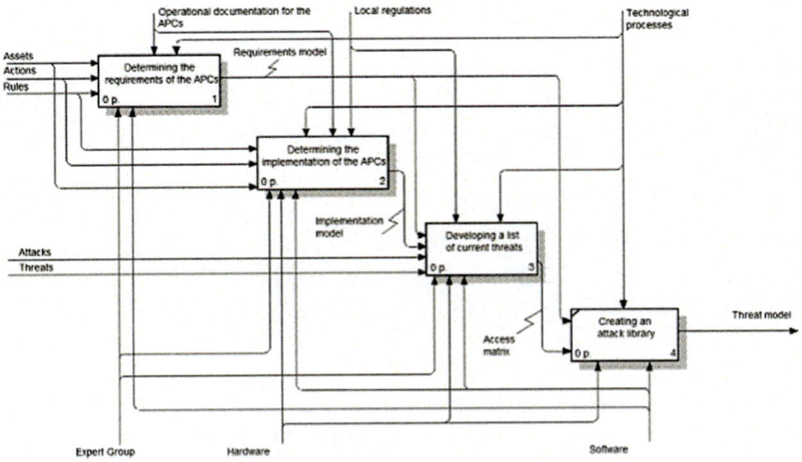

Fig. 3. IDEF0 – Decomposition of OCTAVE.

The obtained by the given algorithm information security threat model TRIKE can be formalized in accordance with the expression (2).

Thus OCTAVE structure is divided into three definite stages related in output to the decomposition presented above, namely: creating threat profiles based on high-level assets, identifying high-risk vulnerabilities of an IT-infrastructure, developing a strategy and security plans based on a threat impact level. Assuming this, the procedure of APCs information security threat modeling based on OCTAVE can be presented as a tuple (1).

$$U_{di} = \sum_{d=1}^{n} \left(U_d(a) + \sum_{i=1}^{3} U_{\overline{d}}(b_i) \right) \tag{2}$$

where U_d - threats to operations $d = \overline{1, n}$; $U_d(a)$ - the function that characterizes a "service denial" threat in relation to operations d; $U_{\overline{d}}(b_i)$ - function describes "privilege elevation" threats b_i, $i = \overline{1, 3}$, for the inverted operation $\overline{d}$ under the following conditions: b_1- a threat of a completely prohibited action; b_2- a threat of a partly prohibited action; b_3- a "social responsibility" threat.

The main disadvantage of OCTAVE methodology of automating information security threat modeling processes is the lack of tools for determining protection measures to compensate for actual threats.

The above-considered advantages and disadvantages of APCs information security threat modeling means were taken into account by the author when developing APCS AS «ISTM».

4 Description and Assessment of APCs as "ISTM" Effectiveness

APCs AS "ISTM", the architecture of which is presented in Fig. 4, is a dynamic expert system, which allows solving the problem of building an APCs information security threat model using the methods and algorithms presented in the work [20].

APCs AS "ISTM" provides reliable and up-to-date information for determining current information security threats at all levels of the APCs; has a single information space for storing consistent knowledge about the security processes of multi-level APCs; allows you to analyze the situation and form recommendations for decision-making on countering current information security threats of the APCs, as well as a well-founded explanation of the recommended solution.

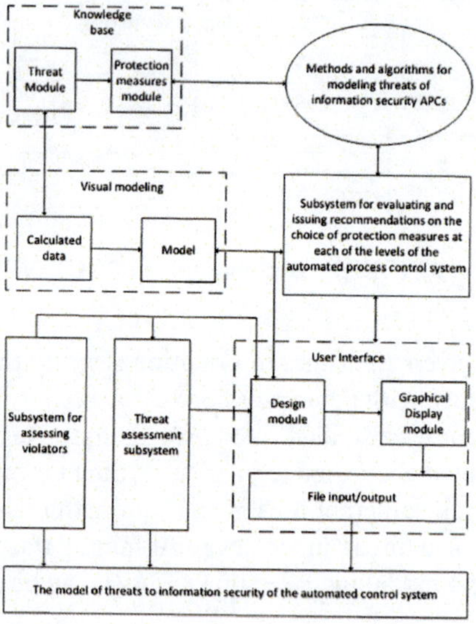

Fig. 4. Architecture of AS "ISTM" APCs.

The main components of APCs AS "ISTM" are subsystems "*Estimation of APCs information security intruders*" and "*Estimation of APCs information security threats*". The connection of the given subsystems in graphic notation IDEF0 is presented in Fig. 5.

The result of the intruder assessment subsystem functioning is so-called a threat model, which contains information about possible attackers with assessment of their potential in relation to the ongoing technological processes and the applied means for APCs information protection. The threat model is built based on expert assessments of intruders' motivation and equipment, possible threats to information security. The decomposition of the subsystem is shown in Fig. 6.

To describe all possible types of violators, when compiling threat models, the process of forming models of information security violators is used. The information security violator model is inextricably linked with the information security threat model, since the information security violator is often both a source of threats and a consequence.

The information security violator model is a set of assumptions about one or more possible information security violators, their qualifications, their technical and material values, etc.

The threat model based on work results of the threat model, the decomposition of which is presented in Fig. 7, is based on the results of the assessment of information

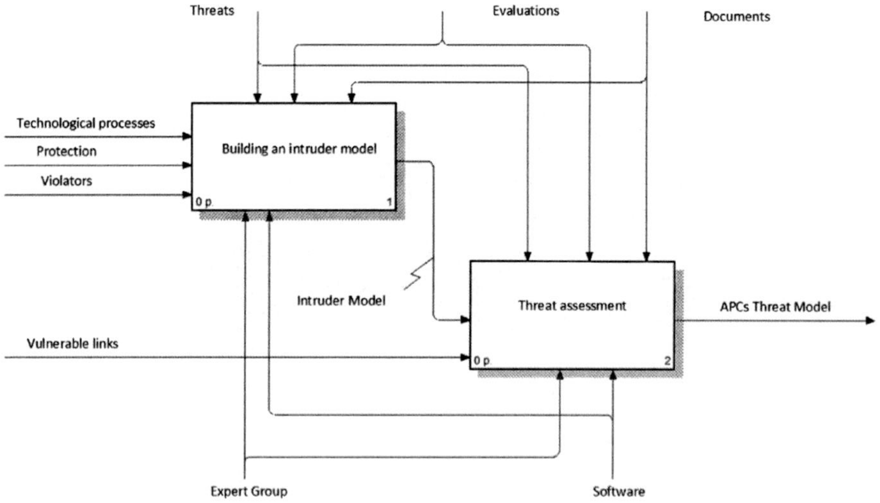

Fig. 5. Functional model of AS "ISTM" APCs.

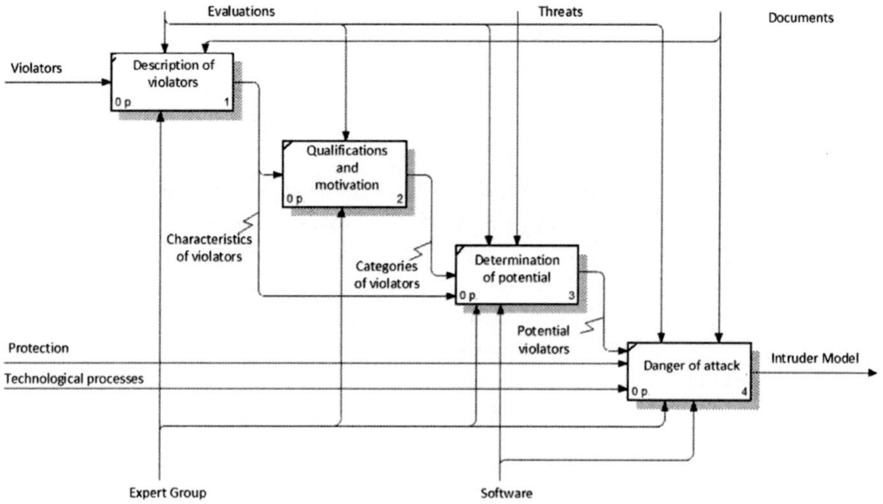

Fig. 6. Decomposition of intruder assessment subsystem AS "ISTM" APCs.

security threats and vulnerable APCs links. The nodes of each system levels, in which vulnerabilities can dynamically occur, act as vulnerable APCs links.

The threat model contains a list of possible intruders, a list of actual threats and protection measures at each APCs level in editable format, which allows using the modeling results in organizational and administrative documentation of APCs information protection system.

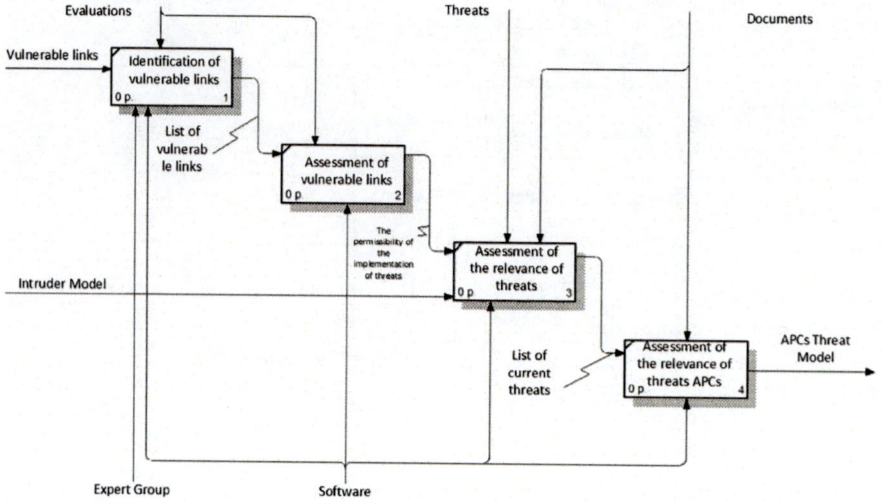

Fig. 7. Decomposition of threat assessment subsystem AS "ISTM" APCs

The formalization of information security threat modeling methods used in APCs AS «ISTM» comes to the expression (3).

$$U_i^k = L_i^k * Z_i^k \tag{3}$$

where vulnerability implementation estimation U_i^k are calculated through identified vulnerable links i at every APCS level k - based on safety estimation Z and the calculation results of confidence degree in case of favorable conditions L_i.

For a comparison assessment of the results of information security threat modeling under the conditions of use the developed APCs AS "ISTM" at objects of informative infrastructure the methodology of estimation the performance indicator V (4), is applied and described in the work in detail [21, 22].

$$E = (1 - \frac{T}{\sum V + \sum P}) * 100\% \tag{4}$$

where $\sum V = V_1 + V_2 + V_3$—is a total number of identified actual threats to information security for each of the three APCs levels: V_1 - actual threats of the upper level; V_2 - actual threats of the middle level; V_3 - actual threats of the field level. $\sum P = P_1 + P_2 + P_3$ – the total number of protective measures offered to minimize the consequences of information security threats implementation at all APCs levels. T – the total number of person-hours spent to obtain the results of APCs threat modeling.

5 Results Discussing

In order to test APCs AS "ISTM", the software was used in the processes of threat modeling for two information infrastructure objects of two different industrial organizations that operate APCs, object 1 and object 2. The important fact is that AS "ISTM" appliance estimation has been carried out by different information security specialists, who are information security administrators at the above-mentioned information infrastructure objects, which allows avoiding subjective estimation of APCs AS "ISTM".

The obtained estimations of APCs AS "ISTM" appliance when modeling information security threats at the two objects of information infrastructure are represented in the Table 1.

Table 1. The comparison estimation results of APCs AS "ISTM" appliance.

No	Evaluation parameters	Estimations					
		Object 1			Object 2		
		OCTAVE	TRIKE	AS ISTM	OCTAVE	TRIKE	AS ISTM
1	Actual threats at the upper level (V_1)	68	48	81	18	17	29
2	Actual threats at the middle level V_2	51	45	65	10	15	23
3	Actual threats at the field level V_3	93	0	14	2	0	5
4	Protective measures at the upper level P_1	0	74	117	0	27	41
5	Protective measures at the middle level P_2	0	19	63	0	7	22
6	Protective measures at the field level P_3	0	3	21	0	0	5
7	Person-hours (T)	72	70	30	14	31	19
8	**Performance indicator (E), %**	**34**	**63**	**92**	**46**	**54**	**85**

On the basis of the foregoing calculations, APCs AS "ISTM" shows higher performance indicators of information security threat modeling for information infrastructure objects, allows to indicate actual threats, and determine compensating protective measures in greater numbers than the considered automated and normative threat modeling means.

The given fact demonstrates that APCs AS "ISTM" can be applied as an additional tool for increasing information security level at information infrastructure objects that operate on the basis of APCs, when building information security threat models in the framework of document requirements [1, 7].

6 Conclusion

Thus APCs information security threat modeling process is a required step when building information protection system at an information infrastructure object, and the developed APCs AS «ISTM» may act as a tool to perform the above-mentioned activities.

References

1. Kozyreva, A.A., Tarasov, D.A.: Sovremennoe sostoyanie gosudarstvennoj politiki v sfere informacionnoj bezopasnosti (The current state of state policy in the field of information security). Vestnik Voronezhskogo instituta MVD Rossii **4**, 243–247 (2018). https://doi.org/10.24412/FdxFWSMj7p8
2. Gryzunov, V.V., Grishechko, A.A., Sipovich, D.E.: Vybor naibolee opasnyh uyazvimostej dlya perspektivnyh informacionnyh sistem kriticheskogo primeneniya (Selection of the most dangerous vulnerabilities for advanced information systems of critical application). Cybersec. Iss. **47**(01), 66–75 (2022). https://doi.org/10.21681/2311-3456-2012-1-66-75
3. Kornienko, A.A., Kornienko, S.V., Karpov, D.: Metodika detalizirovannogo ocenivaniya znachimosti sistem organizacii dvizheniya poezdov kak ob"ektov kriticheskoj informacionnoj infrastruktury (Methodology of detailed assessment of the significance of train traffic management systems as objects of critical information infrastructure). Bull. Sci. Res. Results **2**, 5–19 (2022). https://doi.org/10.20295/2223-9987-2020-2-5-19
4. Kuznecova, N.M., Karlova, T.V., Bekmeshov, A.: Reshenie zadachi avtomatizacii processov zashchity strategicheski vazhnyh resursov predpriyatiya ot kom-pleksnyh kiberatak na osnove analiza taktik zloumyshlennikov (Solving the problem of automating the processes of protecting strategically important enterprise resources from complex cyber-attacks based on the analysis of attackers' tactics). Bull. Bryansk State Tech. Univ. **92**(07), 48–53 (2020). https://doi.org/10.30987/1999-8775-2020-7-48-53
5. Gavrilov, V.E.: Informacionnaya bezopasnost' avtomatizirovannyh sistem upravleniya kriticheski vazhnyh ob"ektov otmechaet yubilej (Information security of automated control systems of critical facilities celebrates the anniversary). CONNECT. Mir informacionnyh tekhnologij **1–2**, 4–22 (2022)
6. Pasechnik, R.M., Tabunkova, M.P., Korolev, I.D.: Matematicheskoe obespechenie optimizacisistemy zashchity informacii v avtomatizirovannyh sistemah upravleniya (Mathematical support for optimization of the information security system in automated control systems). I-Methods **13**(1), 1–10 (2021)
7. Chernov, D.V.: Metodiki ocenki ugroz bezopasnosti informacii avtomatizirovannyh sistem upravleniya tekhnologicheskimi processami (Methods for assessing information security threats of automated process control systems). Sovremennaya nauka: aktual'nye problemy teorii i praktiki. Seriya: Estestvennye i Tekhnicheskie Nauki **9**, 81–87 (2021). https://doi.org/10.37882/2223-2966.2021.09.25
8. Miloslavskaya, N.: Security operations centers for information security incident management. In: IEEE 4th International Conference on Future Internet of Things and Cloud (FiCloud), pp. 131–136 (2016). https://doi.org/10.1109/FiCloud.2016.26
9. Burlakov, M.E.: Dvuhklassifikacionnaya iskusstvennaya immunnaya sistema (Two-classification artificial immune system). Vestnik Samarskogo gosudarstvennogo universiteta **118**(7), 207–220 (2014)
10. Bolshakov, A.S., Rakovskiy, D.I.: Programmnoe obespechenie modelirovaniya ugroz bezopasnosti informacii v informacionnyh sistemah (Software for modeling information security threats in information systems). Pravovaya informatika **1**, 26–39 (2020). https://doi.org/10.21681/1994-1404-2020-1-26-39

11. Boyarov, E.N., Abramova, S.V., Stankevich, P.V.: Analiz informacionnyh riskov v obrazovanii (Analysis of information risks in education). Perspektivy nauki i obrazovaniya **51**(3), 451–464 (2021). https://doi.org/10.32744/pse.2021.3.32
12. Shinakov, K.E., Golembiovskaya, O.M.: Formalizaciya processa ocenki riskov informacionnoj bezopasnosti na osnove metodiki OCTAVE (Formalization of the information security risk assessment process based on the octave methodology). Vestnik Bryanskogo gosudarstvennogo tekhnicheskogo universiteta **47**(3), 175–179 (2015)
13. Chernov, D.V., Sychugov, A.A.: Certificate of registration of the computer program RU 2022611043, 18.01.2022. Modelirovanie ugroz informacionnoj bezopasnosti ASU TP (Modeling of threats to information security of automated control systems) (2021). Application no. 2022610172. Dated 18 Jan 2021
14. Chernov, A.A., Sychugov, D.V.: Analysis of modern requirements and problems of information security of automated process control systems. Neurocomputers **8**, 38–46 (2018). https://doi.org/10.18127/j19998554-201808-04
15. Minyaev, A.A.: Modelirovanie ugroz bezopasnosti informacii v territorial'no-raspredelennyh informacionnyh sistemah (Modeling of information security threats in geographically distributed information systems). Naukoemkie tekhnologii v kosmicheskih issledovaniyah Zemli **13**(2), 52–65 (2021). https://doi.org/10.36724/2409-5419-2021-13-2-52-65
16. Kristanosov, V.B.: Metodologiya ocenki i upravleniya cifrovymi riskami (Methodology of digital risk assessment and management). Vestnik Novosibirskogo gosudarstvennogo universiteta. Seriya: Informacionnye tekhnologii **250**(2), 15–36 (2021). https://doi.org/10.52065/2520-6877-2021-250-2-15-36
17. Grishunin, S.V., Pischalkina, I.Yu., Suloeva, S.B.: Ocenka kiberriskov v proektah interneta veshchej (Evaluation of cyber risks in Internet of Things projects), π-Economy **14**(6), 102–116 (2021). https://doi.org/10.18721/JE.14608
18. Chernov, D.: Application TRIKE methodology when modeling threats to APCs information security. In: Radionov, A.A., Gasiyarov, V.R. (eds.) RusAutoCon 2021. LNEE, vol. 857, pp. 452–461. Springer, Cham (2022). https://doi.org/10.1007/978-3-030-94202-1_42
19. Vitkus, D., Salter, J., Goranin, N., Čeponis, D.: Method for attack tree data transformation and import into it risk analysis expert systems. Appl. Sci. (Switzerland) **10**(23), 1–21 (2020). https://doi.org/10.3390/app10238423.-EDNNWTTOK
20. Chernov, D.V.: Nauchnye rezul'taty vypolneniya proekta "Razrabotka metodov i algoritmov modelirovaniya ugroz IB ASU TP" (Scientific results of the project "Development of methods and algorithms for modeling threats of information security automated control system"). In: Collection of Scientific Papers Based on the Materials of the All-Russian Scientific and Theoretical Conference "Theory and Practice of Information Security, vol. 09, pp.:153–157 (2021)
21. Petrichenko, G.S., Petrichenko, V.G.: Ocenka effektivnosti programmnogo obespecheniya (Software performance evaluation). Ekonomika. Informatika **2308**(9), 108–112 (2016)
22. Ilyina, O.P., Sotavov, A.K.: Arhitekturnoe modelirovanie sistemy informacionnoj bezopasnosti (Architectural modeling of the information security system). Tekhniko-tekhnologicheskie problemy servisa **48**(2), 30–37 (2019)

Problems of Developing User Identification Systems by Keyboard Handwriting

R. Sharipov[1](✉) and O. Panchenko[2]

[1] Kazan National Research Technical University named after A.N. Tupolev–KAI, 10, K. Marx Street, Kazan 420111, Russia
riphat@mail.ru

[2] Kazan National Research Technological University, 68, K. Marx Street, Kazan 420015, Russia

Abstract. This article is devoted to the analysis of problems in the development of user recognition systems by keyboard handwriting. The works of famous scientists in this field are analyzed. Based on the results of the analysis of the works, the problems of developing systems for identifying and authenticating users by keyboard handwriting were identified. It is shown that the use of standard built-in events in modern high-level object-oriented programming languages, pressing and releasing keys, leads to errors in measuring the time of bigrams, where the results shifted to values of 200, 240, 280 and 320 ms. In addition, studies were conducted when working on various keyboards, the results were different related to design features and ergonomics. On the example of three different keyboards, the time distributions of the user's bigrams were obtained. For accuracy, an approximation based on a mixture of the distribution of Gaussian components was used. This paper proposes methods for solving these problems in order to increase the reliability of authentication.

Keywords: Dynamics of keystrokes · Keyboard handwriting · Identification · Authentication · User · Poly-Gaussian distribution

1 Introduction

In recent years, there has been a rapid development of computing systems, more and more people are becoming users of various telecommunication systems, which manifest themselves in many areas of life, for example, in queuing systems for people. Thus, it simplifies the procedures for obtaining or providing services in a particular type of activity. However, on the other hand, issues related to the identification and authentication of users of various systems, as well as the provision of measures to protect confidential data, are being updated. These tasks have non-trivial solutions: today there are a lot of different approaches and methods for identifying and authenticating users of telecommunication systems of the widest range. The question of the effectiveness of such approaches and methods is also important, since, according to various Internet publications, the level of cybercrime is constantly growing every year, and the trend will continue for many years.

There are many approaches to developing identification and authentication systems. In a number of works, a conditional classification of identifiers is presented, which

A. A. Radionov and V. R. Gasiyarov (Eds.): RusAutoCon 2022, LNEE 986, pp. 386–395, 2023.
https://doi.org/10.1007/978-3-031-22311-2_37

suggests that the identifier can be some kind of information in the form of symbols (symbolic passwords, pin codes), it can also be in the form of a device (cards, tokens, touch pads, radio frequency keys, etc.) and biometric user identifiers [1].

Among biometric systems, systems based on the features of typing characters on various keyboard input devices are developing. There are many publications in this direction, we will review some of them, consider approaches and features.

2 Overview of Keyboard Handwriting Identification Systems

In [1–4], the author introduces the concept: a pattern of keyboard handwriting, where a pattern refers to the time intervals between pressing the previous key and the current key. In addition, the author made the assumption that the measured time intervals are subject to the normal Gaussian distribution and developed a mathematical model. To do this, he carried out a practical implementation, described and proposed several algorithms for training a system for recognizing and identifying users, and also developed recommendations for the software implementation of methods. In his work, the author does not provide the results of the implementation of the developed methods, the practical results of using their own algorithms were not presented on the example of developing an integrated identification and authentication system adapted to a specific operating system and implemented in a specific programming language.

In [5], the task was to develop a convenient and reliable software tool for identifying working personnel. To solve this problem, research work was carried out, on the basis of which software tools were created for identifying and authenticating users based on their keyboard handwriting. Two groups of recorded handwriting parameters were distinguished:

1. The speed of the operator's response to detect and correct errors and combinations of characters in which errors are most common when entering a predefined text (typical for this operator).
2. Parameters of temporal characteristics of text input:
   - the speed of learning to enter a constant, individual (password) phrase.
   - parameters that determine the length of the time interval between successive keystrokes, while determining the "slowest" and the "fastest" pairs of keys;
   - the average character set speed (the average number of characters that the operator enters per unit of time);
   - the number of characters that the operator remembers in one continuous input.

A set of software tools was developed for collecting and recording the necessary statistical data and for processing them. In this case, a rather complex mathematical apparatus was used. In the developed system, it was possible to reduce the rate of erroneous decisions to 20% by selecting the response threshold, using a rather complex technique for statistical processing of a data set obtained by entering parameters, and carefully choosing the type of discriminant function. The coefficient of erroneous decisions was calculated as the sum of errors of the first and second kind.

In another work [6], the method of user identification by keyboard handwriting is combined with the method of standard password protection. In this case, despite the fact that the attacker somehow gains access to the password, access to the computer system can be limited due to additional authentication by keyboard handwriting. So, for example, if the user enters an incorrect password, then he is instantly denied access, and if the password is correct, a sample of his keyboard handwriting is measured and entered into the novelty detector. And if the detector recognizes the handwriting sample as approved, access is allowed to it. And if the detector rejects the sample as new (this user is not registered in the system), access to this user will be denied.

In works [7, 8] two novelty detectors were specified to authenticate users of computer systems by handwriting on the keyboard, such as 1-LVQ and SVDD. If attacker samples are not available at the beginning of training, they may be available later after failed intrusion attempts. The average integral errors for 21 users were studied, the values of which did not exceed 2%.

The authors of works [9, 10] proposed to use the apparatus of neural networks when solving the problem of identifying and authenticating users by keyboard handwriting, highlighting this approach as the most relevant in comparison with standard statistical methods, which, as a rule, use averaged values of time parameters that characterize the manner of work. Keyboard users. And also neural networks have some advantages. Among them: the ability to identify hidden patterns of dependence in the data flow, which can be trained on many examples, work with noisy input signals, the ability to parallelize data processing processes leading to high performance, etc.

In a number of works [11, 12] it is proposed to use keyboard handwriting in forensic science to increase the level of efficiency of using forensic records, along with the currently existing interactive information databases, which allow obtaining information about persons by entering an individual information number of citizens.

There is another promising direction - the creation of a keyboard polygraph [13]. The authors of this work define the polygraph as a kind of complete software tool that makes it easy to monitor the psychophysiological reactions of a test person who types certain texts from the keyboard.

At the same time, indicators of working with the keyboard are registered and displayed in the form of some graphical representation of a group of parameters that form a polygamma. And also statistical studies were carried out and the results were obtained according to which the authors made sure that the indicators of aspects of the operator's work with the keyboard have the following properties:

- weak statistical dependence of a large number of indicators;
- slow trend of change in psychophysiological state in the absence of external influences;
- sensitivity to various aspects of the mental state of the operator.

To assess the efficiency of keyboard handwriting recognition systems, error rates are used –false rejection rate (FRR) and false acceptance rate (FAR) where FAR is understood as an event in which an illegal user is recognized as one of the registered ones, and FAR is understood as a case in which a legal user is recognized as not registered [14].

The conducted studies of methods and approaches for recognizing keyboard handwriting showed that there are a number of problems in the development of such systems that limit the creation of fully completed systems for user authentication by keyboard handwriting, where this approach would be the main one, trusted in the development of complete systems. However, in reality, there is not a single ready-made and reliable commercial software product, although in some works the authors achieve handwriting recognition accuracy of up to 90% [15], but usually this result can only be obtained when the operator types one well-known phrase a large number of times during one session or within one business day. Let's look at some of the problems associated with this.

3 Problems of Creating Identification Systems by Keyboard Handwriting

3.1 Deviations in the Parameters of Keyboard Handwriting

In [16–19], the authors studied the parameters of the keyboard handwriting of several users and found that these parameters are subject to deviation depending on the time of day and their psychophysiological state. An analysis of the characteristics of the users' handwriting made it possible to distinguish them into 2 groups:

- A group of users with fairly stable characteristics who have a well-formed style of working with the keyboard;
- A group of users with highly stable characteristics, who ideally own the ten-finger typing method from the keyboard.

When working for some time, users had deviations. Moreover, the users of the first group had deviations of the order of 11–25%, while those of the users of the second group were 7–18%. To solve this problem, many authors propose to register a large number of parameters at the software level [20] to reduce FAR and FRR. However, this approach does not give good results.

3.2 Parameter Measurement Errors

Another significant problem is the accuracy of measuring the parameters of the keyboard handwriting. Mostly this is done only at the program level: in various programming languages, usually high-level object-oriented languages, in which events such as KeyDown (key pressed) and KeyUp (key released) are already supported. By capturing the times of these events and combining these data, you can easily calculate key hold intervals, keystroke intervals, average typing speed, etc.

We have done research with these events. To do this, in the C# programming environment, we measured the time intervals between adjacent keys (bigrams) of 5 random users, while they typed arbitrary texts. To measure the time of digrams, we used the KeyDown and KeyUp events, to measure the time intervals, we used the system timers of the Windows operating system (Fig. 1).

Figure 1 shows that all the measured time values of bigrams are concentrated mainly around the values of 200, 240, 280 and 320 ms. This happens as a result of various delays

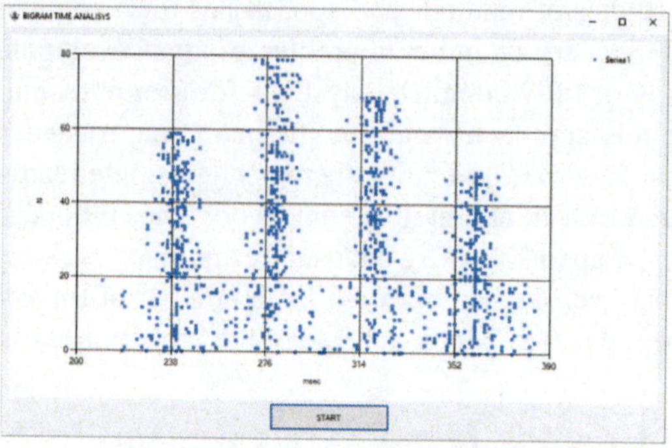

Fig. 1. Distribution of bigram times.

in the operating system, which leads to the approximation of the results to the values of 200, 240, 280 and 320. Some authors [16, 17] have added inserts in assembler codes from the programming environment in order to measure events at a low level of programming in order to reduce errors. This approach improves the measurement accuracy, but not enough.

3.3 Dependence on the Characteristics of Keyboards

And the last problem is the dependence of user recognition systems by keyboard handwriting on the type of keyboard where the data collection process and further user identification are carried out [18]. For example, conventionally keyboards can be divided into "hard" and "soft". Studies have shown that on different types of keyboards, there are some changes in the temporal parameters of the keyboard handwriting when typing characters by the same user. For example, when working on a hard keyboard, the time parameters may decrease slightly, since it is more difficult for fingers to push through hard keys and this leads to slower typing [18, 19].

Studies have been conducted on the influence of various keyboards on the parameters of keyboard handwriting. The following keyboards were selected for the study: Oklick 90M, HP KU-0316 and Genius Slim Star 130 on which the selected user - Operator_1 typed the same text. At the same time, the time of bigrams was fixed without reference to a specific bigram combination. After we collected a sample of the time of the bigrams, we had the task of approximating them and determining the distribution function.

Usually, for example, in the works [20] a Gaussian approximation was used, with known parameters as the mathematical expectation and variance. For a more accurate approximation, we used a poly-Gaussian distribution model representing a mixture of Gaussian distributions based on reference Gaussian signals [21]. For the convenience of approximation, a poly-Gaussian distribution model based on reference Gaussian signals was used [20–22].

The density of the poly-Gaussian distribution is described as [22]:

$$W(t) = \sum_{n=1}^{N} q_n W_n(t), \quad q_n \geq 0, \quad \sum_{n=1}^{N} q_n = 1 \tag{1}$$

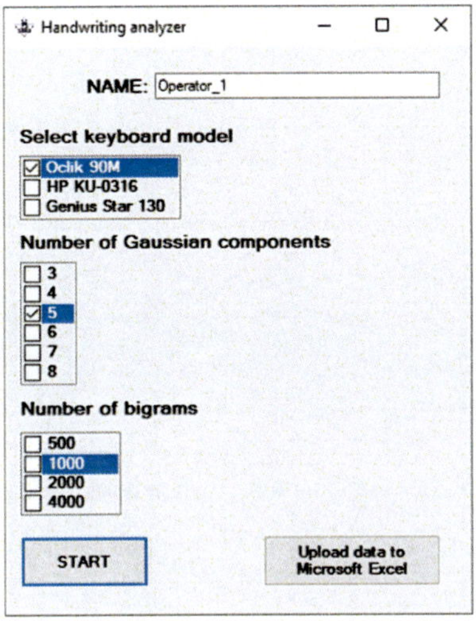

Fig. 2. Interface of the software package "Handwriting analyzer".

where: N – number of Gaussian signals, q – coefficients of each Gaussian component.

For practical research and data analysis of the influence of various keyboards on handwriting, the "Handwriting Analyzer" software package was developed. We set 5 Gaussian components and selected 1000 bigrams for each keyboard (Fig. 2). The choice of 5 components is justified by the sufficiency of overlapping the interval of the most probable values. The number of values 1000 is justified enough to obtain statistical data.

After the user Operator_1 typed a text message, we obtained the time distribution densities of bigrams for three different selected keyboards. From the presented graphical data of Operator_1 (Figs. 3, 4, 5) one can see various approximate distributions of parameters when working on different keyboards. This is due to the design features of the keyboards, the quality of their manufacture. For example, the most ergonomic of these keyboards is the Genius Slim Star 130, the average bigram time is less and the character typing speed was higher (Fig. 5). The Oklick 90M keyboard, on the other hand, has a rigid key structure. Therefore, the bigram time was less, and the average value was within 300 ms, and the typing speed was less (Fig. 3).

In [23], an additional parameter of keyboard handwriting is considered - this is the speed of movement of the keys when they are pressed, where the speed is understood as the rate of change in the capacitance of the contact pad over time, in addition, a technical solution is given, a block diagram is presented, algorithms for the operation of the controller and statistical processing of the measured data are given. There is another approach [24, 25] in which recognition of users of computer systems by keyboard handwriting is carried out taking into account the parameters of pressure on the keys and their vibration. The authors developed a keyboard using special vibration and pressure sensors for data acquisition, proposed feature categorization based on the use of wavelet transforms, revealed the distribution laws of keyboard handwriting features, assessed the correlation dependence of features, and found that additional features, on average,

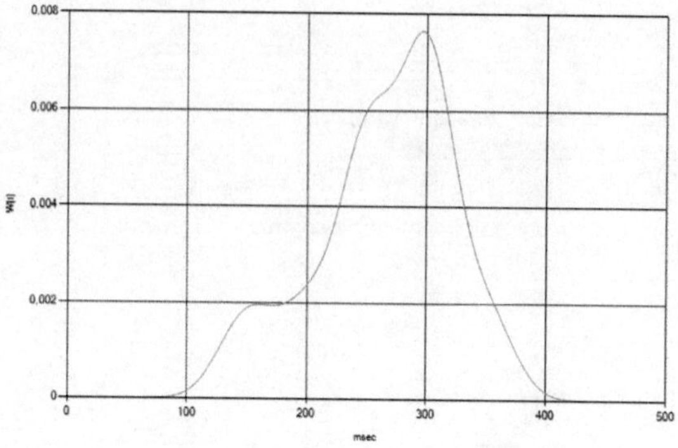

Fig. 3. Time distribution of bigrams of Operator_1 when working at the keyboard "Oklick 90M".

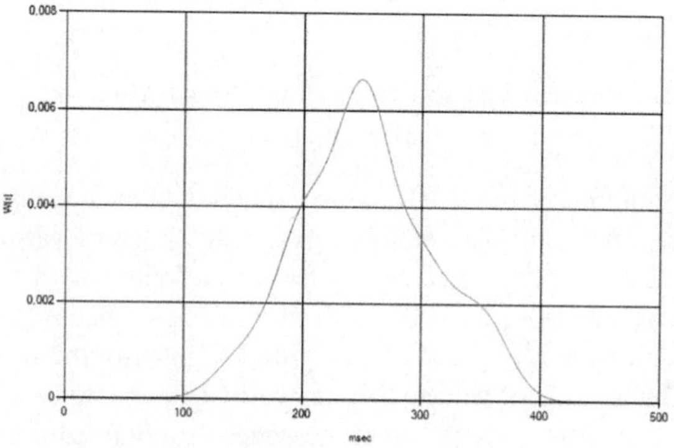

Fig. 4. Time distribution of bigrams of Operator_1 when working at the keyboard "HP KU-0316".

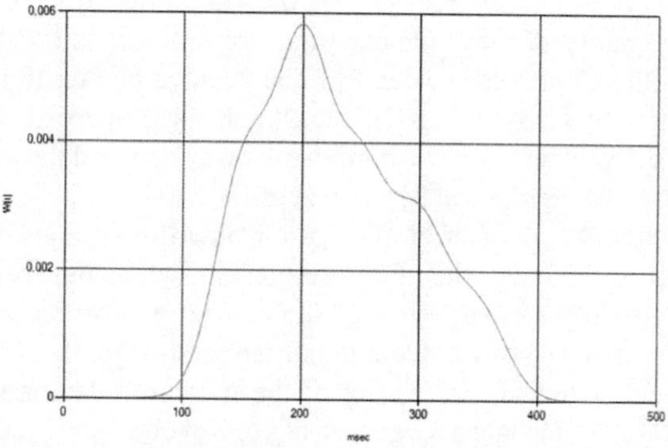

Fig. 5. Time distribution of bigrams of Operator_1 when working at the keyboard "Genius Slim Star 130".

can reduce the number of erroneous solutions up to 2–4 times and thereby improve the accuracy.

4 Conclusion

To improve the accuracy of user identification, it is necessary to:

1. Measure the parameters of keyboard handwriting and fix them at the level of keyboard devices to eliminate measurement errors associated with various delays both at the hardware level during keyboard operation and at the software level associated with processes at the level of operating systems;
2. When testing and developing user identification systems by keyboard handwriting, it is necessary to always use the same keyboard input devices – if the system was trained and standards were formed on certain keyboard devices, then identification procedures should be carried out on the same devices. Accordingly, when transferring user standards to other workstations, it is also necessary to transfer the keyboard or retrain the system again, which is a labor-intensive procedure;
3. Increase the number of recorded keyboard handwriting parameters using hardware support to measure new individual and unique user parameters, for example, vibration and pressing force of keyboard keys or key movement speed, where movement speed is represented as a process of changing the capacitance of the contact pairs of keys.

All of the above approaches lead to the concept of creating keyboard devices with additional built-in modules for measuring, processing and storing user standards, as well as the use of identification mechanisms for protecting access to systems at the level of keyboard devices, where the keyboard is an integral part of the system, having a unique identifier identified by the system on hardware level and thus not giving the opportunity to replace the keyboard device.

References

1. Flior, E., Anaya, T., Moody, C., Beheshti, M., Han, J., Kowalski, K.: A knowledge-based system implementation of intrusion detection rules. In: Abstracts of the Seventh International Conference on Information Technology: New Generations, Las Vegas, NV, USA, 12–14 April 2010
2. Sharipov, R., Tumbinskaya, M., Abzalov, A.: Analysis of Users' Keyboard Handwriting based on Gaussian Reference Signals. In: Abstracts of the International Russian Automation Conference (RusAutoCon), Sochi, Russia, 8–14 September 2019. https://doi.org/10.1109/RUSAUTOCON.2019.8867753M
3. Delattre, M., Genon-Catalot, V., Samson, A.: Mixtures of stochastic differential equations with random effects: application to data clustering. J. Stat. Plan. Infer. **173**, 109–124 (2016). https://doi.org/10.1016/j.jspi.2015.12.003
4. Peralta, D., Saeys, Y.: Distributed, numerically stable distance and covariance computation with MPI for extremely large datasets. In: Abstracts of the International Congress on Big Data (BigDataCongress), Milan, Italy, 8–13 July 2019. https://doi.org/10.1109/BigDataCongress.2019.00023

5. Nelasa, A., Krischuk, V.: Using of the user identification methods on keyboard handwriting at digital signature shaping. In: Abstracts of the Experience of Designing and Applications of CAD Systems in Microelectronics, Lviv-Slavsko, Ukraine, 12–17 February 2001. https://doi.org/10.1109/CADSM.2001.975824
6. Shalagin, S., Nurutdinova, A.: Stochastic identification of the "object-attribute" table based on the modified Rabiner's method. J. Phys. Conf. Ser. **1925**(1), 012014 (2021). https://doi.org/10.1088/1742-6596/1925/1/012014
7. Peralta, D., Triguero, I., García, S., Herrera, F., Benitez, J.M.: DPD-DFF: a dual phase distributed scheme with double fingerprint fusion for fast and accurate identification in large databases. Inf. Fus. **32**, 40–51 (2016). https://doi.org/10.1016/j.inffus.2016.03.002
8. Ibragimov, A., Ilinkova, T., Shafigullin, L., Gabdrakhmanov, A., Sharipov, R., Lakhno, A.: Investigation of mechanical properties of thermal barrier coating by tested on a 4-point bending. Res. J. Pharm., Biol. Chem. Sci. **7**, 2308–2317 (2016)
9. Flior, E., Kowalski, K.: Continuous biometric user authentication in online examinations. In: Abstracts of the Seventh International Conference on Information Technology: New Generations, Las Vegas, NV, USA, 12–14 April 2010. https://doi.org/10.1109/ITNG.2010.250
10. Gibadullin, R., Marushkai, N.: Development of predictive CNN based model for vital signs alerts. In: International Conference on Industrial Engineering, Applications and Manufacturing (ICIEAM), Sochi, Russia, 21 May 2021, pp. 404–409 (2021). https://doi.org/10.1109/ICIEAM51226.2021.9446354
11. D'Lima, N., Mittal, J.: Password authentication using keystroke biometrics. In: Abstracts of the International Conference on Communication, Information & Computing Technology (ICCICT), Mumbai, India, 15–17 January 2015
12. Eremenko, I., Oliunina, I.: Use of machine learning methods for solving problem of user identifying by keyboard handwriting. In: Abstracts of the International Russian Automation Conference (RusAutoCon), Sochi, Russia, 8–14 September 2019. https://doi.org/10.1109/RUSAUTOCON.2019.8867767
13. Iryna, D., Anton, K., Hanna, K., Bohdan, Y.: Corporate system users identification by the keyboard handwriting based on neural networks. Int. J. Innov. Technol. Expl. Eng. **9**(1), 4156–4161 (2019)
14. Gibadullin, R., Nikonorov, V.: Development of the system for automated incident management based on open-source software. In: 2021 International Russian Automation Conference (RusAutoCon), Sochi, Russia, 5 September 2021, pp. 521–525 (2021). https://doi.org/10.1109/RusAutoCon52004.2021.9537385
15. Gibadullin, R., Perukhin, M., Ilin, A.: Speech recognition and machine translation using neural networks. In: 2021 International Conference on Industrial Engineering, Applications and Manufacturing (ICIEAM), pp. 398–403 (2021). https://doi.org/10.1109/ICIEAM51226.2021.9446474
16. Safiullina, L.K, Gabdullin, A.S., Anikin, I.V.: Face recognition in biometric systems using HAAR cascade classification. In: 2021 Dynamics of Systems, Mechanisms and Machines (Dynamics), pp. 1–5 (2021). https://doi.org/10.1109/Dynamics52735.2021.9653460
17. Safiullina, L.K., Maturov, R.R.: Image processing for biometric scanning of the palm vein pattern. In: Kravets, A.G., Bolshakov, A.A., Shcherbakov, M. (eds.) Society 5.0: Cyberspace for Advanced Human-Centered Society. SSDC, vol. 333, pp. 25–34. Springer, Cham (2021). https://doi.org/10.1007/978-3-030-63563-3_3
18. Gizatullin, Z., Shkinderov, M.: Increasing the noise immunity of the access monitoring and control system under the influence of electrostatic discharge. In: 2021 International Conference on Industrial Engineering, Applications and Manufacturing (ICIEAM), p 515–519 (2021). https://doi.org/10.1109/ICIEAM51226.2021.9446434

19. Balagani, K., Phoha, V., Ray, A., Phoha, S.: On the discriminability of keystroke feature vectors used in fixed text keystroke authentication. Pattern Recogn. Lett. US **32**, 1070–1080 (2011)
20. Gizatullin, Z., Shkinderov, M.: Research of electromagnetic interference in the power supply network of the access monitoring and control system. In: 2021 International Conference on Industrial Engineering, Applications and Manufacturing (ICIEAM), pp. 520–524 (2021). https://doi.org/10.1109/ICIEAM51226.2021.9446451
21. Bhatt, Sh., Santhanam, T.: Keystroke dynamics for biometric authentication–a survey. In: Proceedings of the 2013 International Conference on Pattern Recognition, Informatics and Mobile Engineering, PRIME 2013, Tamil Nadu, India, 21–22 February 2013. https://doi.org/10.1109/ICPRIME.2013.6496441
22. Bhattasali, T., Panasiuk, P., Saeed, K., Chaki, N., Chaki, R.: Modular logic of authentication using dynamic keystroke pattern analysis. In: Transl. on Numerical Analysis and Applied Mathematics, United States, pp. 1–4, June 2015. https://doi.org/10.1063/1.4951959
23. Zakharov, V., Shalagin, S., Eminov, B.: Algorithms for computing the sets of stochastic matrices with predefined properties, based on automaton models. In: International Scientific and Practical Conference on Mathematical Modeling, Programming and Applied Mathematics (MMPAM), Veliky Novgorod, Russia, 27–28 June 2019. https://doi.org/10.1088/1742-6596/1352/1/012062
24. Sulavko, A., Fedotov, A., Eremenko, A.: Users' identification through keystroke dynamics based on vibration parameters and keyboard pressure. In: Proceedings of XI International IEEE Scientific and Technical Conference "Dynamics of Systems, Mechanisms and Machines" (Dynamics), November 2017, Omsk, Russia, pp. 1–7 (2017). https://doi.org/10.1109/Dynamics.2017.8239514
25. Lozhnikov, P., Sulavko, A., Eremenko, A., Buraya, E.: Methods of generating key sequences based on keystroke dynamics. In: Proceedings of X International IEEE Scientific and Technical Conference "Dynamics of Systems, Mechanisms and Machines" (Dynamics), November 2016, Omsk, Russia, pp. 1–5 (2016). https://doi.org/10.1109/Dynamics.2016.7819038

Corporate Network for Organizing PDM and PLM Systems Using Wireless Technologies

N. I. Pikuleva, A. Sh. Khafizova, and A. F. Sultanov(✉)

Kazan National Research Technical University named after A.N. Tupolev – KAI, 10, K. Marx Street, Kazan, Tatarstan 420111, Russia
Sultan.of@ya.ru

Abstract. This article is devoted to the development of a solution for the organization of the wireless corporate network access to information databases, PDM and PLM systems in modern industrial enterprise, and the communication between employees within the same department and, if necessary, with the overall corporate network. The functional elements of this solution are completely domestic and have a certificate of compliance in the field of communications. The scientific novelty of this work is to formalize the optimization criteria and principles for the development of typical architecture and the structure of corporate networks using wireless technologies oriented to work with the production processes of high-tech engineering. The wireless corporate information and telecommunications network developed by me for organizing PDM and PLM systems outperforms the traditional wired network in mobility and flexibility of the workplace, payback, ease of installation, and deployment.

Keywords: Wireless technologies · Wi-Fi · Wireless access point · PDM · PLM · Design

1 Introduction

Currently, modern industrial enterprises are already actively using hybrid networks, which consist of wired and wireless technologies.

The subject of the study is the criteria for optimizing the architecture and technological principles of organizing a corporate information and telecommunications network for access to design and technological documentation at a modern industrial enterprise using wireless technologies. The object of research is a corporate wireless information and telecommunications network for access to design and technological documentation for organizing PDM and PLM systems at a modern industrial enterprise.

The master's thesis aims to develop a wireless corporate network in a modern industrial enterprise, which will provide access to information databases of PDM and PLM systems, as well as communication between employees within the same workshop and, if necessary, with a common corporate network. The network will have to provide fast data exchange and maintain an information transfer rate of at least 500 Mbps (for network access to electronic CAD documentation) within a network coverage area of 80,000 m^2

A. A. Radionov and V. R. Gasiyarov (Eds.): RusAutoCon 2022, LNEE 986, pp. 396–407, 2023.
https://doi.org/10.1007/978-3-031-22311-2_38

with about 500 users. In In addition, network elements should predominantly be domestic, for security and reliability. To achieve the goal of the master's thesis, it is necessary to solve the following tasks:

- An overview of wireless technologies for creating a communication network and data exchange in a modern industrial enterprise;
- Substantiation of the advantages of wireless technology for building a corporate information and telecommunications network in modern machine-building production;
- To conduct a comparative analysis of indicators of wired and wireless technologies for building a corporate network at a modern industrial enterprise;
- Explore typical network architecture;
- Develop a typical structural diagram of a wireless corporate network;
- Explore the functional nodes of the network and propose an implementation, taking into account the latest advances in technology;
- Formulate recommendations for designing and building a wireless corporate network;
- Develop the principles of modernization and consider the main directions for the development of information networks, about the corporate network of machine-building production.
- To formulate the basic principles and criteria for optimizing the construction of a corporate information and telecommunications network at a modern industrial enterprise;

The theoretical basis for solving the tasks was the following methods:

- Analysis of literature;
- Analysis of normative documents regulating the development and operation of wireless technologies;
- Study and generalization of domestic and foreign practice.

2 Overview of Wireless Technologies

2.1 Advantages

Currently, wireless technologies are the most relevant for connecting production units of large industrial enterprises that do not have an extensive cable infrastructure. It should be noted that it is more convenient to connect buildings on the territory of the enterprise with each other precisely by a wireless network than by leased communication lines [1].

The physical basis of a wireless network is radio waves. That is, data transmission occurs using radio channels of a certain frequency. The difference between them is only in the width and frequency of the wave. The predominant standard for wireless local area networks is the IEEE 802, standard, which regulates data transmission in the 2.4 and 5 GHz bands.

Wireless communication technologies are by far the most in-demand, so they have great prospects. In the future, people will be able to completely abandon wired communications and switch to wireless communications. And this is not only convenient but also profitable since the cable has a fairly high cost.

The main trends in the development of wireless technologies:

- Increase in speed and stability of data transmission at speeds above 10 Gbit/s;
- Increase in connection capacity and speed, reduction of authentication time;
- Introduction of the 5G standard with a data transfer rate of about 5 Gbit/s;
- Increase in Internet coverage;
- Development of Li-Fi technology, which is based on the transmission of information using light waves, the speed of such transmission will be about 10 Gbps [2];
- Growth of stability and reliability of the data transmission channel;
- Development of energy-saving policy [3];
- Guaranteed safety;
- Construction of self-organizing networks [4].

Wireless technologies and networks must evolve because the number of users who want to have wireless access to Internet services and information databases is growing. Ultimately, advances in the field of wireless technologies should help people easily manage complex modern production, master the accumulated knowledge most completely and rationally and free up the maximum amount of their time for further creativity and comprehensive development.

2.2 Comparative Analysis

According to the terms of reference, the corporate information and telecommunications network should be based on a wireless network. Wireless networks are not inferior to wired networks in data transfer speed, communication quality, and security, in addition, they are superior in mobility and installation speed. In turn, these parameters are decisive when choosing the design of a corporate network.

Based on Table 1 and terms of reference, Wi-Fi technology is the most suitable, as it satisfies the requirements for data transfer speed and range.

To organize a corporate network, commercial Wi-Fi is used, which operates at frequencies of 2.4 GHz and 5 GHz:

- 2.4 GHz frequency channels are used for Wi-Fi 802.11n, 802.11g, 802.11b standards
- 5 GHz is used for 802.11n, 802.11a, 802.11ac.

As can be seen from Table 2, the 802.11n and 802.11ac standards are more promising; moreover, they satisfy the conditions and requirements of the terms of reference for the development of a corporate information and telecommunications network for organizing PDM and PLM systems of modern engineering production. 802.11ac Wi-Fi technology has a whole range of advantages that will ensure rapid adaptation in our region, as it is currently happening around the world. For example, these are such aspects as: several times higher transmission speed (compared to the Wi-Fi 802.11n standard); Saving batteries of mobile and portable devices (2–4 times compared to the Wi-Fi 802.11n standard).

Table 1. Comparative table of wireless technologies [5].

Technology	Bandwidth	Radius of action	Frequency
Wi-Fi	up to 7 Gbps	up to 100 m	2.4–2.5 or 5 GHz
WiMax	up to 1 Gbps (WMAN), up to 100 Mbps (Mobile WMAN)	up to 150 km	1.5–14 GHz
Bluetooth	up to 24 Mbps	up to 100 m	2.4 GHz
UWB	up to 480 Mbps	up to 10 m	10.6 GHz

Table 2. Comparative table of IEEE 802.11 Wi-Fi standards [6].

Standard	Bandwidth	Radius of action	Frequency
802.11a	up to 54 Mbps	up to 120 m	5 GHz
802.11b	up to 11 Mbps	up to 140 m	2.4 GHz
802.11g	up to 54 Mbps	up to 140 m	2.4 GHz
802.11n	up to 600 Mbps	up to 250 m	2.4–2.5 GHz or 5 GHz
802.11ac	up to 7 Gbps	up to 100 m	5 GHz

3 Design

Currently, the design of a Wi-Fi network in a modern industrial enterprise is a very popular direction in design development. The Wi-Fi network is becoming an important tool in daily operations, for example, broadcasting the message of the head of the company to all employees.

3.1 Initial Data Analysis

General concept:

- There are internal segments of the Wi-Fi network (administrative premises);
- There are internal segments of the Wi-Fi network with high ceilings and production purposes (warehouse and production facilities);
- Location of the control unit (telecommunication room, server room, and distribution cabinets);
- Wi-Fi standard network control (wireless controller, specialized software);
- A general approach to the provision of services on the network and requirements for the network: for comfortable work with information databases of PDM and PLM systems, a speed of 100 Mbps per employee is sufficient. For communication between employees within the same workshop and, if necessary, with a common corporate network, a speed of 10 Mbps is sufficient;

- Security principles on Wi-Fi standard networks: division of users into groups and application of authentication methods for different groups, implementation of redundancy of the main network segments;
- The centralized multi-level architecture of building a LAN

Architecture and design features of the environment of the proposed network coverage area:

- Typical office space with plasterboard partitions;
- Warehouse and industrial premises with ceilings 10 m high and walls made of concrete and brick;
- Presence in warehouses and industrial premises of metal structures and equipment that are sources of interference;

Distribution density, number of users, as well as the used user equipment of the Wi-Fi network:

- The deployment density of the radio system depends on the distribution of users, the concentration of users, and the services provided;
- According to the terms of reference, the total number of network users is expected to be about 500 people;
- Zone of maximum concentration of users
- These are administrative premises;
- The percentage of active users out of the total is 50%;
- The percentage of the expected increase in the number of network users in the foreseeable future up to 25%;
- Corporate tablets, laptops, and PCs are used on the network.

Necessary equipment for organizing a Wi-Fi standard wireless network in a modern industrial enterprise:

- In administrative premises, internal office-type access points with integrated antennas are used, in the warehouse and industrial premises, access points with external antennas are used;
- Specialized switches are used to connect access points;
- A wireless network controller is used to manage the network;
- To work with databases and store information, an application server and a database server are used.

3.2 Selecting Wireless Access Points

One of the main components of a corporate wireless network is an access point. In this paper, only enterprise-class access points are considered.

After analyzing foreign and domestic companies in the field of development and production of network equipment, several companies were identified that manufacture network equipment for organizing a wireless corporate information and telecommunications network.

- Cisco
- Ruckus Wireless
- LLC "Enterprise" "Eltex"
- Russian company QTECH

A distinctive feature of wireless access points of foreign companies is the support of the 802.11ac Wave 2 standard, which allows you to use the MU-MIMO multi-user mode of multi-channel reception and transmission. Domestic companies plan to implement support for this standard in their access points by 2018, justifying this by the fact that at present there are few devices and equipment on the Russian market that support the full operation of devices based on the 802.11ac Wave 2 standard. Thanks to the MU-MIMO mode of the access point foreign companies can simultaneously exchange data with several users at once. The use of the MU-MIMO mode means an increase in network bandwidth by 2–3 times with an indispensable increase in the speed of service for all connected devices, which is of decisive importance in the face of a constant increase in information traffic. Some foreign access points, such as the Cisco Aironet 3800 series access point, support data transfer rates up to 2.6 Gb/s per 5 GHz radio module, in addition, this access point supports dual 5 GHz operation, which allows you to increase the transmission speed data up to 5.2 Gbps. Comparing, in general, the wireless access points of foreign and domestic companies, it can be seen that the access points of domestic companies are inferior to foreign ones in terms of functionality and transmission speed. The main criteria for evaluating wireless access points are the latest IEEE 802.11ac standard, PoE technology, simultaneous operation in two bands of 2.4 GHz and 5 GHz, as well as a technological design that allows you to install access points in industrial and office premises. In addition, according to the terms of reference, the network elements should mainly be domestic, so you need to choose among the Russian companies QTECH and ELTEX.

Fig. 1. QTECH QWP-67-AC indoor access point with integrated antennas [7].

In this paper, we use QTECH wireless access points Fig. 1 and Fig. 2, because this company has a full-fledged set of wireless access equipment necessary to build a high-quality corporate network.

The QTECH QWP-67-AC is a dual-band, enterprise-grade 802.11ac access point. It is specifically designed for use in high-density user environments such as offices, universities, hotels, and hospitals.

The QTECH QWO-63-AC is an 802.11ac dual-band access point specifically designed for use by organizations in industrial and harsh outdoor environments. To

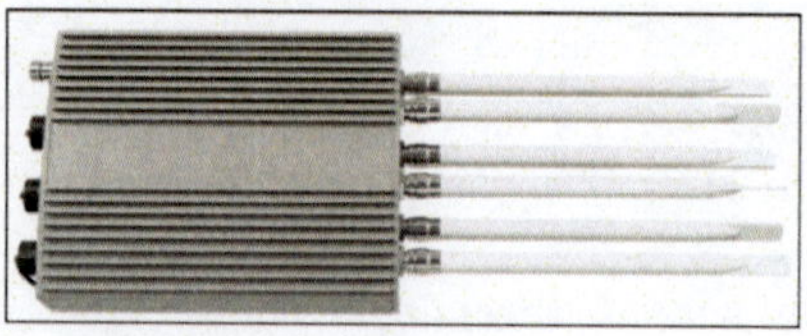

Fig. 2. QTECH QWO-63-AC access point with external antennas [8].

choose the right switches and controller, it is necessary to make an estimated network planning and calculate the number of access points [9, 10] (Fig. 3).

3.3 Estimated Planning

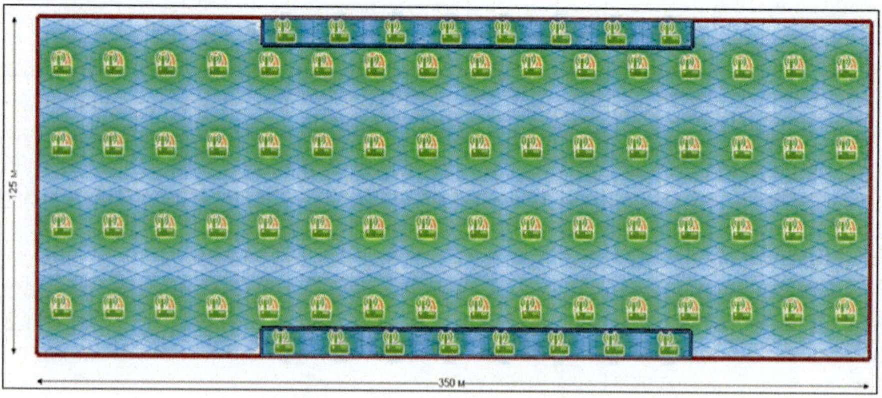

Fig. 3. Wi-Fi network coverage map.

Estimated planning was carried out and a Wi-Fi network coverage map was designed in the 5 GHz frequency range for the entire area of modern engineering production. Wireless access points. Access points are distributed in such a way that their coverage area overlaps with the coverage area of neighboring access points by 10–15%, in addition, access points have a power reserve of 20–30%. This is done for the case when it is necessary to quickly tighten the "hole in the coverage" with neighboring access points when one fails [11, 12].

The ring topology is easier to install, deploy and operate the network. The use of the "ring" topology in conjunction with redundancy protocols will be considered the best, more reliable, and economical option for organizing a corporate information and telecommunications network, since the same fiber is used as a reserve as in the working state but in the opposite direction. To redundant communication lines and network traffic, we use a centralized multi-layer architecture in conjunction with a physical "ring" topology based on the MRP (Multi-layer Ring Protection) protocol – a standardized redundant ring. The ring topology is convenient, primarily because it achieves a certain and guaranteed time to restore communication after a failure, taking into account the number of switches in the ring [13].

4 Optimization Criteria and Development Principles

Wireless Technology Selection Criteria:

- Of all available wireless technologies, Wi-Fi technology is mostly focused on organizing a wireless corporate network, since the IEEE 802.11 standard is supported by all portable personal devices. The latest standards in the IEEE 802.11 family are IEEE 802.11n and IEEE 802.11ac. These standards have a high data transfer rate of up to 7 Gbps and a large coverage area of about 100 m.

Criteria for selecting wireless access points, access points must support:

- IEEE 802.11n and IEEE 802.11ac Wave 1 standard (at present, there are few devices and equipment on the Russian market that support the full operation of devices based on the 802.11ac Wave 2 standard);
- SU-MIMO 3 × 3:3 technology, i.e., 3 antennas for reception and 3 antennas for transmission (at present, there are few devices and equipment on the Russian market that support the full operation of devices with 4 × 4:4 MIMO technology);
- Simultaneous operation in two bands 2.4 GHz and 5 GHz;
- Dynamic bandwidth selection technology from 80 MHz and higher;
- 1000BASE-T interface;
- Centralized control via the controller;
- Powered by Power of Ethernet (PoE) technology;
- Installation of external antennas for using access points in rooms with high ceilings;
- Wide temperature range, for example, from −20 °C to +55 °C for installation in the warehouse and industrial premises;
- Traffic prioritization technology to ensure that bandwidth-intensive applications such as 4K video streaming work flawlessly while meeting stringent voice over data service quality requirements, and time-critical services.

When planning and designing a corporate information and telecommunications network using wireless technologies, the following should be taken into account:

- Architecture and design features of the environment of the proposed network coverage area (plasterboard partitions, concrete and brick walls, doors, windows, ceiling height, the presence of metal structures and equipment that are sources of interference, etc.);
- Zones of maximum concentration of users, the number of active users, the expected increase in users, as well as the used user equipment of the Wi-Fi network and the services provided;
- The corporate network should be the multi-service transmission of all types of traffic should occur through single communication channels;
- The corporate network should be built based on open standards and interfaces to provide the possibility of expanding the network and combining it with other networks;
- Access points are placed in such a way that their coverage area overlaps with the coverage area of neighboring access points by 10–15%, in addition, access points have a power margin of 20–30%. This is done for the case when it is necessary to

quickly tighten the "hole in the coverage" with neighboring access points when one fails;

- Estimated planning and design of a Wi-Fi network coverage map will be performed using specialized software designed for designing wireless networks. The software must have the following functionality:
- The ability to set the scale of the room map to estimate the size of the floor plan;
- The ability to automatically place wireless access points with the function of editing their placement;
- The ability to add an external antenna with gain control;
- The ability to change the parameters of the wireless access point (transmission power, channel selection, range, standard, minimum sensitivity, etc.)
- The ability to specify the coverage area of the wireless network and the area in which access points will not be placed;
- The ability to mark obstacles taking into account the attenuation of the material from which they are made: walls or doors, as well as indicate special areas, for example, isolated office space or a warehouse.
- The possibility of obtaining a two-dimensional color map that demonstrates the theoretical range of the wireless network. This will help the planner design the wireless network more accurately, before its physical construction and active radio survey.

Principles for the development of a structural and wiring diagram of a corporate information and telecommunications network:

- The use of a centralized multi-level architecture based on the principles of hierarchy and modularity as an information network topology that describes the direction of information flows transmitted over the network and includes four levels: the network core, the aggregation level, the access level, and the server level (server farm);
- The use of the physical topology "ring" in conjunction with the MRP (Multi-layer Ring Protection) redundancy protocol to quickly restore communication after a failure, approximately 50 ms;
- The optimal bandwidth for connecting access switches with an aggregation switch and building a backbone communication line, taking into account the future development of wireless technologies and network expansion, is 10 Gbit/s (with 5 Gbit/s of active expected traffic);
- The optimal bandwidth for connecting access switches to wireless access points is 1 Gbps;

Criteria for choosing an access switch and an aggregation switch, switches must support:

- Industry-standard modular compact SFP and SFP + transceivers supporting data rates of 1 Gb/s and 10 Gb/s;
- Router functions (RIP), for the aggregation switch, the routing functions are more advanced (RIP, OSPF, etc.);
- Protection and redundancy protocols (MRP, G.8032, MSTP)
- PoE (Power Over Ethernet) technology.

Wireless controller selection criteria, the controller must support:

- Management of the required number of wireless access points;
- The function of automatic discovery of access points with the subsequent configuration;
- User authentication function;
- The ability to assign policies;
- Traffic shaping control;
- Firewall function;
- Centralized management of the entire wireless network infrastructure of the organization;
- Compatibility with access and aggregation switches.

5 Conclusion and Further Work

In this work, a solution was developed for organizing a wireless corporate network in a modern industrial enterprise, which will provide access to information databases of PDM and PLM systems, as well as communication between employees within the same workshop and, if necessary, with a common corporate network.

The network provides fast data exchange and maintains the data transfer rate necessary to implement network access to electronic CAD documentation within a network coverage area of 80,000 m^2. The functional elements of this solution are completely domestically produced and have a certificate of conformity in the field of communications. Optimization criteria and principles for developing a typical architecture and design structure of a corporate information and telecommunications network for organizing PDM and PLM systems of modern engineering production using wireless technologies are formulated:

- Wireless technology selection criteria;
- Criteria for selecting a wireless network standard;
- Criteria for selecting functional elements of a wireless corporate network;
- Principles of planning and designing a corporate information and telecommunications network using wireless technologies;
- Methods for estimating planning and designing a network coverage map;
- Principles of choosing the architecture for building a local area network using wireless solutions;
- Methods for calculating the optimal network throughput and building a backbone communication line, taking into account the future development of wireless technologies and network expansion;
- Methods and principles of organization of network redundancy.

The wireless corporate information and telecommunications network developed by me for organizing PDM and PLM systems surpasses the traditional wired network in mobility and flexibility of the workplace, payback, ease of installation, and deployment. This solution can be used as a guide and recommendation for the design and planning of a corporate information and telecommunications network for access to design and technological documentation for organizing PDM and PLM systems in a modern industrial enterprise. This solution will be in demand, because the installation of a wireless network in an enterprise will help to combine all processes into one workflow and make it more convenient, optimizing organizational and production processes in the organization. In addition, the use of wireless technologies in the organization of a corporate information and telecommunications network will increase labor efficiency through more efficient use of labor, office space, and the latest technologies.

References

1. Kalfas, G., Pleros, N., Alonso, L., Verikoukis, C.: Network planning for 802.11ad and MT-MAC 60 GHz fiber-wireless gigabit wireless local area networks over passive optical networks. J. Opt. Commun. Netw. **8**(4), 206–220 (2016). https://doi.org/10.1364/JOCN.8.000206
2. Xu, L., Xie, J., Xu, X., Wang, S.: Enterprise LTE and WiFi interworking system and a proposed network selection solution. In: 2016 ACM/IEEE Symposium on Architectures for Networking and Communications Systems (ANCS), pp. 137–138 (2016). https://doi.org/10.1145/2881025.2889476
3. Coronado, E., Villalon, J., Garrido, A.: Wi–balance: SDN–based load–balancing in enterprise WLANs. In: 2017 IEEE Conference on Network Softwarization (NetSoft), pp. 1–2 (2017). https://doi.org/10.1109/NETSOFT.2017.8004257
4. Li, Z., Li, X., Chen, L., Yu, S., Xiao, N.: Localization system of underground mine trackless facilities based on wireless sensor networks. In: 2008 IEEE International Conference on Mechatronics and Automation, pp. 347–351 (2018). https://doi.org/10.1109/ICMA.2008.4798778
5. Hu, Guan, Y.L., Law, C.L., Yan, L.S., Li. S.: Development and performance evaluation of mobile WiMAX testbed. In: 2007 IEEE Mobile WiMAX Symposium, pp. 104–107. https://doi.org/10.1109/WIMAX.2007.348688
6. Gibadullin, R.F., Nikonorov, V.V.: Development of the system for automated incident management based on open-source software. In: 2021 International Russian Automation Conference (RusAutoCon), pp. 521–525 (2021). https://doi.org/10.1109/RusAutoCon52004.2021.9537385
7. Raikhlin, V.A., et al.: Reliable recognition of masked cartographic scenes during transmission over the network. In: 2016 International Siberian Conference on Control and Communications (SIBCON), pp. 1–5 (2016). https://doi.org/10.1109/SIBCON.2016.7491657
8. Vershinin, I.S., et al.: Associative steganography. Durability of associative protection of information. Lobachevskii J. Math. **41**, 440–450 (2020). https://doi.org/10.1134/S1995080220030191
9. Gibadullin, R.F., Zakirov, R.R.: Mobile application for neural network analysis of human functional state. In: Radionov, A.A., Gasiyarov, V.R. (eds.) RusAutoCon 2020. LNEE, vol. 729, pp. 745–755. Springer, Cham (2021). https://doi.org/10.1007/978-3-030-71119-1_73
10. Raikhlin, V.A., et al.: Reliable recognition of masked binary matrices. Connection to information security in map systems. Lobachevskii J. Math. **34**, 319–325 (2013). https://doi.org/10.1134/S1995080213040112
11. Raikhlin, V.A., et al.: The elements of associative stegnanography theory. Mosc. Univ. Comput. Math. Cybern. **43**(1), 40–46 (2019)

12. Vershinin, I.S., Gibadullin, R.F., Pystogov, S.V., Raikhlin, V.A.: Associative steganography of text messages. Mosc. Univ. Comput. Math. Cybern. **45**(1), 1–11 (2021). https://doi.org/10.3103/S0278641921010076
13. Gibadullin, R.F., Perukhin, M.Y., Ilin, A.V.: Speech recognition and machine translation using neural networks. In: 2021 International Conference on Industrial Engineering, Applications and Manufacturing (ICIEAM), pp. 398–403 (2021). https://doi.org/10.1109/ICIEAM51226.2021.9446474

Diagnostics and Reliability of Automatic Control Systems

Development of Weighting Coefficients for Assessing the Quality of Security Measures Implementation for Significant Objects of Critical Information Infrastructure

A. Sitskaya[1(✉)], V. Selifanov[1], and M. Purgina[2]

[1] Siberian State University of Geosystems and Technologies, 10, Plahotny Street, Novosibirsk 630108, Russia
AnSits@yandex.ru

[2] Novosibirsk State University of Economics and Management, 56, Kamenskaya Street, Novosibirsk 630099, Russia

Abstract. One of the main tasks of the state is the proper functioning of internal systems that provide a decent standard of living for citizens. But what actually ensures the functioning of the activities of the state in different areas? Critical information infrastructures and their significant objects, as well as telecommunication networks used to organize the interaction of such objects, have become one of the main support systems. Order number 239 of Federal Service of Technical and Export Control (FSTEC) contains protection measures, but does not describe the methods for assessing the effectiveness of the protection of significant objects of critical information infrastructure (SO CII), which in turn casts doubt on the quality of the assessment of the built protection systems. This article discusses the ranking of threats and the development of a method for evaluating the effectiveness of the built protection system based on weighting factors. The developed method is based on the methodology for assessing the level of security protection of the information system of the Central Bank of Russia.

Keywords: Critical infrastructure · Significant objects · Information security · Weighting coefficients · Ranking · Security assessment methodology of information systems

1 Introduction

2021 was characterized by the beginning of scheduled inspections significant objects of critical information infrastructure (SO CII). A little more than three years have passed since the beginning of the federal law No. 187 «On the Security of the Critical Information Infrastructure of Russian Federation» and entering information about the first objects in the relevant register [1].

In accordance with the Decree of Russian Federation Government no. 162 [2], the subject of a scheduled inspection is the compliance of security requirements by the subject of critical information infrastructure, that is, as far as possible, the stable functioning of SO CII during computer attacks against him.

A. A. Radionov and V. R. Gasiyarov (Eds.): RusAutoCon 2022, LNEE 986, pp. 411–420, 2023.
https://doi.org/10.1007/978-3-031-22311-2_39

Thus, in the course of control, it is necessary to establish that a hypothetical intruder will not be able to make unauthorized access and/or impacts that lead to an unacceptable risk for the critical information infrastructure subject.

That is, there are three main control issues:

- assessment of the correctness of determining the unacceptable risk for the subject CII - category CII;
- assessment of the correctness of the definition of the intruder and his capabilities (intruder model) and actual threats - threat models;
- assessment of the compliance of the implemented organizational and technical protection measures with the violator's capabilities to carry out computer attacks.

At the same time, the order of FSTEC in Russia №. 127 established the provisions of regulatory legal acts, the implementation of which is subject of verification. Their analysis shows that there are only two gradations: corresponds and does not correspond [3]. And no criteria are given by which it is possible to judge not only the official implementation or non-compliance with the provisions of Russian Federation Government decrees and the orders of the regulator, but it is possible to draw a conclusion about the possibility or impossibility of the offender committing actions leading to unacceptable risks (consequences).

An analysis of practice of FSTEC control activities in Russia in related areas (monitoring of information security in state information systems, state personal data information systems, key information infrastructure systems) and scientific research (publications) in these areas showed a similar picture.

2 Compliance Assessment Methods of the Central Bank's of Russia Measure

The documents of the Central Bank of Russian Federation, interstate standard GOST R 57580 [4, 5], have a strict evaluation system that contains qualitative and quantitative assessments, unlike the documents of FSTEC in Russia.

In 2017 and 2018, the Central Bank of Russia developed a series of GOSTs 57580 [4, 5]. GOST R 57580.1 - 2017 which contains the basic composition of organizational and technical measures related to information security [4]. This set of measures turned out to be flexible and adaptable to any information systems. But GOST R 57580.2 was no less important development, which contains methods for assessing the effectiveness of protection, according to the selected measures of GOST 57580.1 [4, 5]. The assessment method in use proved to be simple and effective, as evidenced by the effective protection of banks information system. That is why the method of assessing the Central Bank of Russia should be taken as the basis for the method for assessing compliance with measures of the 239 decree of FSTEC in Russia [6].

The Central Bank's of Russia method is based on weight coefficients, which, in turn, reflect the degree of significance of a particular characteristic in relation to others. It is the weighting factor that has the greatest influence on the final result. The weighting factor can be determined in many ways, but in the case of CII, it is advisable to turn to the graph method. It is the graph that most clearly and consistently reflects the relationship of objects. After its construction, it will be possible to describe a matrix that shows the connections of elements in numerical form.

3 Building a Measure Ranking Graph

Decree №. 239 of FSTEC in Russia contains 17 groups of measures that regulate both the technical part and the organizational part. Let's build a graph related to the security audit of SO CII (Fig. 1) [6]:

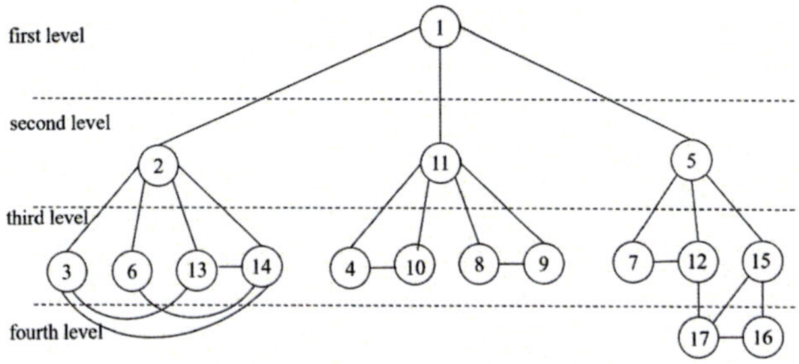

Fig. 1. SO CII security audit graph.

The nodes of the graph are designated in accordance with the groups of measures 239 order:

1. Identification and authentication;
2. Access control;
3. Restriction of the software environment;
4. Protection of machine storage media;
5. Security audit;
6. Antivirus protection;
7. Prevention of intrusions (computer attacks);
8. Ensuring integrity;
9. Ensuring accessibility;
10. Protection of technical means and systems;
11. Protection of the information (automated) system and its components;
12. Response to computer incidents;
13. Configuration management;
14. Management of software updates;
15. Planning of measures to ensure security;
16. Ensuring actions in emergency situations;
17. Information and training of personnel.

The graph is built on the basis of the analysis and systematization of the SANS CIS Controls 8 rating [1] and the analysis of thematic literature [7–13], but the greatest contribution to the ranking of measures was made by the practical experience of the FSTEC of Russia in the Siberian Federal District in conducting inspections of significant CII objects.

In the column, the measures are divided into levels according to their weight in the protection system, as well as the sequence of verification. The graph clearly demonstrates all possible connections between groups of measures, which, in turn, are necessary when building an effective protection system. Three branches of the graph are equivalent to each other, because talk about three parallel checks.

There are two main levels in the graph: the first and the second. The first level of the graph contains one node - a group of measures that relates to identification, authentication. This group of measures stands above the rest, because it is the most significant. If identification and authentication are not organized, then absolutely any user, intruder will have free access to CII, which, in turn, makes further protection absolutely meaningless and ineffective.

The second level of the graph includes three nodes: access control, security audit, protection of the information system and its components. This level has a close relationship with subsequent ones, to some extent it can be said that it generalizes all subsequent groups of measures related to a particular node. The second node, or access control node, contains groups of measures directly related to the organization of the technical part, namely the operation system and software, node five, or the node of the security audit, in one way or another contains, for the most part, organizational measures related to the detection of threats and response in accordance with them, as well as personnel training. The eleventh node, or the node of the protection of the information system and its components, also contains a mixture of technical and organizational parts. The top combines groups of measures related to ensuring the safety of equipment and access to it, and the information system itself.

4 Building a Measure Ranking Matrix

We have built a matrix based on graph for most clear demonstration the connections of the graph elements (Fig. 2):

All connections in the matrix between elements are two-way, except for the connection of the first and second element. The connection is carried out from the first element (identification/authentication) to the second element (access control), this is due to the fact that the very first step both in checking and when logging into the system is precisely identification and authentication, and only then comes the access control process, which in turn consists in determining the access rights of the object after receiving the identification code, which is why the relationship between the first and second elements is one-way.

5 Weighting Coefficients

One of the most important audit processes is the determination of information system security assessment. But for CII objects there are no criteria and requirements for determining assessment, unlike in GOST R 57580.2 [14, 15].

	1	2	3	4	5	6	7	8	9	10	11	12	13	14	15	16	17
1	0	1	0	0	1	0	0	0	0	0	1	0	0	0	0	0	0
2	1	0	1	0	0	1	0	0	0	0	0	0	1	1	0	0	0
3	0	1	0	0	0	0	0	0	0	0	0	0	1	1	0	0	0
4	0	0	0	0	0	0	0	0	0	1	1	0	0	0	0	0	0
5	1	0	0	0	0	0	1	0	0	0	0	1	0	0	1	0	0
6	0	1	0	0	0	0	0	0	0	0	0	0	0	1	0	0	0
7	0	0	0	0	1	0	0	0	0	0	0	1	0	0	0	0	0
8	0	0	0	0	0	0	0	0	1	0	1	0	0	0	0	0	0
9	0	0	0	0	0	0	0	1	0	0	1	0	0	0	0	0	0
10	0	0	0	1	0	0	0	0	0	0	1	0	0	0	0	0	0
11	1	0	0	1	0	0	0	1	1	1	0	0	0	0	0	0	0
12	0	0	0	0	1	0	1	0	0	0	0	0	0	0	0	0	1
13	0	1	1	0	0	0	0	0	0	0	0	0	0	1	0	0	0
14	0	1	1	0	0	1	0	0	0	0	0	0	1	0	0	0	0
15	0	0	0	0	1	0	0	0	0	0	0	0	0	0	0	1	1
16	0	0	0	0	0	0	0	0	0	0	0	0	0	0	1	0	1
17	0	0	0	0	0	0	0	0	0	0	0	1	0	0	1	1	0

Fig. 2. Graph matrix.

A method of effectiveness evaluating of security measures compliance was developed according to order №239 of FSTEC in Russia, based on the approach of the quantitative assessment calculating method of the Central Bank of Russia [6].

It is necessary first of all to rank all the measures to calculate the weight coefficients. The method of developing weighting coefficients is partially based on the method described in the article [16]. The ranking was partially done during the graph and matrix building, but it requires destabilization. 3 branches are explicitly represented in the graph in Fig. 1, but in fact there are 4 of them. The first branch consists of elements from 1 to 17, the second – 2, 3, 6, 13, 14 elements, third – 11, 4, 10, 8, 9 elements, fourth – 5, 7, 12, 15, 16, 17 elements. As you can see, the first branch includes all the elements, this is due to the fact that without identification and authentication, information system protection cannot be effective, because it becomes open to anyone. That is why it can be considered a separate branch on which all other elements of the graph depend.

To calculate the weight coefficient of each branch of the graph r_i, it is necessary to find out the sum of the values of the elements of each branch of the graph hi (Table 1).

The value of the entire graph is 92.

The Table 1 shows that the value of the first branch is 49, the second is 15, the third is 12, and the fourth is 16.

Next, we calculate the weighting coefficient of each branch using formula 1:

$$r_i = \frac{h_i}{\sum_{i=1}^{n} h_i} \tag{1}$$

where n – amount of branches in the graph, i – the number of the branch of the graph.

The results are presented in Table 2.

The obtained weighting coefficients indicate that the measures regulating identification/authentication have the greatest weight in the evaluation of information system.

Table 1. The values of each graph branch.

1 branch		2 branch		3 branch		4 branch	
Element	Value	Element	Value	Element	Value	Element	Value
1	3	2	3	4	2	5	3
2	4	3	3	8	2	7	2
3	3	6	2	9	2	12	3
4	2	13	3	10	2	15	3
5	4	14	4	11	4	16	2
6	2	–	–	–	–	17	3
7	2	–	–	–	–	–	–
8	2	–	–	–	–	–	–
9	2	–	–	–	–	–	–
10	2	–	–	–	–	–	–
11	5	–	–	–	–	–	–
12	3	–	–	–	–	–	–
13	3	–	–	–	–	–	–
14	4	–	–	–	–	–	–
15	3	–	–	–	–	–	–
16	2	–	–	–	–	–	–
17	3	–	–	–	–	–	–
Total:	49	Total:	15	Total:	12	Total:	16

Table 2. Weight of each branch in the graph.

	Branch 1	Branch 2	Branch 3	Branch 4
Weight	0.5	0.2	0.1	0.2

6 Measure Conformity Assessments

In order to use weighting coefficients when calculating the evaluation of the effectiveness of the measures implemented to protect the system, it is first necessary to evaluate these measures.

An effective evaluating measures system is offered by mentioned above GOST R 57580.2, which proposes the following [5]:

- a score of 1 is given if the measure is fully implemented;
- a score of 0.5 is given if the measure is not fully implemented;
- a score of 0 is given if the measure is not implemented.

But these estimates are valid for the main group of measures, however, there are measures, non-compliance with which can be critical, for example, measures related to identification and authentication, etc. In this case, it is rational to use the following estimates:

- a score of 1 is given if the measure is fully implemented;
- a score of 0 is given if the measure is not implemented or only partially implemented.

These groups of measures are the first and second levels of the graph, i.e., measures related to identification/ authentication, access control, security auditing and protection of information system and its components.

It is also necessary to remember that each system is individual and requires its own, different approach. Therefore, a common reference protection system does not exist, which means that each protection system needs its own ideal. Such an ideal is a protection system in which all developed protection measures are fully implemented.

Next, you need to calculate the score values for each of the measure groups:

$$E_k = \frac{\sum_{i=1}^{n} M_i}{N} \tag{2}$$

where Ek -the numerical value of the intermediate assessment of the groups of measures selected by the organization; Mi – evaluation of i- measure; N - the number of measures required in the protection system; n -the number of implemented measures.

After receiving intermediate assessments of the implemented groups of measures, it is necessary to calculate the effective assessment of information system security. To do this, it is necessary to calculate intermediate values for each branch of the graph (formula 3):

$$E_{Bk} = \sum_{i=1}^{n} E_i \tag{3}$$

where EBk - numerical value of the intermediate estimate of the selected branch of measures; Ei - intermediate score of i-group of measure relating to identification and authentication; n -the number of measure groups in branch.

In order to obtain a weighted assessment of the effectiveness of information system protection, weighting factors are used. There are 4 branches in the protection system, therefore, in order for the assessment to be objective and complete, it is necessary to calculate the arithmetic mean of all the assessments received. Based on the foregoing, formula 4 is obtained:

$$E = \frac{0.5 \times E_{B1} + 0.2 \times E_{B2} + 0.2 \times E_{B3} + 0.1 \times E_{B4}}{4} \tag{4}$$

where E - weighted assessment of the measures chosen by the organization; EB1 - intermediate assessment of groups of measures of the 1st branch; EB2 - intermediate assessment of groups of measures of the 2nd branch; EB3 - intermediate assessment of groups of measures of the 3rd branch; EB4 - intermediate assessment of groups of measures of the 4th branch.

Using formula 4, a weighted estimate is obtained, because each group of measures is ranked according to its weight (value) in the implementation of the IP protection system.

Having received an assessment of measures effectiveness taken, it is necessary to understand at what numerical value the developed protection system is effective. This requires criteria, or a range of efficiency values, which are not currently available in any document of FSTEC of Russia, but are prescribed in the documents of the Central Bank of Russian Federation, namely GOST R 57580.2 [5].

The information systems of Russian banks are of great value to the state, since violations of their functioning can lead to the collapse of the country's financial system and, as a result, a difficult situation in the state, described in more detail in the article [17]. This suggests that the criteria for their assessment are the most stringent, which in turn is also necessary for SO CII, because a violation of their functioning can lead to significant "negative social, political, economic, environmental consequences, consequences for the country's defense, state security and law and order" [18], which means that it is advisable to apply the criteria of the Central Bank of Russian Federation as criteria for assessing SO CII. However, in contrast to the categories of SO CII, the Central Bank has developed five levels of compliance, presented in Table 3.

Table 3. Qualitative assessment of security information system processes compliance level.

Evaluation	Compliance level
$E = 0$	zero
$0 < E \leq 0.5$	first
$0.5 < E \leq 0.7$	second
$0.7 < E \leq 0.85$	third
$0.85 < E \leq 0.9$	forth
$0.9 < E \leq 1$	fifth

Thus, it becomes necessary to adapt the compliance levels of protection developed by the Bank of Russia under CA CII. At the moment it is advisable for CA CII to focus on the three highest levels of compliance: the fifth, fourth and third.

7 Conclusions

This article presents the development of information system effectiveness evaluating method of CA CII. Weighting coefficients have been developed based on groups measures ranking of the order №. 239 of FSTEC in Russia [17].

A quality quantitative assessment of information system protection is calculated on a mentioned above weighting coefficients.

A correctly calculated information system protection assessment allows timely detect the vulnerabilities in the built security system. Early vulnerabilities detection allows resolve them in time, which means to ensure the necessary level of SO CII security.

References

1. Federal Law of 26 July 2017 N 187-FZ. On the Security of the Critical Information Infrastructure of Russian Federation of 19 June, 2017. http://www.consultant.ru/document/cons_doc_LAW_220885/
2. Decree of Government of Russian Federation of February 17, 2018 N 162. On approval of the Rules for the implementation of state control in the field of ensuring the security of significant objects of critical information infrastructure of the Russian Federation
3. Decree of the FSTEC of Russia dated 26.06.2019 No. 127. On Approval of the Review of the Law Enforcement Practice of the FSTEC of Russia in the Framework of monitoring Compliance by Russian Participants in Foreign Economic Activity with the Legislation of the Russian Federation in the field of export control for 2018. https://www.alta.ru/tamdoc/19a00127/
4. GOST R 57580.1-2017 Security of financial (banking) transactions. Protection of information of financial organizations. Basic composition of organizational and technical measures. input. 1 Jan. 2018. http://docs.cntd.ru/document/1200146534
5. GOST R 57580.2-2018 Security of financial (banking) transactions. Protection of information of financial organizations. Conformity assessment methodology. input. 1 Sept. 2018. http://docs.cntd.ru/document/1200158801
6. Decree of the FSTEC of Russia dated December 25, 2017 N 239. Federal Service for Technical and Export Control. On approval of the requirements for ensuring the security of significant objects of the critical information infrastructure of the Russian Federation/DIALOG: https://fstec.ru/en/53-normotvorcheskaya/akty/prikazy/1592-prikaz-fstek-rossii-ot-25-dekabrya-2017-g-n-239
7. Vasiliev, V.I., Khisamutdinov, T.Z., Krasko, A.S., Matveev, P.V.: Expert decision support system in the process of information security audit. Inf. Counteraction Terror. Threat. **4**, 98–102 (2005)
8. Vantseva, I.O., Zyryanova, T.Y., Medvedeva, O.O.: The influence of the federal law "on the security of the critical information infrastructure of the Russian Federation" on the owners of critical information infrastructures. Bull. Ural. Fed. District. Secur. Inf. Sphere **1**(27), 71–76 (2018)
9. Belyakov, M.I.: Model of the process of functioning of the system for ensuring the information security of an object of critical information infrastructure in the task of evaluating its effectiveness. Questions of defense technology. Series 16: Technical Means of Countering Terrorism, vol. 11–12, no. 149–150, pp. 71–75 (2020)
10. Elekova, A.S.: Development of the regulatory framework in the field of security in critical information infrastructures. In: MNSK-2017: Information technologies: Proceedings of the 55th International Scientific Student Conference, Novosibirsk, April 17–20 2017, Novosibirsk National Research State University, Novosibirsk, p. 36 (2017)
11. Krasnov, A.E., Mosolov, A.S., Feoktistova, N.A.: Assessing the sustainability of critical information infrastructures to information security threats. Secur. Inf. Technol. **28**(1), 106–120 (2021). https://doi.org/10.26583/bit.2021.1.09
12. Averchenkov, V.I., Rytov, M.Y., Kuvylkin, A.V., Rudanovsky, M.V.: Information Security Audit of Executive Authorities: Textbook. Limited Liability Company "FLINTA", Moscow, p. 100 (2011)

13. GOST R ISO/IEC TO 18044-2007 Information technology (IT). Methods and means of ensuring security. Information Security Incident Management (Reissue). http://docs.cntd.ru/document/1200068822
14. Sitskaya, A.V., Tabakaeva, V.A., Selifanov, V.V.: Issues of automation of the audit in accordance with GOST R 57580.2-2018. Interexpo Geo-Siberia **6**, 268–275 (2021). https://doi.org/10.33764/2618-981X-2021-6-
15. Zavedenskaya, A.A., Zyryanova, T.Y.: Correlation of the information protection measures specified in GOST R 57580.1-2017 with the protection measures from the order of the FSTEC of Russia dated February 18, 2013 No. 21. Security of the information space: Proceedings of the XIX All-Russian scientific and practical conference of students, graduate students and young scientists, Yekaterinburg, December 08–11, 2020. Ural State University of Economics, Yekaterinburg, pp. 93–97 (2021)
16. Grigoriev, A.V., Kozin, P.A., Ostapchuk, A.V.: Methods for determining the values of weight coefficients. Property relations in Russian Federation, no. 8. https://cyberleninka.ru/article/n/metodika-opredeleniya-znacheniy-vesovyh-koeffitsientov (2004)
17. Zabegalin, E.V.: Logical model of activities for complex technical diagnostics of information security of organizations and significant objects of critical information infrastructure. Control. Syst. Commun. Secur. **3**, 145–178 (2019). https://doi.org/10.24411/2410-9916-2019-10308
18. Federal Law "On information, information technologies and information protection" 27 July 2006 N 149-FZ (last edition). http://www.consultant.ru/document/cons_doc_LAW_61798/

Development of Electric Drive Control Scheme for Biogas Plant

I. V. Bochkarev[1], R. R. Khramshin[2(✉)], A. R. Sandybaeva[1], and L. A. Usacheva[1]

[1] Kyrgyz State Technical University Named After I. Razzakov, 66, Ch. Ajtmatov Avenue, Bishkek 720044, Kyrgyzstan

[2] Nosov Magnitogorsk State Technical University, 38, Lenin Avenue, Magnitogorsk 455000, Russia

hramshinrr@mail.ru

Abstract. At present, biogas plants are widely used to process organic waste materials and to produce biogas from organic wastes. To carry out various technological operations, such biogas plants contain a lot of individual electric drives based on asynchronous motors. The efficiency and reliability of the biogas plant operation are obviously determined by the asynchronous motor operational reliability. It should be taken into account that when an asynchronous motor operates as a part of a biogas plant, and on the whole, when electrical equipment is used in farming, there are certain specific factors concerning both operation conditions and maintenance characteristics. It is shown that most of the factors mentioned above result in stator winding overheating and, consequently, to the electric motor failure, which, in its turn, causes biogas plant malfunction. This requires asynchronous motor thermal state monitoring directly during operating modes and this monitoring makes it possible to provide well-timed thermal protection of the asynchronous motor. The analysis shows that as small dimension low-power asynchronous motors are applied in biogas plant electric drives, it is reasonable to carry out the biogas plant electric drive thermal control and protection on the basis of the asynchronous motor copper resistance instantaneous value, which depends uniquely on the temperature. A thermal protection system was proposed, which, unlike other common schemes provides insignificant influence of the direct measuring current additionally formed in the stator winding on the asynchronous motor output parameters. Besides, the proposed scheme provides protection from the open-phase operating conditions of the asynchronous motor.

Keywords: Biogas plant · Asynchronous motor · Armature resistance · Thermal protection · Control scheme

1 Introduction

By now, the technology has been developed which makes it possible to convert almost any kind of organic material into biogas and fertilizers. This organic material is processed by the methane fermentation method in biogas plants [1, 2], which make it possible to

A. A. Radionov and V. R. Gasiyarov (Eds.): RusAutoCon 2022, LNEE 986, pp. 421–429, 2023.
https://doi.org/10.1007/978-3-031-22311-2_40

efficiently recycle animal wastes and to convert them into two useful products: biogas for electric and heat energy production and eco-friendly organic fertilizers [3, 4].

The general design and process flow diagram of the biogas plant for organic raw material recycling is given in Fig. 1.

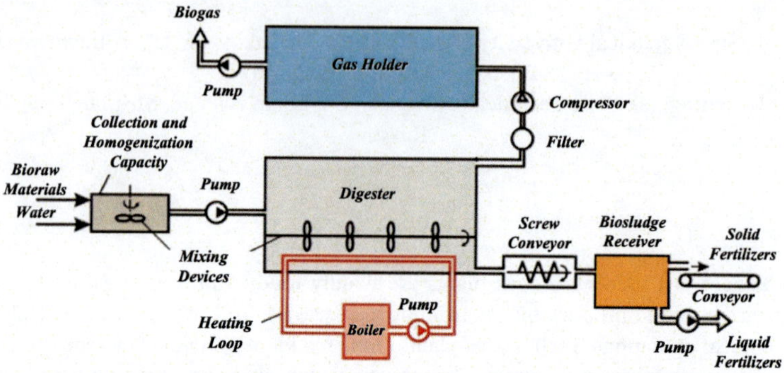

Fig. 1. General design and process flow diagram of the biogas plant.

Depending on the raw material and the final product volume, the biogas plant structural layout and the plant specifications can vary. Nevertheless, in order to provide the production process of biomass conversion, any biogas plant must contain a lot of various electrical equipment. For example, raw material treatment and its feeding into the homogenizer, further transportation of the obtained mixture into the fermenting box, biomass constant stirring in these facilities, biogas feeding into the cogeneration plant and biosludge transportation into the receiving tank and other technological operations are carried out by various shredding machines, conveyors, pumps and fans equipped with individual electric drives [5]. It is obvious that to improve the biogas plant characteristics, it is necessary to apply automatic control in these electric drives and use diagnostic tools of their operating conditions. This makes it possible both to improve biogas plant production efficiency and operation quality and also to increase its service life and operation reliability.

2 Operational Characteristics of Biogas Plant Electrical Equipment

The statistics shows that all biogas plant electric drives are based on asynchronous squirrel-cage electric drives and are usually quite low-power. The biogas plant reliability and the performance efficiency on the whole depend on the reliable operation of these electric drives. It should be taken into account that when an asynchronous motor operates as a part of a biogas plant, and on the whole, when electrical equipment is used in farming, there are certain specific factors concerning both operation conditions and maintenance characteristics as compared with electric drive operation at industrial enterprises, such as:

- Possible supply line parameter fluctuations (voltage and supply frequency value deviations as well as deviation of harmonicity from the specified parameters, supply

asymmetry), phase failure, etc. Besides, breaks in power supply are possible, which require asynchronous motor restarts, which are commonly accompanied by high-ampere starting currents. However, the statistics shows that the most frequent cause of asynchronous motor malfunctions and faults at agricultural enterprises is the phase loss;
- The workload is not constant and can vary in a wide range;
- Various foreign objects can get into the equipment kinematic elements and cause the asynchronous motor short-circuit condition due to the shaft seizure;
- Adverse weather conditions (high humidity, significant temperature fluctuations, high dust level, subversive gas in livestock buildings, etc.);
- Particular operating conditions including possible ventilation duct clogging between the heat radiating ribs and the fan breakdown caused by dirt and foreign objects, which results in cooling failure; possible operation under direct Sun exposure, which causes additional overheating, etc.

Besides, it should be noted that the maintenance level can be quite low, the choice of start-control devices can be incorrect and the start-control devices themselves can be quite worn-out.

Most of these factors cause stator winding overheating, and, as a consequence, electric drive breakdown and, in its turn, biogas plant faulty operation. At the same time, it is obvious that if the asynchronous motor thermal state monitoring is carried out directly during the operating conditions and the proper asynchronous motor thermal protection is based on such monitoring, then even in spite of biogas plant electrical equipment hard operating conditions, it is possible to provide high reliability of its operation. Thus, development and application of new efficient electric drive thermal protection means is an important way to improve the biogas plant performance level.

3 Thermal Protection Method Analysis and Research Problem Setting

Two different synchronous motor thermal control methods have been developed in recent years. They can be divided into the two classes:

- Methods of direct winding thermal state control;
- Methods of indirect parameter change control whose values depend on or influence the winding heating value.

It should be noted that not all of the temperature control methods can be used to develop thermal control systems. In particular, direct temperature control methods using thermometers, temperature-sensitive inks or infrared image converters can't be used for this purpose. That is why let us analyze only those methods, which can be used in practice for thermal protection purposes.

The first class includes thermal protection systems using various thermal sensors (thermoelectrical, thermistor, semiconductor), which convert temperature into electric signal [6]. These sensors are usually installed into stator end winding. However, such systems can control only the winding surface temperature and only in the place where they are installed. Thus, reliable thermal protection systems can't be developed on their basis. Besides, installation of such sensors into a mass-produced electric motor, especially low-power one, can be quite difficult and not always possible.

The second class thermal protection systems based on the indirect temperature control can be divided into the following types [7–9]: 1) based on asynchronous motor current control; 2) based on asynchronous motor thermal model; 3) based on copper resistance control.

However, the first two methods mentioned above do not take into account the changes in the cooling conditions during the asynchronous motor operation or the changes in the weather conditions, which are characteristic for a biogas plant operation. That is why the method based on the asynchronous motor copper resistance control, whose value depends on the temperature, seems to be the most suitable for thermal protection of biogas plant electric drives. In this way, the winding itself acts as a thermal sensor [10–13]. The efficiency and reliability of this approach are defined by the fact that as biogas asynchronous motors are relatively low-power, and, consequently, small size, the temperature in different winding parts is almost the same and the average temperature controlled by the thermal sensor, represents the actual temperature in any winding part. In order to implement this method, small direct current is introduced in the asynchronous motor supply circuit. These direct current values and the corresponding voltage drops are used to calculate stator winding resistance by Ohm's law. One possible circuit design described above is shown in Fig. 2 [14].

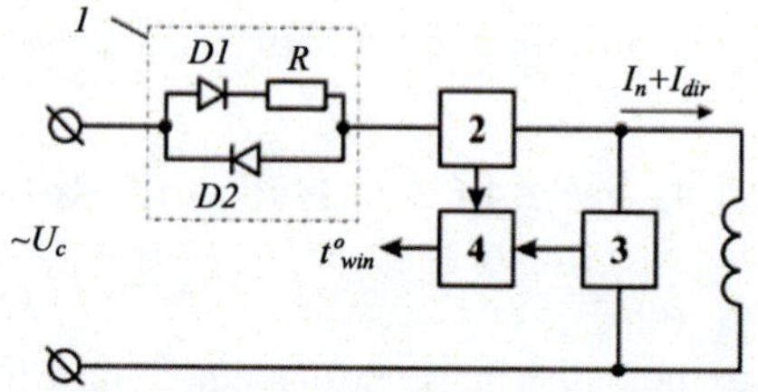

Fig. 2. Asynchronous motor stator winding temperature control system.

Direct current introduction is implemented by its growth from the working alternating current by adding to the winding circuit a device 1 with a reversible conductivity, which has different resistance values for the opposite directions of current flow thanks to the resistor *R*. Due to this, a small form asymmetry is developed for different alternating currents wave half-cycles, which results in development of a small DC component I_{dir}, whose value in accordance with the requirements given in [10] can't exceed 0.5% of the rated alternating current value I_n. The DC value and the voltage drops in the asynchronous motor winding from this current are measured by blocks 2 and 3 respectively, while the winding heating value t^o_{win} is calculated in block 4 whose output signal can be used to develop the thermal protection system [15].

A significant disadvantage of this method is the fact that the introduced direct current component causes changes in the asynchronous motor output parameters, in particular, it causes changes in the mechanism heating conditions, torque pulsation and negative torque also appear. That is why thermal protection system development, which minimizes the disadvantages described above, is of great practical importance.

4 Research Findings

The investigations described above were used to develop an automatic control system of biogas plant electric drives, which provides asynchronous motor thermal protection and protection from the open-phase operating conditions. The basic diagram of this system is given in Fig. 3.

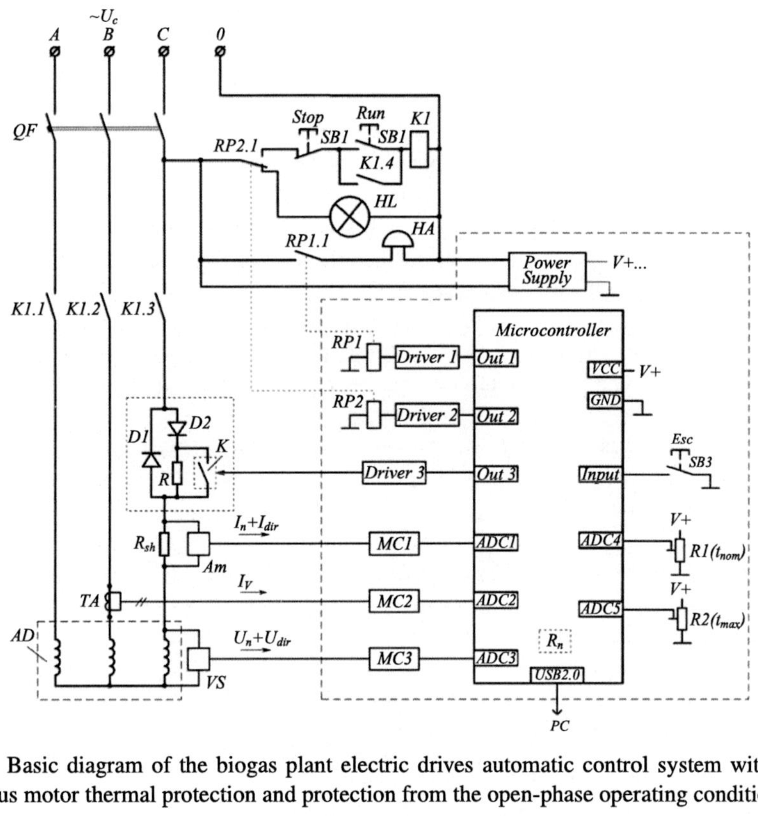

Fig. 3. Basic diagram of the biogas plant electric drives automatic control system with asynchronous motor thermal protection and protection from the open-phase operating conditions.

The control system is based on the 32-bit ARM microcontroller, which performs logical functions, zero-detection and filtering functions and calculates the values of voltage, current and provides electronic switch control, relay switch control through the corresponding drivers, etc. The power supply unit (PSU) provides all the necessary voltage levels for the microcontroller supply, the reference voltage for the analog-to-digital converter, measuring converter power supply, supplementary relay power supply and driver power supply. Key K can be implemented as a thyristor or as a transistor electronic key. Interface *USB*2.0 is used to set the mathematical model constants using the *PC*.

Current transformer *TA* switched in phase "*B*" is necessary to control the phase loss. The microcontroller uses it to define phase "*C*" loss, if current is present in phase "B" while there is no current in "*C*" (which is registered by the current in the shunt circuit R_{sh}). In case of phase "*A*" loss, currents will increase in "*B*" and "*C*" causing winding heating and thermal protection will come into action. In case of phase "*B*" loss, current will be present in phase "*C*" but there will be no current in phase "*B*". In case of two or more phase losses, the motor will not start, while in case of "*B*" and "*C*" phase loss, the motor will shut down immediately, while in case of "*A*" phase loss, the motor shutdown will be provided by the thermal protection.

The system works in the following way. After switching on the power supply by the automatic *QF* switch, the microcontroller takes the readings from the preset resistors *R*1 and *R*2 (set points t_{nom} and T_{max}) and checks if the protection operation has been reset after the previous operation cycle. During the initialization, the supplementary relays *RP*1 and *RP*2 are switched off and the indicator lamp *HL* is on, which might indicate the initialization process, protection operation or malfunction of the thermal protection module. In this case, the motor startup is impossible.

If at the previous asynchronous motor operating stage the protection system was not initiated, then after fulfilling all the actions listed above, the microcontroller uses driver 2 to switch on the supplementary relay *RP*2. The indicator lamp *HL* goes down and now it is possible to start the motor by pushing *SB*2 "*Start*". If the error has not been corrected, then, after eliminating the cause, the *SB*3 button must be pushed to reset it, after that the *RP*2 relay will switch on and make it possible to start the asynchronous motor.

When the motor is switched off, after presetting the *K* key is open and the microcontroller periodically uses readings from the voltage sensor *VS* on the asynchronous motor winding to assess if the motor is switched on. At this stage, the temperature is not measured.

During the start-up period, the electronic key *K* is open and the direct measuring current I_{dir} is introduced into the asynchronous motor stator winding circuit. The copper resistance value *R* is calculated by the Ohm's law

$$R = U_{dir} \big/ I_{dir}, \tag{1}$$

where I_{dir} and U_{dir} are the direct measuring current and the voltage drop from it on the asynchronous motor stator winding, respectively.

Change in *R* value due to the temperature change makes it possible to calculate the winding average temperature t_{win} in operating condition

$$t_{\text{win}} = (R_R - R_{\text{N}})/(\alpha \cdot R_{\text{N}}) \tag{2}$$

where R_N is the initial winding resistance at the temperature of 0 °C; where R_R is resistance in operating condition; α is the temperature resistance coefficient.

The numerical value R_N is fed into the microcontroller at the stage of its setting-up. In order to implement the expression (1), two independent measuring channels were designed in the scheme; these channels measure active current and voltage components of the winding. To measure the current, the current sensor in the form of an instrument shunt R_{sh} with reference active resistance was introduced into the winding circuit. The signal from the shunt is amplified by the amplifier *Am* and is fed into the microcontroller analog-to-digital converter $ADC1$ through the measuring converter $MC1$. The second measuring channel contains a voltage drop measuring module on the *VS* winding whose output signal is supplied to the microcontroller analog-to-digital converter $ADC3$ through the measuring converter $MC3$.

During the motor start-up, the microcontroller defines the motor start-up, constantly measures the winding temperature t_{win}, compares it with the preset value t_{nom} and checks the phase loss. However, if some emergency situation takes place in the asynchronous motor operation, for example, if its shaft is restrained due to a foreign object present in the biogas plant equipment kinematic elements or the conveyor chain is frozen down in the winter, then the asynchronous motor winding temperature will increase too quickly and the heating curve will be too steep, the microcontroller will identify this and use driver 1 through the supplementary relay *RP*1 to activate the alarm *HA* informing the biogas plant operator that the asynchronous motor start-up does not take place in the standard mode. At the same time, the motor itself does not switch off because the problem can repair itself (for example, conveyor flights can overcome the freezing strength due to the asynchronous motor pulling force), in this case, the alarm will switch off. However, if the synchronous motor winding heating reaches the value $t_{win} \geq T_{max}$, the microcontroller will switch off the asynchronous motor power supply by switching off the supplementary relay *RP*2, and the starting switch $K1$. The restart will be impossible, the signal lamp will burn and the alarm *HA* will provide audible signal.

After elimination of the overheating cause or the phase loss, the resetting for the asynchronous motor start-up suppression is carried out by double pressing of the reset button (the first one switches on the alarm, while the second one allows the restart operation by switching on the *RP*2).

In order to reduce the impact of the measuring current on the synchronous motor, the current can be introduced at intervals. Depending on the motor type and its operating conditions, the interval length can be about 3–5 min. For this, after each cycle of direct current signal sending and the procedure of winding temperature measurement, this signal can be removed and supply the winding only with alternating operating voltage. The asynchronous motor thermal protection does not suffer much due to the machine heat retention. That is why in steady-load conditions of the asynchronous motor, if $t_{win} \leq t_{nom}$, the key *K* should be closed. Closing and opening of the key *K* takes place at the moment when current starts flowing through *D*2, which is provided by the program zero detector inside the microcontroller and the beginning is defined by the instantaneous current value from the shunt. Further opening/closing of the key *K* takes place periodically every 3–5 min (the time interval is set during adjustment or depending on the temperature change rate). If the temperature approaches t_{win} the $T_{\max}$ value, this is recorded by the microcontroller and the key *K* will remain open thus providing constant control of the winding temperature.

In case of protection operation, that is if $t_{win} \geq T_{max}$ and/or in case of the phase loss, the asynchronous motor switches off and the alarm *HA* and the signal lamp *HL* switch on. At the same time, the fact of protection operation is recorded in the non-volatile memory. This is necessary in order to inform about switching off power supply after protection operation during the further start-up. In this case, in order to switch off the audible alarm, it is necessary to press the reset button *SB*3 once. The second push on the button *SB*3 will switch on the *RP*2, which will make it possible to start up the asynchronous motor and remove the data about protection operation from the memory.

When the asynchronous motor is switched off by pushing the "*Stop*" *SB*2 button, the microcontroller defines the shutdown by the signals from the voltage sensor and opens the electronic key *K* if it was closed.

5 Conclusion

Thus, the proposed electric drive automatic control system integrated into the biogas plant provides monitoring of the asynchronous motor stator winding thermal state taking into account its operating characteristics, which makes it possible to obtain reliable thermal protection for the asynchronous motor. This system can be correlated with the other biogas plant structural components and in this way develop their generalized control system.

The experimental investigations based on the asynchronous motor of 4A80B2 type and the ARM microcontroller STM32F103C8T6 proved high efficiency of the proposed control system.

In conclusion, it should be noted that the developed automatic control system can be used for both biogas plant electric drive thermal protection and for control of temperature protection from winding overheating for other devices supplied from alternating voltage mains and applied in various industrial electrical installations and electric drives.

References

1. Liangwei, D., Yi, L., Wenguo, W.: Biogas plant. In: Biogas Technology, pp. 109–156 (2020). Springer, Singapore. https://doi.org/10.1007/978-981-15-4940-3_4
2. Eder, B., Shults, K.: Biogas plants. ZorgBiogas, Germany (2008)
3. Braun, R.: Anaerobic digestion: a multi-faceted process for energy, environmental management and rural development. In: Improvement of Crop Plants for Industrial End Uses, pp. 335–416. Springer, Dordrecht (2007). https://doi.org/10.1007/978-1-4020-5486-0_13
4. Deublein, D., Steinhauser, A.: Biogas from Waste and Renewable Resources: An Introduction, 2nd, Revised and Expanded Edition. Wiley-VCH (2010). https://doi.org/10.1002/9783527632794
5. Bochkarev, I.V., Sandybaeva, A.R., Bagiev, K.: Control system of a bioenergetic complex electric drive. Autom. Control. Probl. **1**(38), 33–42 (2020)
6. Sachenko, A.A.: Temperature Measurement by Sensors with Built-In Calibrators. Energoatomizdat, Moscow (1986)
7. Kramarukhin, Y.: Temperature Measuring Devices. Machine building, Moscow (1990)
8. Soloviev, A.L.: Protection of 0.4 kV Asynchronous Electric Motors. Energoprogress, Moscow (2007)
9. Makarov, A.V., Vecherkin, M.V., Zavyalov, A.S.: Review of asynchronous motor thermal models. Electrotech. Syst. Complexes **21**, 75–84 (2013)

10. State Standard 27222-91: Electric rotating machines. Winding resistance measuring in alternating current machines without switching off from the mains. Publishing house of standards, Moscow (1991)
11. Dostov, L.I.: State and prospects for further development of winding temperature measuring devices in electric machines in operating modes. Electrical engineering industry. Series 01. Electric machines: Survey information 30:44 (1990)
12. Sabaghi, M., Farahani, H.F.: Monitoring of induction motor temperature under unbalanced supplying by stator resistance estimation. Indian J. Sci. Technol. **5**(3), 2354–2359 (2012). https://doi.org/10.17485/ijst/2012/v5i3.6
13. Bochkarev, I.V., Bryakin, I.V., Khramshin, V.R., Sandybaeva, A.R., Litsin, K.V.: Developing new thermal protection method for AC electric motors. Machines **9**(3):51. https://doi.org/10.3390/machines9030051
14. Bochkarev, I.V., Ashmarin, A.V.: Temperature measuring device in an electric machine winding. USSR Patent 998,873, 23 Feb 1983 (1983)
15. Bochkarev, I.V., Ashmarin, A.V.: Thermal protection device for an electric machine. USSR Patent 1,187,234, 23 Nov 1985 (1985)

Check for updates

Experimental Time Difference Effect Studies of the Vacuum Circuit Breakers in Electric Arc Furnaces on Switching Overvoltages

A. Nikolaev(✉), E. Smirnov, and S. Morozov

Nosov Magnitogorsk State Technical University, 38, Lenin Avenue, Magnitogorsk 455000, Russia

alexniko@inbox.ru

Abstract. To date, the furnace transformers of high-capacity arc furnaces and ladle-furnaces are switched on and off quickly by means of vacuum circuit breakers. The advantages of modern vacuum circuit breakers are the best possible arc extinguishing efficiency as well as a long switching life of up to 30,000 operating cycles and more. The main drawback of these vacuum circuit breakers is the switching overvoltage. They are caused by the phenomena of current chopping and virtual current chopping during unloaded and loaded furnace transformer switching. This overvoltage can have a negative effect on the insulation of the furnace transformer and cable lines. This paper reflects the results of experimental overvoltage studies during switching the vacuum circuit breakers of electric arc furnaces at Cherepovets Steel Mill, PJSC "Severstal". The measures are proposed to limit switching overvoltage by setting optimal timing of closing and opening of vacuum circuit breakers contacts.

Keywords: Vacuum circuit breaker · Arc steelmaking furnace · Ladle-furnace installation · High-voltage transformer · Switching overvoltages · Contact switching differences

1 Introduction

At present, the on- and off-switching of furnace transformers in arc furnaces (EAFs) and ladle-furnaces (LFs) is realized by means of vacuum circuit breakers (VCB). The advantages of modern vacuum circuit breakers are their high arc quenching efficiency and long switching life [1–3]. The main disadvantages of VCB are current and virtual current chopping which lead to high switching overvoltages when the furnace transformers are switched on and off [4–6]. These overvoltages have a negative effect on the insulation of the transformer and cable lines. Specialized RC circuits and surge arresters are used to protect the electrical equipment against switching overvoltages, installed after the VCB and in some cases also before the VCB [7–9]. The operation efficiency of protective RC circuits and surge arresters depends on the correct choice of their parameters, as well as the current state and characteristics of the explosive, such as: 1) the actual timing of closing and opening of the switch contacts, 2) the contact divergence speed, 3) the

A. A. Radionov and V. R. Gasiyarov (Eds.): RusAutoCon 2022, LNEE 986, pp. 430–437, 2023.
https://doi.org/10.1007/978-3-031-22311-2_41

contact bounce presence/absence, 4) the actual value of contact transient resistance, etc. [10, 11]. The operation experience of modern VCB has shown that difference change of contacts closing and opening the switches because of its mechanical part deterioration at a considerable quantity of operation cycles can lead to decrease in use efficiency of protective RC-circuits and surge arresters. This poses a danger to the serviceable operation of the electrical EAF and LF equipment and requires the application of additional measures to monitor the VCB condition.

Within the framework of this article the experimental research results of voltages and currents transients on the primary sides of furnace transformers at electric furnace melting units, operating at Cherepovets metallurgical plant of PJSC "Severstal" during switching of operative explosives are presented. On the basis of the obtained results the main boundary values of contacts switching difference have been determined and it ensures the risk absence for switching overvoltages.

2 The Description of the Experimental Research Methodology for Overvoltages During Switching Vcb in Electric Arc Furnaces as Exemplified by Chermk Pjsc "Severstal"

During the experimental studies the main device to assess the nature of high-frequency switching overvoltages was the NI-USB 6251 recorder with a high sampling frequency $f$rec. = 300 kHz per channel, equipped with the wide-bandwidth galvanic isolation units and resistive voltage dividers whose inputs were connected with measuring voltage transformers 35000/100 V, placed in the cells of the operating vacuum circuit breakers [12]. A duplicating device was a multi-channel recorder RES-3 (electrical signals recorder), with which a similar recording of instantaneous values of phase voltages, as well as linear currents on the primary side of the furnace transformers was performed. The sampling rate of RES-3 was $f$rec. = 18 kHz per channel. For the record of instantaneous values of currents the measuring shunts were used which were included in the series measuring circuits of current transformers placed in breaker cells Fig. 1. The oscillograms recorded by two recorders were synchronized in the mathematical package Matlab-Simulink and as a result a full picture of instantaneous values of currents and voltages was reproduced during the switching of operational vacuum circuit breakers of the investigated electric steel-smelting units Fig. 2.

Tables 1 and 2 show the following parameters: Ur.v - rated line voltage of VI; Ir - rated current; I rated sh.circ. Break. Current - rated short-circuit breaking current; Ncycle.Irated = 30000 - allowed number of switching cycles during interrupting of rated Irated current; Ncycle. Actual. - actual value of switching cycles. Note: REQU, CEQU - equivalent one phase active resistance and capacitance of protective RC- circuits; Uph.rated - rated voltage of phase group varistor connected in a star; Ul.rated - rated voltage of the linear group varistor connected in delta; Uph - long-term operating voltage of phase group varistor; Ulo - long-term operating voltage of the linear group varistor; Imeas.rated – the rated varistor voltage of both groups; Il.rated - long-term rated value of pulse current (through-current) at timp. = 2000 μs; Ua.rated is the rated voltage of arrester; Ua.break. is the arrester breakdown voltage in dry state at f = 50 Hz; Ua.imp.bre. - the arrester pulse breakdown voltage at pre-discharge time from 2 to 10 μs. It should be

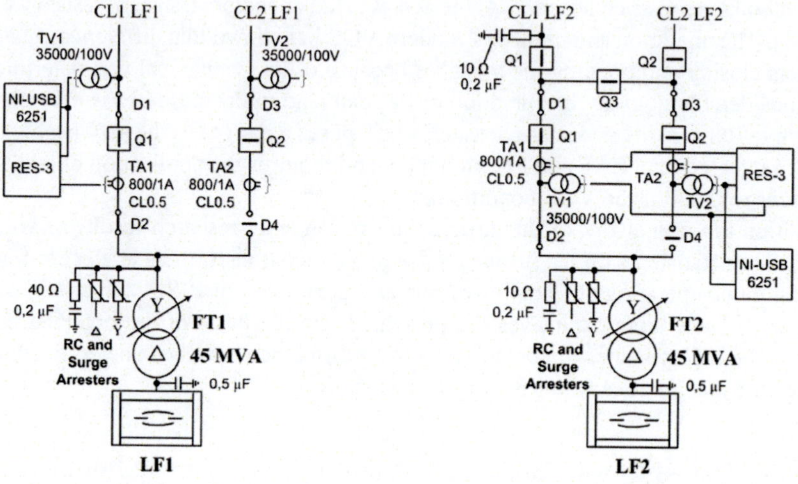

a) b)

Fig. 1. The connection diagrams of measuring complex based on NI-6251 and RES-3 recorders to voltage (TV) and current (TA) transformers of the vacuum circuit breaker cells of the studied ladle-furnace units (375 t, 45 MVA).

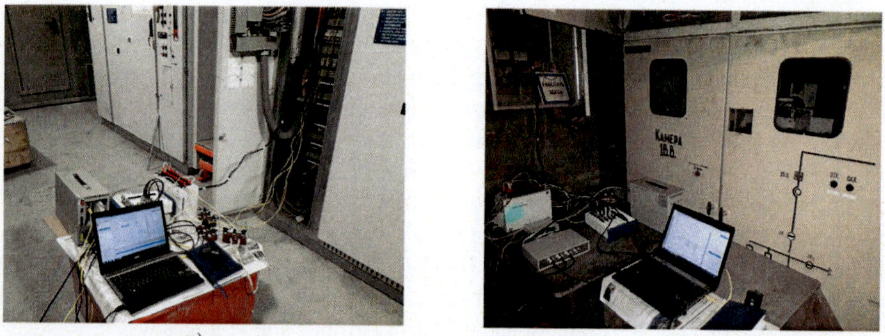

a) b)

Fig. 2. The photos of the measuring system based on NI USB-6251 and RES-3 recorders, connected to the voltage and current measurement circuits of the operational vacuum circuit breaker cells.

noted, that the voltage measuring transformers are installed at LF1 before the operational VCB. In this case during the experimental studies the phase voltages from the supply side were recorded. At the electric steel-melting unit LF2 the measuring transformers are placed after operative VCB, that made it possible to record transient processes of phase voltages from the electric load side. These peculiarities of measuring transformers location were taken into account in the analysis of the recorded transients [13, 14].

Table 1. Technical data of the operational vacuum circuit breakers of the investigated electric steelmaking units.

Electric furnace unit	Input Number	Vacuum circuit breaker type	Rated data	Number of switches at the time of measurement
LF 1	1	Siemens 3AH4 305–2	$U_{r.v} = 36$ kV; $I_r = 1250$ A;	$N_{\text{cycle. Act}} = 50$
	2		$I_{\text{rated sh.circ. Break. Current}} =$ 31.5 kA; $N_{\text{cycle.Irated}} = 30000$	$N_{\text{cycle. Act.}} =$ 45255
LF 2	1	Siemens 3AH4 305–2	$U_{r.v} = 36$ kV; $I_r = 2000$ A;	$N_{\text{cycle. Act}} = 67$
	2		$I_{\text{rated sh.circ. Break. Current}} =$ 31.5 kA; $N_{\text{cycle. Irated}} = 30000$	$N_{\text{cycle. Act}} =$ 46578

Table 2. Technical data of the main rc-circuits and surge arresters of the electric arc furnace.

Electric furnace unit	Parameters RC circuits after VCB	Surge Arresters' Parameters after VCB	Parameters RC circuits before VCB
LF 1	$R_{EQU} = 40\ \Omega$; $C_{EQU} = 0.20\ \mu F$	SPD-P/PDP-35/40.5/10/1000-IV NF1 (phase), SPD - P/PDP -35/43.0/10/1000-IV NF 1 (linear) $U_{pho} = 40.5$ kV, $U_{rated} = 43.0$ kV, $I_{meas.rated} = 10$ kA, $I_{l.rated} = 1.00$ kA	Missing
LF 2	$R_{EQU} = 20\ \Omega$; $C_{EQU} = 0.30\ \mu F$	Type: TRIDELTA SBK-IV-48/10.3 (phase), TRIDELTA SBK-IV-48/10.3 (linear) $U_{ph.rated} = 48.0$ kV, $U_{l.rated} = 54.0$ kV, $U_{pho} = 38.4$ kV, $U_{lo} = 38.4$ kV, $I_{meas.rated} = 10$ kA, $I_{l.rated} =$ 0.50 kA	$R_{EQU} = 10\ \Omega$; $C_{EQU} = 0.20\ \mu F$ (before Q1)

3 The Analysis of Different Types of Switching Overvoltages Vi in Electric Furnace Systems

A number of basic types of overvoltages have been identified from all the obtained oscillograms. The first type relates to overvoltages occurring during arc faults when the poles of the VCB are shorted. The duration of individual voltage pulses is a few microseconds and the amplitude reaches 4.2 rated values. In a number of cases after preliminary breakdowns, multiple re-ignitions of arc gaps of up to 2.5 ms with high overvoltages (up to 2 rated values) occur, which persist until complete closure of the explosive poles

Fig. 3.d,g). With VI interrupting, the overvoltages arise due to the phenomena of current shearing and virtual current shearing. Depending on the interrupter condition, the current to be tripped (no-load current or rated load current) and the operating modes of the RC circuits and arrester, strong overvoltages with the increasing amplitude may occur due to multiple re-ignitions of the arc during the separation of the VI poles. For a number of units, e.g. VCB1 LF2, the overvoltages with an amplitude up to 4,6 Um.rated and duration of 3 ms were observed when the rated load was disconnected Fig. 3.f,i). It should be noted that these overvoltages represent a serious danger for the insulation of 35 kV cable lines and the windings of furnace transformers, which can lead to an emergency situation [15, 16].

In some disconnection cases of FT under load was accompanied by monotonous increase of instantaneous value in phase voltage up to great value (2 Um.rated) without clear re-ignitions in VCB with great timing of closing and opening of contacts Fig. 3.e,h). In some cases on LF2, 1 input tripping of VCB under load without RC circuits and surge protection devices was accompanied by lower overvoltages, than when using protective elements. It testifies to the necessity of check of a condition of VCB mechanics and vacuum chambers, and also check of correctness of a choice of parameters of RC circuits

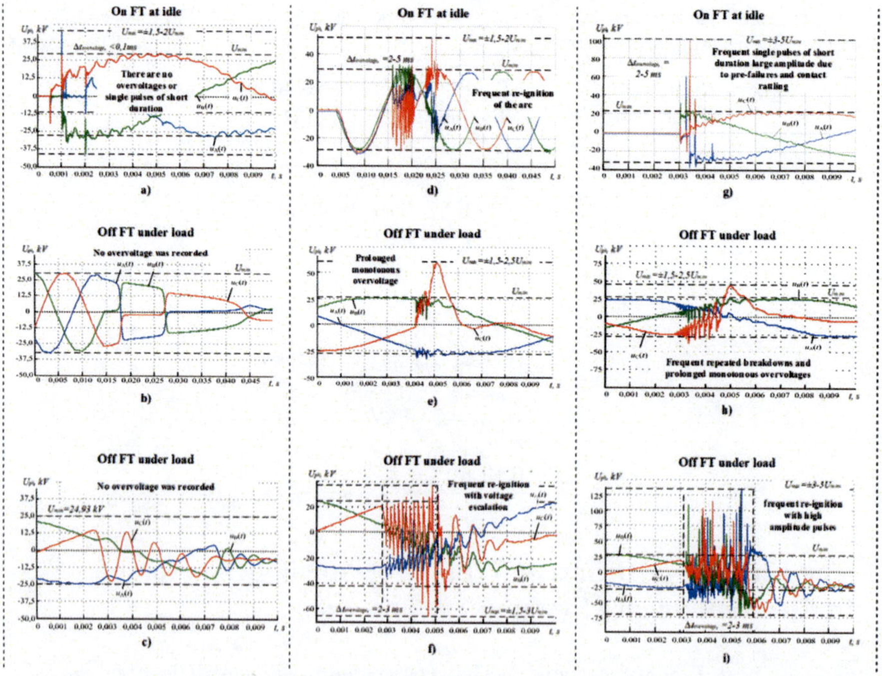

Fig. 3. The sample voltage waveforms used to assess the condition of vacuum interrupters, based on the nature of the phase voltage transients at the moment of switching the electric steelmaking units during a series of test actions: a), b), c)- the nature of the transients is good/acceptable (observation of characteristic processes for all three series of experiments of each type); d), e), f), g), h), i)- the unacceptable transition (observation of characteristic changes for at least one experience of one type of series).

taking into account a mode of power supply of the unit which defines parallel work of the investigated LF with other electric steel-smelting units [17, 18].

The series of oscillograms showed: a), d), g) - switching on the unloaded furnace transformer; b), e), h) - switching off the unloaded furnace transformer; c), f), i) - switching off the furnace transformer under load.

4 The Analysis of Differential Switching Effect of VCB Poles on Overvoltage Processes

The measurements of overvoltage levels for LF2 VCB1 and VCB2 were carried out under identical modes of power supply of the unit with included RC circuits 10 Ω/0,2 μF before and after the circuit breakers provided that no overvoltages were present during switching of VCB1 and they were present at the new switch VCB2 with number of trips 8000. The most probable cause of their occurrence is the non-optimal timing of the pole opening. By comparing the data on the timing difference in tripping for VCB1 of the LF2 the limit value Δt equal to 0.6 ms has been substantiated. The influence research results of a difference timing of switching the interrupter poles on the level of overvoltages are resulted on Fig. 4 and show that thanks to reduction of a difference timing at switching-off the interrupter poles 1 input at the loaded LF2 FT, the strong overvoltages with the amplitude to 4.0 Um.rated. And duration 2.5 ms, observed within the limits of the first series experiments disappeared. In contrast, in the case of the new LF2 VCB2, strong overvoltages with amplitude of 4.6 Um.rated. And duration of 2.0 ms. Were observed twice out of 3 experiments.

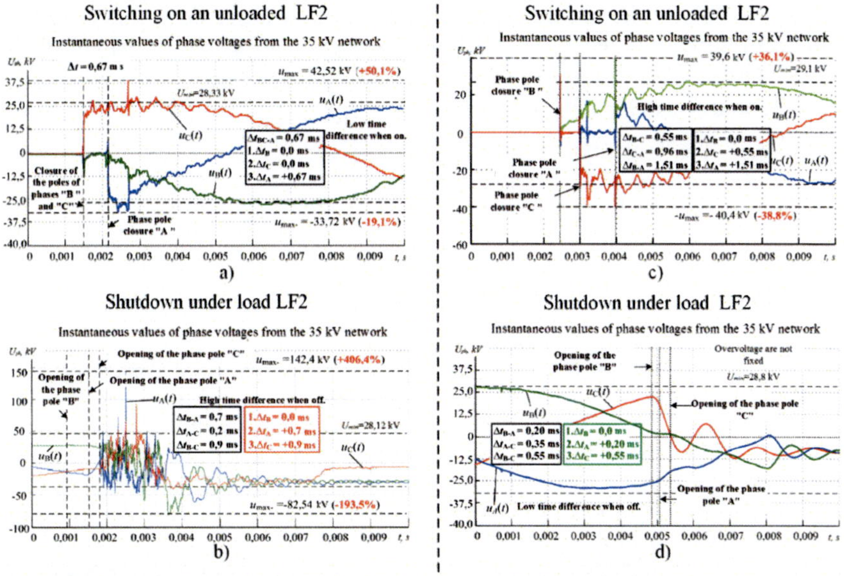

Fig. 4. Comparative analysis of voltage transients during switching VCBs of LF2 with different diversity settings at the same power supply modes and parameters of RC circuits and surge protection devices.

5 Conclusion

1. Based on the analysis of the experimental data, the overvoltages of two types have been identified and investigated: 1) the overvoltages arising at arc gap pre-faults closing the VCB poles; 2) the overvoltages due to the phenomena of virtual and conventional current slices at the unloaded and loaded furnace transformer disconnection. For each case the maximum values of voltage pulses, duration and frequency of arc re-ignitions in vacuum chambers have been determined.
2. In order to exclude dangerous overvoltages, the estimation technique for the operative state of vacuum circuit breakers at electric furnace melting units has been developed with the use of the specialized tests and measurements on the basis of a high-speed measuring complex. The methodology implies routine experimental studies of recording instantaneous values of voltages and currents in the furnace switch gear 35kV of all operating electric furnace melting units under special test effects: 1) three switching-on of the unloaded furnace transformer; 2) three switching-off of the furnace transformer operating under no-load; 3) three switching-off of the FT under load. On the basis of the received data the experts carry out the comparative analysis of experimental oscillograms with reference samples then the conclusion on a serviceable condition and quality of settings of vacuum switches diversity is made. If changes are required, the diversity correction of switching the interrupter poles with the subsequent carrying out repeated measurements and recording of voltages and currents is carried out.
3. The results showed that by reducing the difference in the timing of switching interrupter poles 1 input with a loaded LF2 FT, strong overvoltages with an amplitude up to 4,0 $U_{m.rated.}$ And duration of 2.5 ms, which were observed in the first series of experiments have disappeared. By comparing the trip diversity data, a limit value of Δt equal to 0.6 ms was justified for 1 input of LF2. Therefore, it is recommended to adhere to the condition $\Delta t \leq 0.6$ ms when carrying out the settings for the timing difference of the Siemens 3AH vacuum circuit breakers.

References

1. Knuppel, G.: Deoxidation and Vacuum Treatment of Steel, p. 312. Metallurgy, Moscow (1973)
2. Guidelines for the protection of electrical networks 6–1150 kV from lightning and internal surges: RD 153–34.3–35.125–99: under the scientific editorship of N.N. Tikhodeev. 2nd ed. PEIPK Mintopenergo RF, p. 355, St. Petersburg (2001)
3. Electrician's Bible: PUE, MPOT, PTE, p.752. Eksmo, Moscow (2012)
4. Lyskov, Y., Antonova, N.P., Demina, O.Y., Zuev, A.V.: Guidelines for the use of non-linear surge arresters in electrical networks 6–35 Kv, p. 45. RAO "UES of Russia", Moscow (2001)
5. Ling, Y., Lin, S., Huaren, W.: Simulation of fault arc using conventional arc models. Sch. Electr. Autom. Eng. Energy Power Eng. **5**, 833–837 (2013)
6. Nitu, S., Nitu, C., Mihalache, C., Anghelita, P., Pavelescu, D.: Comparison between model and experiment in studying the electric arc. University "POLITEHNICA" from Bucharest. J. Optoelectron. Adv. Mater. **10**, 1192–1196 (2008)
7. Goncharov, A.F., Epstein, I.Y., Popov, Y.N., et al.: Overvoltage protection of high-voltage motors switched by vacuum switches. Ind. Energy **6**, 21–24 (2000)

8. Model VBU Fault Interrupert. Instructions. Joslyn, pp. 11–14 (2004)
9. Kornilov, G.P., Nikolaev, A.A., Khramshin, T.R.: Modeling of Electrotechnical Complexes of Industrial Enterprises: Textbook Allowance, p. 239. Magnitogorsk Publishing House. Nosov Magnitogorsk State tech. un-ty, Magnitogorsk (2014)
10. Shilov, I.G., Sinyukov, T.V.: Overvoltage in Distribution Electrical Networks, p. 6. Lipetsk Publishing House. state tech. un-ty, Lipetsk (2010)
11. Lavrinovich, V.A., Hoang, T.A.: Investigation of Arc Properties in a Vacuum Arc Chute at Small Distances Between Contacts, p. 5. Publishing house of Tomsk Polytechnic. un-ty, Tomsk (2010)
12. Goncharov, A.F., Pavlov, V.V., Yazev, V.N., et al.: Selection of optimal parameters for active-capacitive surge suppressors. Ind. Energy **2**, 26–29 (1995)
13. Evdokunin, G.A., Tiler, G.: Modern Vacuum Switching Technology for Medium Voltage Networks, p. 136. Publishing house M.P, Sizova (2002)
14. Bushuev, S.A., Volkov, A.Y.: Study of switching overvoltages generated by vacuum generator interrupters at a power facility. Department of TEVN. Soyuz Corporation, MPEI, p. 210 (2004)
15. Shilin, N.V., Nikuev, Y.: Restoring strength in high voltage switches. Electricity **2**, 31–36 (1984)
16. Shevtsov, D.E.: Basic principles of synchronous switching of medium voltage electrical networks. In: XX International Scientific-Practical Conference "Modern Equipment and Technologies", Novosibirsk, pp. 55–56 (2010)
17. Lebedev, I.A.: Study of mathematical models and development of requirements for the operation of synchronous vacuum circuit breakers for 6–10 kV networks from the point of view of generating a minimum overvoltage level. In: Proceedings of the XI All-Russian Student Scientific and Technical Seminar "Energy: Ecology, Reliability, Safety", Tomsk, vol. 1, pp. 22–26 (2009)
18. Bizjak, G., Zunko, P., Pov, D.: Combined model of SF6 circuit breaker for use in digital simulation programs. IEEE Trans. Power Deliv. **19**, 174–181 (2004)

Author Index

A. A. Radionov and V. R. Gasiyarov (Eds.): RusAutoCon 2022, LNEE 986, pp. 439–440, 2023.
https://doi.org/10.1007/978-3-031-22311-2